Preface

Welcome to the International Conference an Image and Video Retrieval, CIVR 2002. Our conference is a snapshot of the current world-wide research in image and video retrieval from digital libraries, databases, and multimedia collections. Topics range from the state of the art in semantic visual retrieval to video summarization to new features and modeling paradigms.

This year 82 papers from 24 countries were submitted and 39 were accepted for presentation at the conference after being reviewed by at least 3 members of the Program Committee.

We would like to thank all members of the Program Committee, as well as the additional referees listed below, for their help in ensuring the quality of the papers accepted for publication. We would also like to thank the Organizing Committee for all their efforts in making the conference happen, as well as our two keynote speakers, Arnold Smeulders from the University of Amsterdam and Alex Hauptmann from Carnegie-Mellon University. Finally, we are grateful to our sponsors, the British Computer Society Information Retrieval Specialist Group, the British Machine Vision Association (BMVA), the Institute for Image Data Research, University of Northumbria, the Institution of Electrical Engineers (IEE), and the Leiden Institute of Advanced Computer Science (LIACS), Leiden Universiy.

May 2002

Michael S. Lew
Nicu Sebe
John P. Eakins

Lecture Notes in Computer Science

Vol. 245: H.F. de Groote, Lectures on the Complexity of Bilinear Problems. V, 135 pages. 1987.

Vol. 246: Graph-Theoretic Concepts in Computer Science. Proceedings, 1986. Edited by G. Tinhofer and G. Schmidt. VII, 307 pages. 1987.

Vol. 247: STACS 87. Proceedings, 1987. Edited by F.J. Brandenburg, G. Vidal-Naquet and M. Wirsing. X, 484 pages. 1987.

Vol. 248: Networking in Open Systems. Proceedings, 1986. Edited by G. Müller and R.P. Blanc. VI, 441 pages. 1987.

Vol. 249: TAPSOFT '87. Volume 1. Proceedings, 1987. Edited by H. Ehrig, R. Kowalski, G. Levi and U. Montanari. XIV, 289 pages. 1987.

Vol. 250: TAPSOFT '87. Volume 2. Proceedings, 1987. Edited by H. Ehrig, R. Kowalski, G. Levi and U. Montanari. XIV, 336 pages. 1987.

Vol. 251: V. Akman, Unobstructed Shortest Paths in Polyhedral Environments. VII, 103 pages. 1987.

Vol. 252: VDM '87. VDM – A Formal Method at Work. Proceedings, 1987. Edited by D. Bjørner, C.B. Jones, M. Mac an Airchinnigh and E.J. Neuhold. IX, 422 pages. 1987.

Vol. 253: J.D. Becker, I. Eisele (Eds.), WOPPLOT 86. Parallel Processing: Logic, Organization, and Technology. Proceedings, 1986. V, 226 pages. 1987.

Vol. 254: Petri Nets: Central Models and Their Properties. Advances in Petri Nets 1986, Part I. Proceedings, 1986. Edited by W. Brauer, W. Reisig and G. Rozenberg. X, 480 pages. 1987.

Vol. 255: Petri Nets: Applications and Relationships to Other Models of Concurrency. Advances in Petri Nets 1986, Part II. Proceedings, 1986. Edited by W. Brauer, W. Reisig and G. Rozenberg. X, 516 pages. 1987.

Vol. 256: Rewriting Techniques and Applications. Proceedings, 1987. Edited by P. Lescanne. VI, 285 pages. 1987.

Vol. 257: Database Machine Performance: Modeling Methodologies and Evaluation Strategies. Edited by F. Cesarini and S. Salza. X, 250 pages. 1987.

Vol. 258: PARLE, Parallel Architectures and Languages Europe. Volume I. Proceedings, 1987. Edited by J.W. de Bakker, A.J. Nijman and P.C. Treleaven. XII, 480 pages. 1987.

Vol. 259: PARLE, Parallel Architectures and Languages Europe. Volume II. Proceedings, 1987. Edited by J.W. de Bakker, A.J. Nijman and P.C. Treleaven. XII, 464 pages. 1987.

Vol. 260: D.C. Luckham, F.W. von Henke, B. Krieg-Brückner, O. Owe, ANNA, A Language for Annotating Ada Programs. V, 143 pages. 1987.

Vol. 261: J. Ch. Freytag, Translating Relational Queries into Iterative Programs. XI, 131 pages. 1987.

Vol. 262: A. Burns, A.M. Lister, A.J. Wellings, A Review of Ada Tasking. VIII, 141 pages. 1987.

Vol. 263: A.M. Odlyzko (Ed.), Advances in Cryptology – CRYPTO '86. Proceedings. XI, 489 pages. 1987.

Vol. 264: E. Wada (Ed.), Logic Programming '86. Proceedings, 1986. VI, 179 pages. 1987.

Vol. 265: K.P. Jantke (Ed.), Analogical and Inductive Inference. Proceedings, 1986. VI, 227 pages. 1987.

Vol. 266: G. Rozenberg (Ed.), Advances in Petri Nets 1987. VI, 451 pages. 1987.

Vol. 267: Th. Ottmann (Ed.), Automata, Languages and Programming. Proceedings, 1987. X, 565 pages. 1987.

Vol. 268: P.M. Pardalos, J.B. Rosen, Constrained Global Optimization: Algorithms and Applications. VII, 143 pages. 1987.

Vol. 269: A. Albrecht, H. Jung, K. Mehlhorn (Eds.), Parallel Algorithms and Architectures. Proceedings, 1987. Approx. 205 pages. 1987.

Vol. 270: E. Börger (Ed.), Computation Theory and Logic. IX, 442 pages. 1987.

Vol. 271: D. Snyers, A. Thayse, From Logic Design to Logic Programming. IV, 125 pages. 1987.

Vol. 272: P. Treleaven, M. Vanneschi (Eds.), Future Parallel Computers. Proceedings, 1986. V, 492 pages. 1987.

Vol. 273: J.S. Royer, A Connotational Theory of Program Structure. V, 186 pages. 1987.

Vol. 274: G. Kahn (Ed.), Functional Programming Languages and Computer Architecture. Proceedings. VI, 470 pages. 1987.

Vol. 275: A.N. Habermann, U. Montanari (Eds.), System Development and Ada. Proceedings, 1986. V, 305 pages. 1987.

Vol. 276: J. Bézivin, J.-M. Hullot, P. Cointe, H. Lieberman (Eds.), ECOOP '87. European Conference on Object-Oriented Programming. Proceedings. VI, 273 pages. 1987.

Vol. 277: B. Benninghofen, S. Kemmerich, M.M. Richter, Systems of Reductions. X, 265 pages. 1987.

Vol. 278: L. Budach, R.G. Bukharajev, O.B. Lupanov (Eds.), Fundamentals of Computation Theory. Proceedings, 1987. XIV, 505 pages. 1987.

Vol. 279: J.H. Fasel, R.M. Keller (Eds.), Graph Reduction. Proceedings, 1986. XVI, 450 pages. 1987.

Vol. 280: M. Venturini Zilli (Ed.), Mathematical Models for the Semantics of Parallelism. Proceedings, 1986. V, 231 pages. 1987.

Vol. 281: A. Kelemenová, J. Kelemen (Eds.), Trends, Techniques, and Problems in Theoretical Computer Science. Proceedings, 1986. VI, 213 pages. 1987.

Vol. 282: P. Gorny, M.J. Tauber (Eds.), Visualization in Programming. Proceedings, 1986. VII, 210 pages. 1987.

Vol. 283: D.H. Pitt, A. Poigné, D.E. Rydeheard (Eds.), Category Theory and Computer Science. Proceedings, 1987. V, 300 pages. 1987.

Vol. 284: A. Kündig, R.E. Bührer, J. Dähler (Eds.), Embedded Systems. Proceedings, 1986. V, 207 pages. 1987.

Vol. 285: C. Delgado Kloos, Semantics of Digital Circuits. IX, 124 pages. 1987.

Vol. 286: B. Bouchon, R.R. Yager (Eds.), Uncertainty in Knowledge-Based Systems. Proceedings, 1986. VII, 405 pages. 1987.

Vol. 287: K.V. Nori (Ed.), Foundations of Software Technology and Theoretical Computer Science. Proceedings, 1987. IX, 540 pages. 1987.

Vol. 288: A. Blikle, MetaSoft Primer. XIII, 140 pages. 1987.

Vol. 289: H.K. Nichols, D. Simpson (Eds.), ESEC '87. 1st European Software Engineering Conference. Proceedings, 1987. XII, 404 pages. 1987.

Vol. 290: T.X. Bui, Co-oP A Group Decision Support System for Cooperative Multiple Criteria Group Decision Making. XIII, 250 pages. 1987.

Vol. 291: H. Ehrig, M. Nagl, G. Rozenberg, A. Rosenfeld (Eds.), Graph-Grammars and Their Application to Computer Science. VIII, 609 pages. 1987.

Vol. 292: The Munich Project CIP. Volume II: The Program Transformation System CIP-S. By the CIP System Group. VIII, 522 pages. 1987.

Vol. 293: C. Pomerance (Ed.), Advances in Cryptology — CRYPTO '87. Proceedings. X, 463 pages. 1988.

Vol. 294: R. Cori, M. Wirsing (Eds.), STACS 88. Proceedings, 1988. IX, 404 pages. 1988.

Vol. 295: R. Dierstein, D. Müller-Wichards, H.-M. Wacker (Eds.), Parallel Computing in Science and Engineering. Proceedings, 1987. V, 185 pages. 1988.

Lecture Notes in Computer Science

Vol. 245: H.F. de Groote, Lectures on the Complexity of Bilinear Problems. V, 135 pages. 1987.

Vol. 246: Graph-Theoretic Concepts in Computer Science. Proceedings, 1986. Edited by G. Tinhofer and G. Schmidt. VII, 307 pages. 1987.

Vol. 247: STACS 87. Proceedings, 1987. Edited by F.J. Brandenburg, G. Vidal-Naquet and M. Wirsing. X, 484 pages. 1987.

Vol. 248: Networking in Open Systems. Proceedings, 1986. Edited by G. Müller and R.P. Blanc. VI, 441 pages. 1987.

Vol. 249: TAPSOFT '87. Volume 1. Proceedings, 1987. Edited by H. Ehrig, R. Kowalski, G. Levi and U. Montanari. XIV, 289 pages. 1987.

Vol. 250: TAPSOFT '87. Volume 2. Proceedings, 1987. Edited by H. Ehrig, R. Kowalski, G. Levi and U. Montanari. XIV, 336 pages. 1987.

Vol. 251: V. Akman, Unobstructed Shortest Paths in Polyhedral Environments. VII, 103 pages. 1987.

Vol. 252: VDM '87. VDM – A Formal Method at Work. Proceedings, 1987. Edited by D. Bjørner, C.B. Jones, M. Mac an Airchinnigh and E.J. Neuhold. IX, 422 pages. 1987.

Vol. 253: J.D. Becker, I. Eisele (Eds.), WOPPLOT 86. Parallel Processing: Logic, Organization, and Technology. Proceedings, 1986. V, 226 pages. 1987.

Vol. 254: Petri Nets: Central Models and Their Properties. Advances in Petri Nets 1986, Part I. Proceedings, 1986. Edited by W. Brauer, W. Reisig and G. Rozenberg. X, 480 pages. 1987.

Vol. 255: Petri Nets: Applications and Relationships to Other Models of Concurrency. Advances in Petri Nets 1986, Part II. Proceedings, 1986. Edited by W. Brauer, W. Reisig and G. Rozenberg. X, 516 pages. 1987.

Vol. 256: Rewriting Techniques and Applications. Proceedings, 1987. Edited by P. Lescanne. VI, 285 pages. 1987.

Vol. 257: Database Machine Performance: Modeling Methodologies and Evaluation Strategies. Edited by F. Cesarini and S. Salza. X, 250 pages. 1987.

Vol. 258: PARLE, Parallel Architectures and Languages Europe. Volume I. Proceedings, 1987. Edited by J.W. de Bakker, A.J. Nijman and P.C. Treleaven. XII, 480 pages. 1987.

Vol. 259: PARLE, Parallel Architectures and Languages Europe. Volume II. Proceedings, 1987. Edited by J.W. de Bakker, A.J. Nijman and P.C. Treleaven. XII, 464 pages. 1987.

Vol. 260: D.C. Luckham, F.W. von Henke, B. Krieg-Brückner, O. Owe, ANNA, A Language for Annotating Ada Programs. V, 143 pages. 1987.

Vol. 261: J. Ch. Freytag, Translating Relational Queries into Iterative Programs. XI, 131 pages. 1987.

Vol. 262: A. Burns, A.M. Lister, A.J. Wellings, A Review of Ada Tasking. VIII, 141 pages. 1987.

Vol. 263: A.M. Odlyzko (Ed.), Advances in Cryptology – CRYPTO '86. Proceedings. XI, 489 pages. 1987.

Vol. 264: E. Wada (Ed.), Logic Programming '86. Proceedings, 1986. VI, 179 pages. 1987.

Vol. 265: K.P. Jantke (Ed.), Analogical and Inductive Inference. Proceedings, 1986. VI, 227 pages. 1987.

Vol. 266: G. Rozenberg (Ed.), Advances in Petri Nets 1987. VI, 451 pages. 1987.

Vol. 267: Th. Ottmann (Ed.), Automata, Languages and Programming. Proceedings, 1987. X, 565 pages. 1987.

Vol. 268: P.M. Pardalos, J.B. Rosen, Constrained Global Optimization: Algorithms and Applications. VII, 143 pages. 1987.

Vol. 269: A. Albrecht, H. Jung, K. Mehlhorn (Eds.), Parallel Algorithms and Architectures. Proceedings, 1987. Approx. 205 pages. 1987.

Vol. 270: E. Börger (Ed.), Computation Theory and Logic. IX, 442 pages. 1987.

Vol. 271: D. Snyers, A. Thayse, From Logic Design to Logic Programming. IV, 125 pages. 1987.

Vol. 272: P. Treleaven, M. Vanneschi (Eds.), Future Parallel Computers. Proceedings, 1986. V, 492 pages. 1987.

Vol. 273: J.S. Royer, A Connotational Theory of Program Structure. V, 186 pages. 1987.

Vol. 274: G. Kahn (Ed.), Functional Programming Languages and Computer Architecture. Proceedings. VI, 470 pages. 1987.

Vol. 275: A.N. Habermann, U. Montanari (Eds.), System Development and Ada. Proceedings, 1986. V, 305 pages. 1987.

Vol. 276: J. Bézivin, J.-M. Hullot, P. Cointe, H. Lieberman (Eds.), ECOOP '87. European Conference on Object-Oriented Programming. Proceedings. VI, 273 pages. 1987.

Vol. 277: B. Benninghofen, S. Kemmerich, M.M. Richter, Systems of Reductions. X, 265 pages. 1987.

Vol. 278: L. Budach, R.G. Bukharajev, O.B. Lupanov (Eds.), Fundamentals of Computation Theory. Proceedings, 1987. XIV, 505 pages. 1987.

Vol. 279: J.H. Fasel, R.M. Keller (Eds.), Graph Reduction. Proceedings, 1986. XVI, 450 pages. 1987.

Vol. 280: M. Venturini Zilli (Ed.), Mathematical Models for the Semantics of Parallelism. Proceedings, 1986. V, 231 pages. 1987.

Vol. 281: A. Kelemenová, J. Kelemen (Eds.), Trends, Techniques, and Problems in Theoretical Computer Science. Proceedings, 1986. VI, 213 pages. 1987.

Vol. 282: P. Gorny, M.J. Tauber (Eds.), Visualization in Programming. Proceedings, 1986. VII, 210 pages. 1987.

Vol. 283: D.H. Pitt, A. Poigné, D.E. Rydeheard (Eds.), Category Theory and Computer Science. Proceedings, 1987. V, 300 pages. 1987.

Vol. 284: A. Kündig, R.E. Bührer, J. Dähler (Eds.), Embedded Systems. Proceedings, 1986. V, 207 pages. 1987.

Vol. 285: C. Delgado Kloos, Semantics of Digital Circuits. IX, 124 pages. 1987.

Vol. 286: B. Bouchon, R.R. Yager (Eds.), Uncertainty in Knowledge-Based Systems. Proceedings, 1986. VII, 405 pages. 1987.

Vol. 287: K.V. Nori (Ed.), Foundations of Software Technology and Theoretical Computer Science. Proceedings, 1987. IX, 540 pages. 1987.

Vol. 288: A. Blikle, MetaSoft Primer. XIII, 140 pages. 1987.

Vol. 289: H.K. Nichols, D. Simpson (Eds.), ESEC '87. 1st European Software Engineering Conference. Proceedings, 1987. XII, 404 pages. 1987.

Vol. 290: T.X. Bui, Co-oP A Group Decision Support System for Cooperative Multiple Criteria Group Decision Making. XIII, 250 pages. 1987.

Vol. 291: H. Ehrig, M. Nagl, G. Rozenberg, A. Rosenfeld (Eds.), Graph-Grammars and Their Application to Computer Science. VIII, 609 pages. 1987.

Vol. 292: The Munich Project CIP. Volume II: The Program Transformation System CIP-S. By the CIP System Group. VIII, 522 pages. 1987.

Vol. 293: C. Pomerance (Ed.), Advances in Cryptology — CRYPTO '87. Proceedings. X, 463 pages. 1988.

Vol. 294: R. Cori, M. Wirsing (Eds.), STACS 88. Proceedings, 1988. IX, 404 pages. 1988.

Vol. 295: R. Dierstein, D. Müller-Wichards, H.-M. Wacker (Eds.), Parallel Computing in Science and Engineering. Proceedings, 1987. V, 185 pages. 1988.

Lecture Notes in Computer Science

Edited by G. Goos and J. Hartmanis

333

Hartmut Noltemeier (Ed.)

Computational Geometry and its Applications

CG'88, International Workshop on Computational Geometry
Würzburg, FRG, March 24–25, 1988
Proceedings

Springer-Verlag

Editor

Hartmut Noltemeier
Lehrstuhl für Informatik I, Universität Würzburg
Am Hubland, D-8700 Würzburg, Federal Republic of Germany

CR Subject Classification (1987): B.6–7, C.3, E.1–2, F.1–2, F.4.1, G.2, H.1, H.2.8, H.3.3, I.2–6

ISBN 3-540-50335-8 Springer-Verlag Berlin Heidelberg New York
ISBN 0-387-50335-8 Springer-Verlag New York Berlin Heidelberg

PREFACE

The International Workshop CG'88 on "Computational Geometry" was held at the University of Würzburg (West Germany), March 24-25, 1988.

Previous meetings of a smaller and much more regional kind were organized 1983 in Zürich (J. Nievergelt), 1984 in Bern (W. Nef, H. Bieri) and 1985 in Karlsruhe (A. Schmitt, H. Müller).

As the interest in the fascinating field of Computational Geometry and its Applications has grown very quickly in recent years we felt the need to have a workshop, where a suitable number of invited participants could concentrate their efforts in this field to cover a broad spectrum of topics and to communicate in a stimulating atmosphere.

This workshop was attended by some fifty invited scientists. The scientific program consisted of 22 contributions, of which 18 papers with one additional paper (M. Reichling) are contained in the present volume.

The contributions covered important areas not only of fundamental aspects of Computational Geometry but a lot of interesting and most promising applications:

Algorithmic Aspects of Geometry
Arrangements
Nearest-Neighbor-Problems and Abstract Voronoi-Diagrams
Data Structures for Geometric Objects
Geo-Relational Algebra
Geometric Modeling
Clustering and Visualizing Geometric Objects
Finite Elements Methods, Triangulating in Parallel,
Animation and Ray Tracing
Robotics: Motion Planning, Collision Avoidance
Visibility
Smooth Surfaces
Basic Models of Geometric Computations
Automatizing Geometric Proofs and Constructions.

The scientific program was accompanied by presentations of some Graphics- and CAD-software systems (University of Karlsruhe and Würzburg) which attracted similar interest to the demonstration of a computer assisted course on "Geometric Algorithms" (by Th. Ottmann and his coauthors, University of Freiburg).

I am very indebted to all participants of the workshop and especially to all contributors to this volume.

Thanks are also due to Hugo Heusinger, who carried much of the burden of preparing the workshop, and to all my fellows and staff members of Lehrstuhl für Informatik I, who did their best for a smooth local organization.

Last but not least I owe thanks to Springer-Verlag and to Dr. Hans Wössner for their support.

Würzburg, July 1988 Hartmut Noltemeier

CONTENTS

Using graphical information from a grid file's directory to visualize patterns in Cartesian product spaces

ETH Zürich
Institut für Informatik
Fachgruppe Wissenschaftliches Rechnen
H. Hinterberger

Abstract

Data management and data visualization - each an extensively explored field - are rarely considered in conjunction with each other. This is not surprising as they apparently lack a common ground. By example of a region directory, a particular directory for the grid file, we show how an elementary measure, namely data density, can usefully serve data management and data visualization by providing the basis for a common structure. The region directory is a data structure that can - at no extra cost - provide graphical information useful to visualize structures in multivariate data sets with approximate reproductions of the data's variably dense regions. If the centers of these regions are visualized with parallel coordinates, regions with higher or lower than average density (clusters and voids) can be approximately localized. Graphical information for these visualizations is readily available in the region directory, giving the grid file the potential for rapid abstractions from large amounts of data. It is especially noteworthy that the data themselves are not accessed for these displays.

Keywords: File structures, computational geometry, multidimensional data, data visualization, multivariate data analysis, graphical data analysis, data density.

Content

1. Parallel coordinates: A novel method for higher-dimensional computational geometry

Motivations to obtain geometrical models of multivariate relationships come from many directions. Based on requirements in *data analysis,* we discuss geometric models (planar diagrams) of relationships in *N* variables, introduced by Alfred Inselberg [Ins 85] who called the underlying method *parallel coordinates.* For *N*-dimensional Euclidean space $\mathbf{R}^N$ a coordinate system is constructed as follows. On the plane with *xy*-coordinates, *N* real lines, labelled $x_1, x_2, \ldots, x_N$, are placed equidistant and perpendicular to the *x*-axis. They are the axes of the parallel coordinate system and have the same positive orientation as the *y*-axis (see Fig. 1). A point *C* with coordinates $(c_1, c_2, \ldots, c_n)$ is represented by the polygonal line whose *N* vertices are at $(i-1, c_i)$ on the x_i-axis for $i = 1, \ldots, N$. In effect, a one-to-one correspondence between points in $\mathbf{R}^N$ and planar polygonal lines with vertices on $x_1, x_2, \ldots, x_N$ is established.

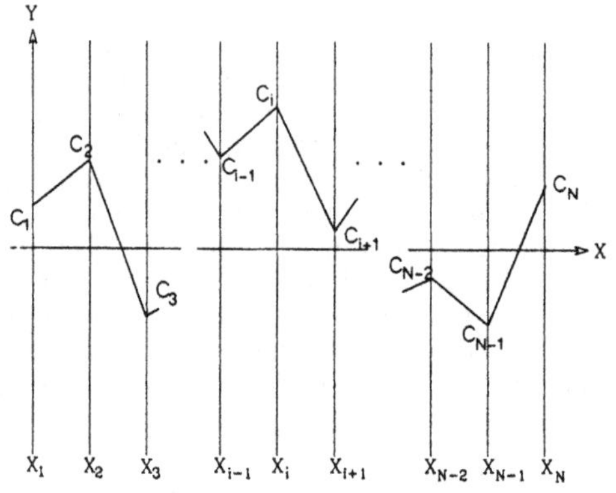

Figure 1 Parallel axes for $\mathbf{R}^N$ (from [Ins 85]).

Points and lines in the plane: The fundamental duality

Points in Cartesian coordinates map into *lines* in parallel coordinates while *lines* in Cartesian coordinates map into *points* in parallel coordinates. Consider the x_1x_2-plane with parallel as well as Cartesian coordinates (see Fig. 2) and the line $l: x_2 = mx_1 + b, \; m < \infty$. The points on *l* as represented in parallel coordinates form an infinite family of lines. When $m \neq 1$, any two of these lines intersect at the point $\bar{l}: (1/1-m, b/1-m)$. The distance between the parallel axes is one and the coordinates of $\bar{l}$ are given with respect to the *xy*-Cartesian coordinates. The line *l* is therefore represented by the point $\bar{l}$ in parallel coordinates. The correspondence for vertical lines is $l: x_1 = c$ <—> $\bar{l}$: (0,c). Lines with $m = 1$ do not have a corresponding point representation in the Euclidean plane. However, considering *xy* and x_1x_2 as two copies of the *projective plane,* points on the line $l: x_2 = x_1 + b$ correspond to parallel lines intersecting at the *ideal point* $\bar{l}$ with slope $y/x = b$.

Parallel lines and the ideal line

In Fig. 3 we see that $\bar{l}$ is to the right of the x_2 parallel coordinate axis for $0 < m(l) < 1$, on the strip between the axes for $m(l) < 0$, and to the left of the x_1-axis when $m(l) > 1$. Horizontal and vertical lines are represented by points on the x_2-axis and the x_1-axis respectively. Let P_m^∞ denote the ideal point where all lines with slope *m* meet. Then, its dual $\overline{P_m^\infty}$ is a vertical line in the *xy*-plane at $x = 1/1-m$, as shown in Fig. 3.

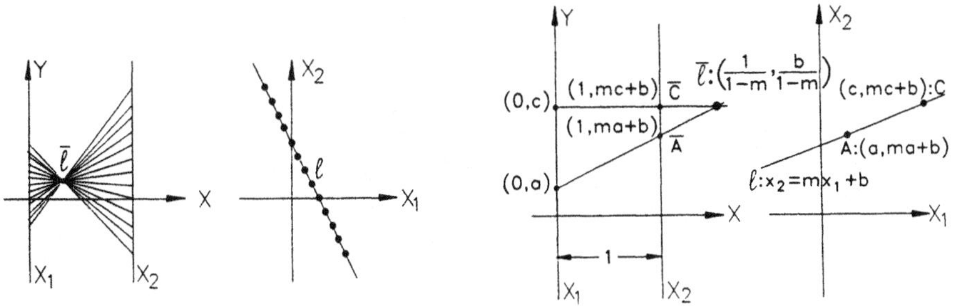

Figure 2 Parallel coordinates induce dualisms in the plane (from [Ins 85]).

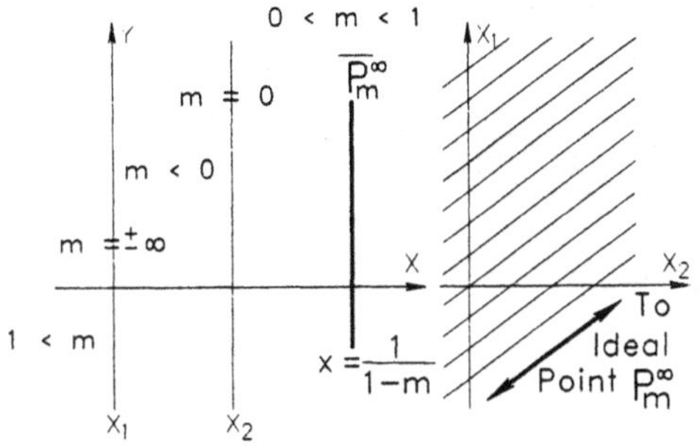

Figure 3 The slope m of l determines the x-coordinate of $\overline{l}$ (from [Ins 85]).

Segments, intersections, and interior points

As shown in Fig. 4, for $x_1(A) < x_1(B)$ the interval $[A, B] \subset l$ in parallel coordinates looks like a "fan" caused by the two lines $\overline{A}$ and $\overline{B}$ intersecting at the point $\overline{l}$. The point $P = l_1 \cap l_2$ in parallel coordinates is represented by the segment $\overline{P}$, between the parallel axes, on the line joining the points l_1 and l_2.

Interior points of convex polygons can be found in parallel coordinates based on the observation shown in Fig. 5. Let the lines l_i , i = 1,2, and the vertical line $v : x_1 = a$ be given. The segments $\overline{P_i}$, where $P_i = v \cap l_i$, are found in the way shown in Fig. 4. Consequently, a point P between l_2 and l_1, with $x_1(P) = a$, is represented by the segment $\overline{P}$ on $\overline{v}$ with $x_2(P_1) <= (P) <= x_2(P_2)$.

Further dualities

If the point conic in the original Cartesian coordinate plane is an ellipse, the image in the parallel coordinate plane is a line hyperbola with a point hyperbola as envelope. Rotations in Cartesian coordinates become translations in parallel coordinates and vice versa. Points of inflection in Cartesian space become cusps in parallel coordinate space (see Fig. 6) and vice versa, making a relatively hard to detect property of a function more manageable with parallel coordinates.

The illustrations in this section have been restricted to the 2-dimensional case for reasons of simplicity. Results from extensions of the method to $\mathbf{R}^N$ are discussed and illustrated in [Ins 85].

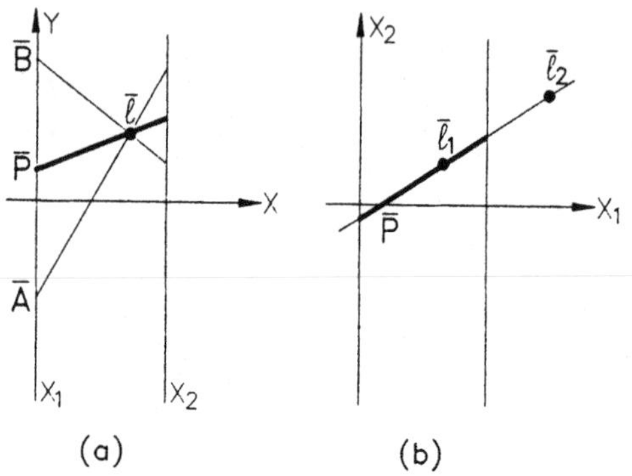

Figure 4 Segments (a) and intersecting lines (b) in parallel coordinates (from [Ins 85]).

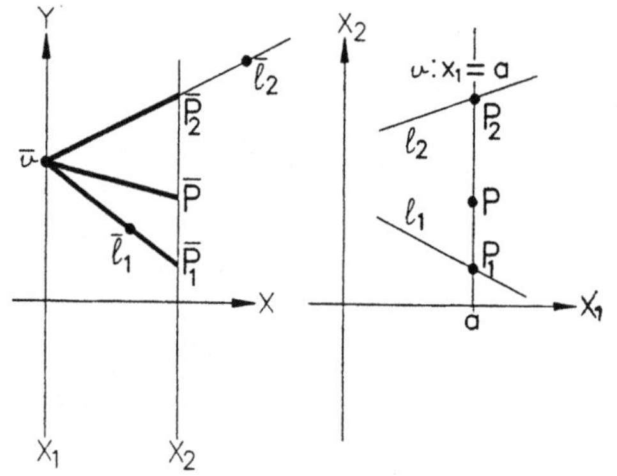

Figure 5 Point between two lines in parallel coordinates (from [Ins 85]).

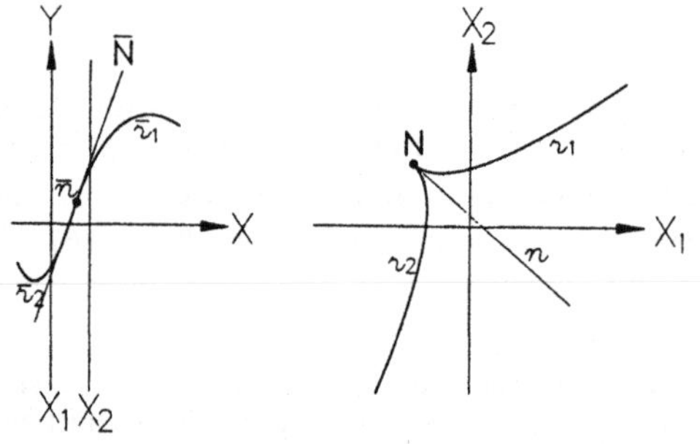

Figure 6 Cusps map into inflection points (from [Ins 85]).

2. Multivariate exploratory data analysis with parallel coordinates

The *exploratory* data analysis paradigm favors graphical representations to visualize data because, contrary to *confirmatory* data analysis (what most books call statistics), the recognition of the data's structural characteristics is considered more important than the presentation of accurate detail.

Exploratory data analysis based on projection techniques, including 2- and 3-dimensional scatter diagrams, is potentially misleading (e.g. projecting the hypervolume of a thin shell onto a 2-plane will not reveal its true structure). The cause of the failure of the standard Cartesian coordinate representation is the requirement for orthogonal coordinate axes. It would be highly desirable to have a simultaneous representation of all coordinates of a data vector. The parallel coordinates, introduced in the previous section, provide such a simultaneous representation. Furthermore, this method treats all components in a similar manner. E.J. Wegman [Weg 86] shows how parallel coordinates can be used for exploratory data analysis. The following diagrams illustrate the interpretation of parallel coordinates in a data analytical setting (all from [Weg 86]).

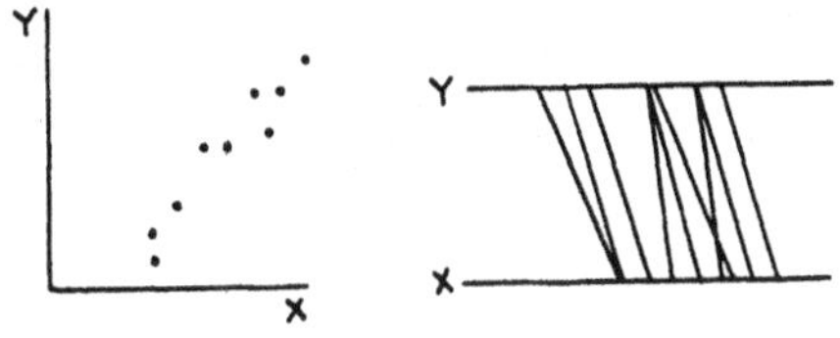

Data points with large positive correlation

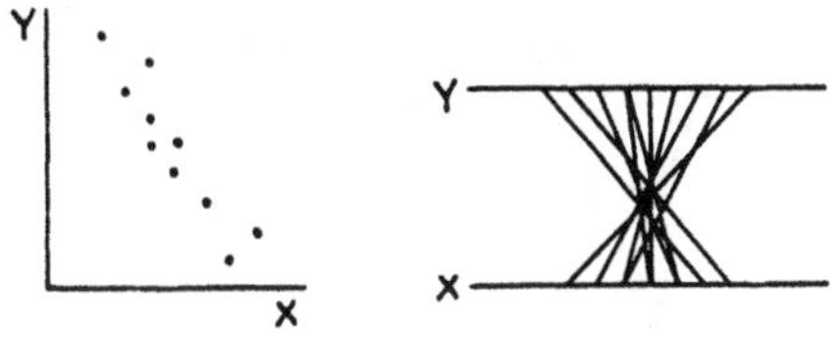

... with large negative correlation

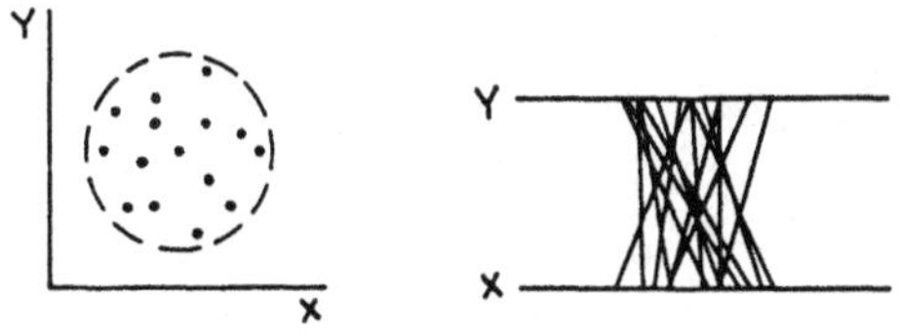

Uncorrelated data tend to an approximately circular convex ball

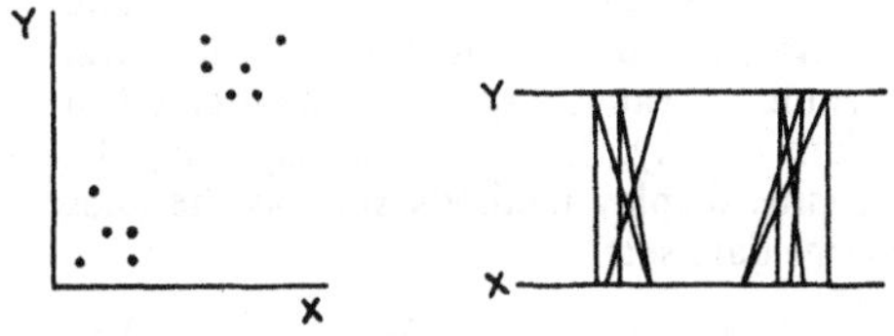

Clustered data separated in both *x* and *y*

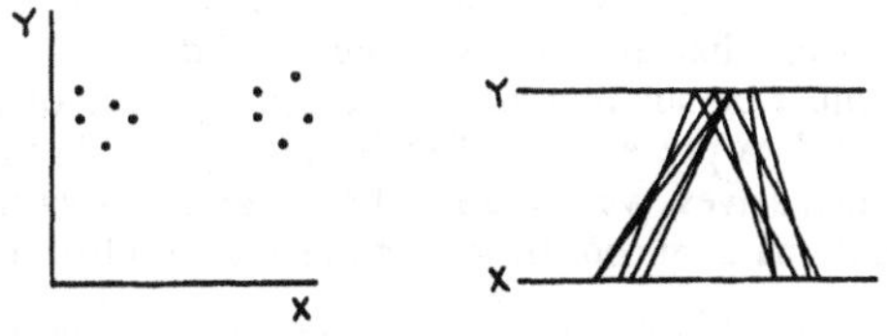

... separated in *x* but not in *y*

Figure 7 illustrates how parallel coordinates can be used to visualize relationships among five automobile data variables. The shaded area highlights the group of cars imported to the US.

The geometric symmetry inherent in this display method readily supports permutation of the axes to provide adjacency of different variables for easy comparison. Although there are $n!$ permutations, many of these duplicate adjacencies and in [Weg 86] E. Wegman shows that $^{n+1}/_2$ permutations are needed to insure adjacency of every pair of axes. This compares favorably with the $^{n!}/_{2!(n-2)!}$ presentations needed to make pairwise comparisons using scatterplots

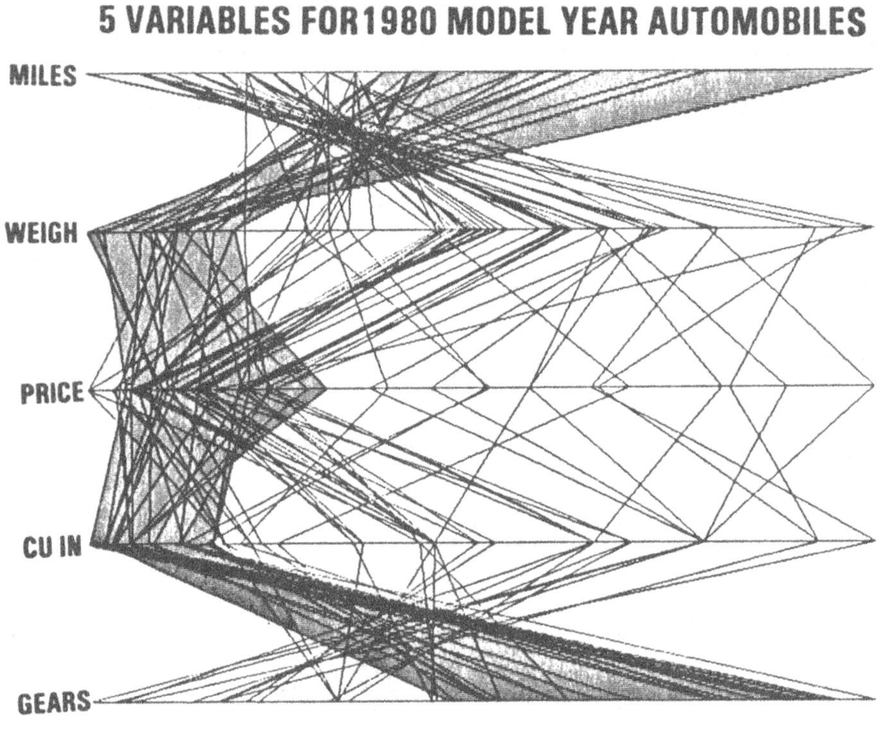

Figure 7 A parallel coordinate plot of five-dimensional automobile data (from [Weg 86]).

This method obviously restricts the number of data items that can be displayed in one diagram. This means that the large volumes of today's data collections, particularly scientific and statistical ones, must be suitably abstracted before these display techniques can be applied. One method to abstract data presents itself with the measure known as *data density*. If, instead of displaying the data themselves, we visualize their regions of variable density, display methods such as the ones based on parallel coordinates can also be applied to very large data sets.

The next section discusses ways to visualize data densities. Then, in the last section, we show that, if the data is suitably managed, information about the data's regions of variable density can be obtained directly from the file directory. This means that a first evaluation of the data, examining it for possible structures for example, can be made without accessing the data themselves!

3. Analyzing multivariate data based on variably dense data regions

It is well known that research institutions are generating and collecting data at a rate far outstripping their capabilities of representing it. According to a Wall Street Journal report (13. January 1988), scientists have looked at only 10% of the data that spacecraft have sent back to earth. What is more disturbing, they have closely analyzed only 1% of the mountain of tape. This state of affairs leads to the frequently heard conclusion that improved data visualization is needed to maintain effective R&D capabilities in engineering, space exploration, medicine and other areas of science. We suggest, that an efficient management of large volumes of multivariate, dynamic data and their successful exploratory analysis depends on suitable order- and structure-preserving abstractions. We show one way how this can be achieved.

Traditional display methods, based on visualizing individual data items, are limited by graphical resolution. Scatterplots, for example, can become problematic when more than 6 attributes and more than 1000 data items must be visualized. Some iconic methods can handle up to 16 or 20 attributes reasonably well but difficulties arise when more than 100 such data items must be displayed simultaneously. The need for data abstraction is evident, however many known data compression techniques cannot provide useful graphical information because they destroy the data's inherent order or its structure or possibly both. Methods based on data density are different, because they "compress" large multivariate data sets symmetrically, thus preserving information in each dimension.

Data point clouds considered as regions of variable density

We generalize density $d = m/v$, to let the mass m represent any number of points in a given Cartesian product space, and volume v represent contiguous regions in a geometric interpretation of this same space. This generalization of *data (point) density* is analogous to the concept of *probability density* in mathematical statistics.

Densities can be investigated in two ways:

1) Method **M** holds volume constant and observes *changes in mass*.
2) Method **V** observes *changes in volume* for a given constant mass.

Method **M** is frequently used when data distributions are approximated with visualizations of variably dense regions (e.g. discretized pictures). We introduce new data visualization techniques based on method **V** by exploiting its close relation to graphical information available in the *region directory* of a grid file (described in [Hin 87]). Using two variables (sepal length and sepal width) from Anderson's [And 36] four-dimensional iris data shown in Fig. 8, the two methods can be illustrated as shown in Fig. 9 where the small regions produced by method **V** to represent high densities are enhanced with bold outlines (see [Hin 87] for more detail).

Method **V** has an inherent simplicity, useful for graphical presentations, because only one type of information must be maintained, namely volume. Method **V**'s difficulty lies in the problem of finding a space partitioning such that each resulting region contains unit mass. Method **M** is simple by comparison because it reduces to counting. However, both mass and volume must be maintained for visualizations using constant volume.

Data density information in matrix directories

With the term "*matrix directory*" we refer to file directories which partition the data's embedding space in an orthogonal fashion, and maintain this partition with a data structure having only one logical level. If, in addition, a correspondence between regions of this grid-like partition and fixed-sized data buckets exists, i.e. records in a given bucket come from the same convex, box-shaped, region (bucket region) in the data space, then a matrix directory approximates data density in the way shown in Fig. 10.

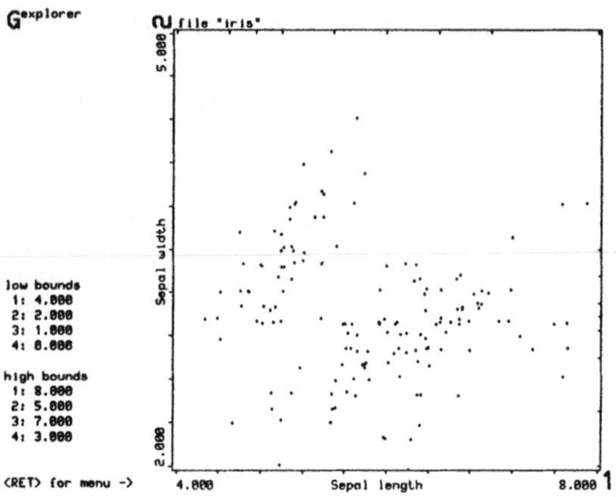

Figure 8 Scatterplot showing the relationship between sepal length and sepal width in a sample of three iris species observed by E. Anderson.

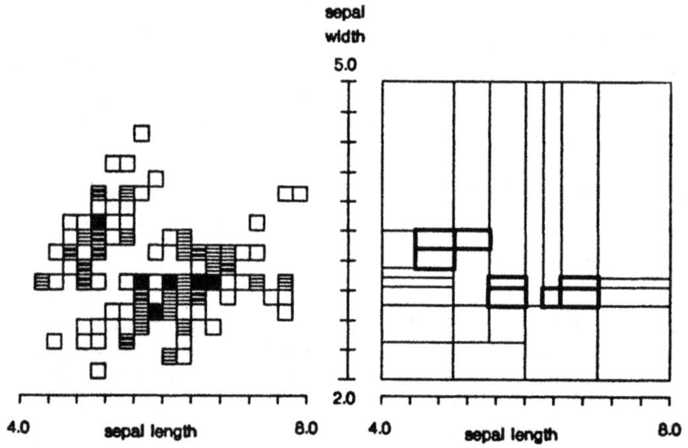

Method **M**

$d = \mathbf{m} / v$

Method **V**

$d = m / \mathbf{v}$

Figure 9 Representation of data densities based on observation of mass (left) or volume (right).

The representation of data density with regional information from a matrix directory, where unit mass is represented by a data bucket, is an approximation, requiring some necessary idealizing assumptions with respect to the representation of mass and volume.

Mass

The unit of measurement for *mass* is the grid file's bucket capacity *c*, with $1 <= c <= m$; where m is the maximum number of records, typically $1 < c << m$. Bucket capacity *c* is a theoretical representation of mass. In practice we have to deal with a variable bucket occupancy *b* with $1 <= b <= c$, where *c* is bucket capacity. Therefore, we consider bucket occupancy *b* a random variable and let its mean, the average bucket occupancy, represent unit mass.

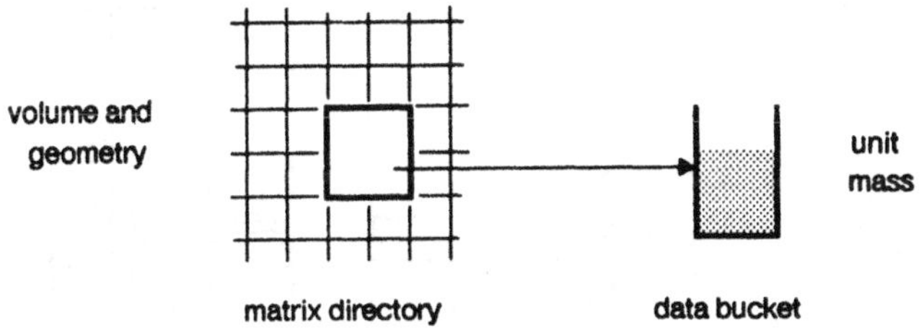

Figure 10 Data density *d* as represented by a matrix directory, using method **V**.

Volume

Method **V** must be conditioned to restrict the geometry of its regions. One method of conditioning can be found in the form of search structures (directories) for data files. These data structures partition the search space in an algorithmically tractable way into regions whose geometries are characteristic for each type of directory. Many multidimensional directories (quad-trees, oct-trees, grid or cell directories) partition the search space suitably to create regions for method **V**. Three constraints, however, must be met to make this approach sensible for *practicable* applications:

1) In the interest of storage efficiency only directories without empty buckets, maintaining average bucket occupancies typically > 50% are considered.
2) Rapid processing requires that the directory provides information about the volume and the location of a region with a single directory entry.
3) In order to accept data sets without restrictions it is mandatory that the directory adapts itself gracefully to any data distribution.

With the region directory of a grid file, a multidimensional, general purpose data structure satisfying these constraints has been introduced. The grid file is a dynamic data structure, modifying its grid partition while the data are being inserted. This means that the $n!$ different orders in which we can insert n records in a grid file with bucket capacity c, can produce potentially $^{n!}/_{c!(n-c)!}$ different bucket region configurations. The 117 unique (two-dimensional) records of the iris data stored in our grid file with bucket capacity 7 could therefore produce approximately 5×10^{10} different representations. Six of these are shown in Fig. 11. One would expect, however, that a given splitting policy reduces this number of possible configurations substantially.

The geometric interpretation of a Cartesian product space's partition is not restricted to two-dimensional data as illustrated so far, but extends to any dimension because Euclidean vector spaces do so by definition. Consequently, method **V** can be applied to k-dimensional data, for $k >$ 2. However, extending the data space's dimensionality beyond two, has serious consequences for data displays. In [Hin 87], the author discusses four methods to visualize convex, box-shaped regions from applications of method **V** to higher-dimensional data. One of them, described in the next section, is based on parallel coordinates and the observation that less detail must be displayed if a bucket region is reduced to its geometric center: A point in the Euclidean vector space $\mathbf{R}^k$. This corresponds to a geometric interpretation of a k-dimensional grid directory.

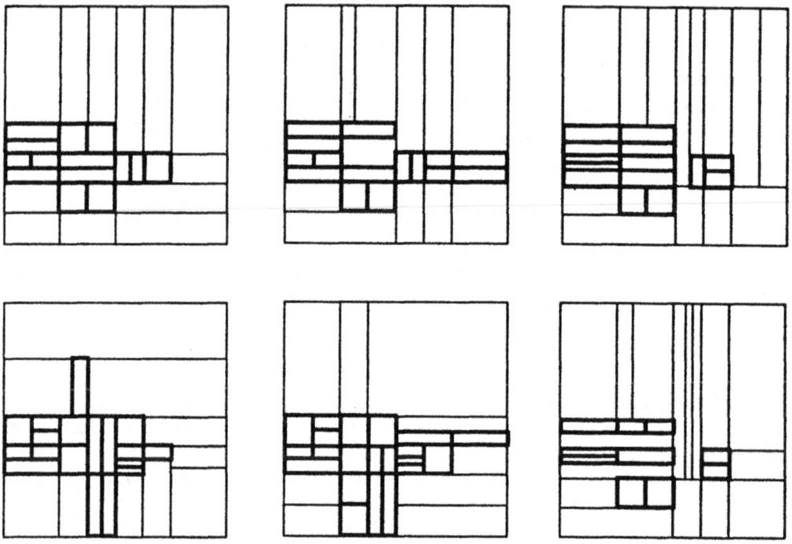

Figure 11 The result of inserting the iris data in a two-dimensional grid file in six different orders. All dense regions (bold outlines) approximate the data clouds, but do it in different ways.

4. Visualizing centers of variably dense data regions without accessing the data

The grid file, as introduced in [Nie 84], is often being accused of providing adaptability only at the cost of superlinear directory growth. With the region directory [Hin 87], however, we have found a data structure to implement the grid file's grid array so that its growth linearly depends on the number of data items stored, whatever the data's distribution. The region directory is different in that it maintains coordinates of the bucket regions instead of the grid arrays' elements.

Each bucket region, by virtue of its definition, can be uniquely identified with two of its vertices as long as they are diagonally opposite. In a k-dimensional bucket region two vertices are diagonally opposite when each vertex is defined as the intersection of k interval boundaries and no interval boundary is common to both vertices. We call them the *near vertex* and the *far vertex*, and adopt the convention that from two parallel interval boundaries, the one closer to the origin of the search space will be assigned to the near vertex. Figure 12 illustrates the two-dimensional case. An entry of the region directory then consists (at least) of a bucket region's near and far vertex and the address of the bucket corresponding to this region. The coordinates of the near and far vertex provide precisely the graphical information needed to visualize the regions required for method **V**, furthermore, they are available in the directory which means that we can obtain information about the data's distribution fast, without accessing the data themselves!

A grid directory based on bucket regions cannot be managed efficiently with a multidimensional array, the data structure suggested in the original grid file report. As alternative we choose to organize the grid directory as a table of bucket regions ordered, say, by near vertex. During a search, the grid block identified by a given scale interval combination is now *not* used to calculate the position of an element in a multidimensional array, but serves as search argument when passing through the region directory to find this grid block's bucket region.

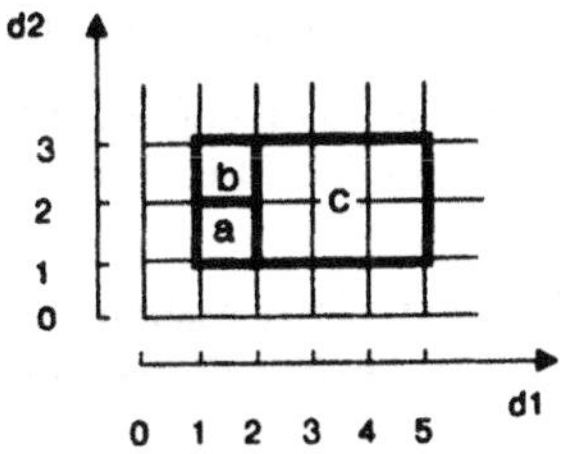

bucket region	near vertex	far vertex
a	1, 1	2, 2
b	1, 2	2, 3
c	2, 1	5, 3

Figure 12 The bucket regions can be identified with near and far vertices.

Clusters and voids

Clusters of data points can be looked at as contiguous regions in the data space with higher than average density and, ideally, any method used in cluster analysis should tell us something about the geometry of these regions. This requirement makes methods visualizing variably dense regions in the data space based on graphical information about these regions interesting alternatives. Small regions of method **V**, for instance, can indicate clusters or parts thereof. Furthermore, traditional cluster analysis methods are ill-suited to find regions in the data space with lower than average density (holes). We call such sparsely occupied regions *voids*. Large bucket regions in the region directory point to voids, making the method to approximate data distributions based on data density nicely symmetrical.

To approximately locate clusters or voids in higher-dimensional Cartesian product spaces, it suffices to visualize centers of bucket regions representing either high or low data densities. We achieve this in the following way using parallel coordinates. Each axis of the parallel coordinate system represents a key attribute domain of the grid file. The axis' partition is identical to the corresponding scale's partition. Figure 13 illustrates the method with two hypothetical bucket regions. At the left we list the near and far vertices, in the middle their centers expressed in Cartesian coordinates, and at the right these two centers are plotted with parallel coordinates.

Cartesian coordinates

	near vertex	[0, 1, 1, 4]
A	far vertex	[3, 3, 5, 5]
	region center	[1.5, 2, 3, 4.5]
	near vertex	[3, 2, 1, 1]
B	far vertex	[7, 3, 2, 4]
	region center	[5, 2.5, 1.5, 2.5]

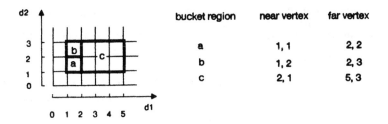

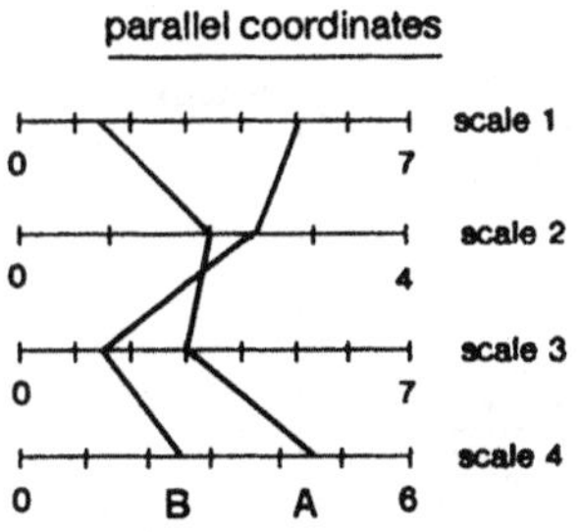

Figure 13 Visualizing the center of two hypothetical, four-dimensional bucket regions.

The difficulty of separating different region–centers meeting at the same point on one or more coordinate axes can be overcome by various means. First, different permutations of the coordinate axes' order often "disentangle" a display. Second, the coordinate points can be plotted "jittered" – a technique often used to avoid overplotting in scatter diagrams. And third, colored lines can be used to visually separate individual region–centers.

Figure 14 illustrates region–centers with data from the iris file. Upon closer examination, the two major clusters (one drawn with solid, the other with dotted lines) inherent in the data show up distinctly.

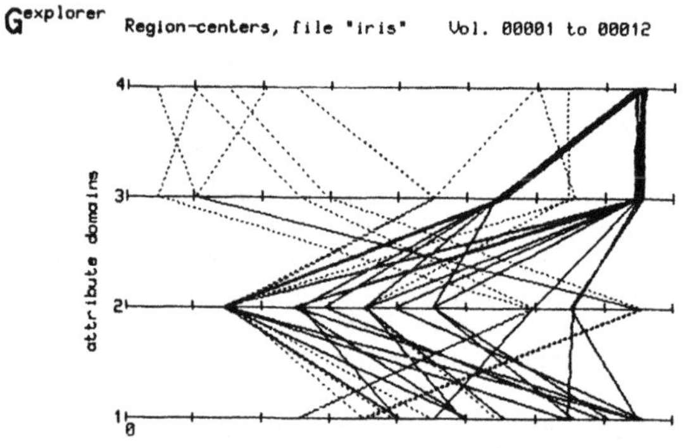

Figure 14 Centers of bucket regions with higher than average density from the grid file storing the iris data, depicted with parallel coordinates.

One additional, representative example shall illustrate the method's usefulness to analyze large sets of multivariate data. Further case study results can be found in [Hin 87]. Figure 15 shows a scatter diagram, plotting the relationship between *CO*–concentrations and *NO*–concentrations obtained during experiments with different types of wood-burning furnaces. These are "raw" experimental data (not "cleaned" or scaled) of which we would like to get a first impression.

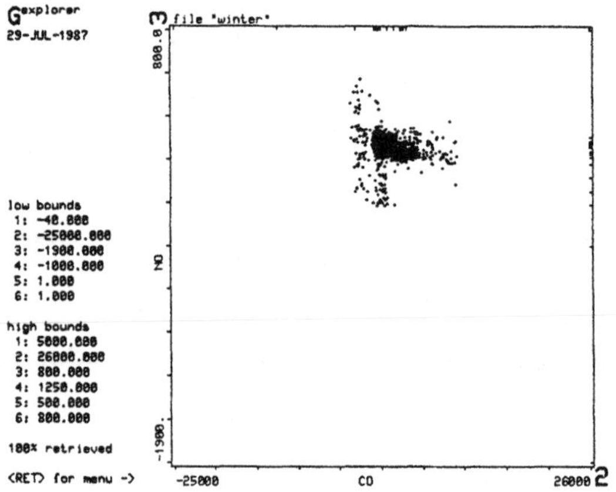

Figure 15 Scatterplot of "raw" emmission control data.

As a first step we display approximately 7% of the bucket regions with the highest density using the region-centers display shown in Fig. 16. The crossing line segments connecting dimensions 2 and 3 - a strong indicator for diagonally opposite clusters - justify a revised scatterplot of *CO*- vs. *NO*-concentrations with lower and upper bounds different from the ones used for the plot shown in Fig. 15. The result is shown in Fig. 17. The two predicted clusters show up in fact, demonstrating the proposed method's usefulness to explore multivariate data fast, based on "free" information (automatically) abstracted from the data.

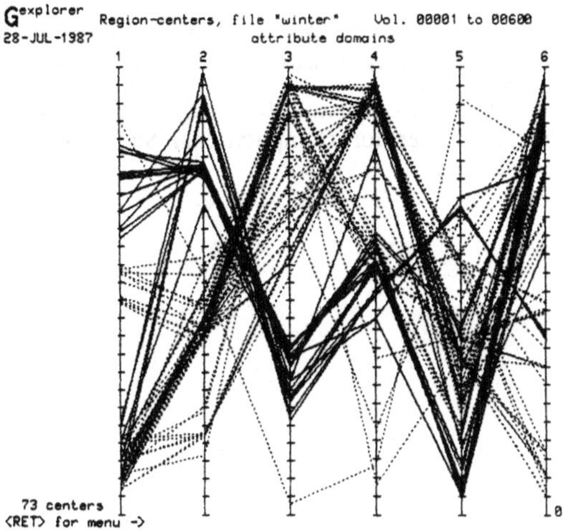

Figure 16 Region-centers display based on information in the region directory of the grid file storing the emission control data's 2908 records shown in Fig. 15.

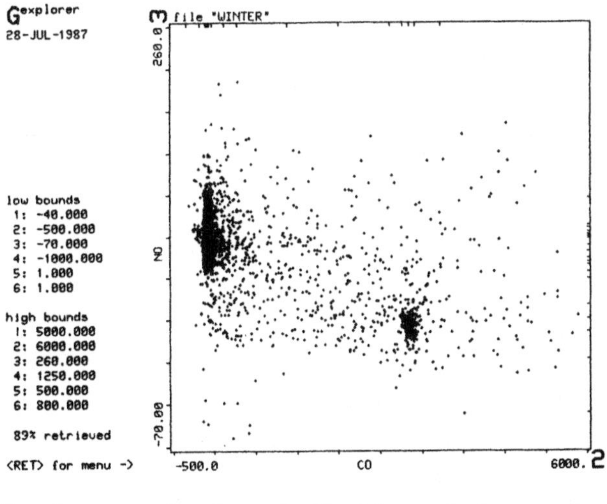

Figure 17 Expanding the scatterplot shown in Fig. 15 reveals the two clusters expected from a characteristic pattern, observable in the region-centers display shown in Fig. 16.

References

[And 36] E. Anderson (1936) The species problem in Iris. *Ann. Mo. bot. Gdn.*, Vol. 23, pp. 511–525.

[Hin 87] H. Hinterberger (1987) *Data Density: A Powerful Abstraction to Manage and Analyze Multivariate Data*. Diss. ETH Nr.: 8330, Verlag der Fachvereine, Zurich.

[Ins 85] A. Inselberg (1985) The plane with parallel coordinates. *The Visual Computer*, Vol. 1, pp. 69–91.

[Nie 84] J. Nievergelt, H. Hinterberger, K.C. Sevcik (1984) The Grid File: An adaptable, symmetric multi-key file structure. *ACM Trans. on Database Systems*, Vol. 9, No. 1, pp. 38–71.

[Weg 86] E.J. Wegman (1986) *Hyperdimensional Data Analysis Using Parallel Coordinates*. Technical Report No. 1, Center for Computational Statistics and Probability, George Mason University, Fairfax, VA.

Twin Grid Files: A Performance Evaluation *

Andreas Hutflesz
Universität Karlsruhe
Postfach 6980
D - 7500 Karlsruhe

Hans-Werner Six
FernUniversität Hagen
Postfach 940
D - 5800 Hagen

Peter Widmayer
Universität Karlsruhe
Postfach 6980
D - 7500 Karlsruhe

Abstract

Data structures for the physical organization of multidimensional points on secondary storage usually suffer from a fairly low average storage space utilization, even for independently and uniformly distributed points. We evaluate the performance of two types of space optimizing access schemes, the optimal static and the suboptimal dynamic twin grid files. It turns out that in comparison with the (standard) grid file, twin grid files achieve considerable savings in storage space, without losing efficiency in any other relevant aspect. This is shown to hold for typical operations in a variety of practical situations. To better grasp the inherent power of twin grid files, variations of crucial parameter settings are studied in detail.

1 Introduction

In non-standard databases for geometric objects, efficient external storage access schemes are needed. For sets of multidimensional points, insertion, deletion, range query, and exact match are the predominant operations. A variety of access structures supporting these operations has been proposed [1,2,3,6,7,9,10,11]. They suffer, however, from a fairly low average storage space utilization of roughly 69%, like many other schemes based on recursive halving [8]. Storage space utilization is of major concern especially for an almost static set of points, like e.g. in geographic databases. Recently, we have proposed a general technique, the twin principle, for deriving storage space efficient static structures, by pairing known access schemes for points, and explained it on the basis of the grid file structure [5]. The optimal static twin grid file consists of two grid files, both spanning the entire data space; the points to be stored are partitioned among the two grid files so as to minimize the total number of buckets used. To support insertions and deletions as well, we have presented a dynamic variant of the twin principle on a grid file basis, where insertions and deletions trigger the local (and therefore suboptimal) redistribution of points among a pair of grid files [4].

A performance evaluation for static and dynamic twin grid files has been carried out to an extent that allows to judge whether the twin principle merits further study. Before actually using twin access schemes in non-standard databases, several issues remain to be clarified. Among these,

* This work was partially supported by grants Si 374/1 and Wi 810/2 from the Deutsche Forschungsgemeinschaft

a detailed study of the influence of parameter variations on twin structure behavior (or its robustness against such variations) is needed; this is the topic of this paper.

For the purpose of reminding the reader of twin grid files, we briefly review the major properties of optimal static twin grid files in the next section, and of suboptimal dynamic twin grid files in Section 3. In Section 4, we present our performance evaluation in some detail. Beforehand, however, let us recall the basic twin structure idea. Whenever convenient, we deliberately quote literally from [4] and [5].

Let us restrict our attention in this paper to two-dimensional points. Consider a set of points in the plane to be stored in a grid file. The storage space for each point as well as the bucket size are fixed. The bucket capacity b is the maximum number of points that can be stored in one bucket; in our example, let $b = 3$. Empty buckets are not stored explicitly on secondary storage; instead, an empty region is represented by a special directory entry, the so-called dummy.

Figure 1.1 (a) shows five points, stored in a standard grid file using three buckets. This is clearly more than the minimum of two buckets, required to store five points. To save storage space, we distribute the points among two grid files, the *primary grid file* P and the *secondary grid file* S, both spanning the entire data space. The primary and the secondary grid file together constitute the twin grid file. By using only one bucket for each grid file, the five example points can be stored in two buckets (see Figure 1.2 (a)).

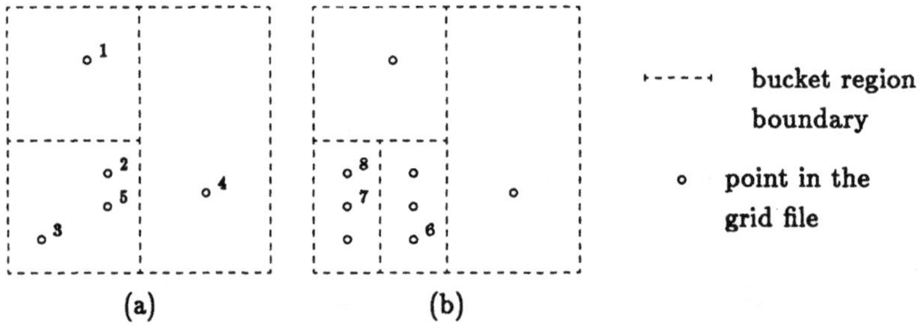

Figure 1.1: A standard grid file ($b = 3$)

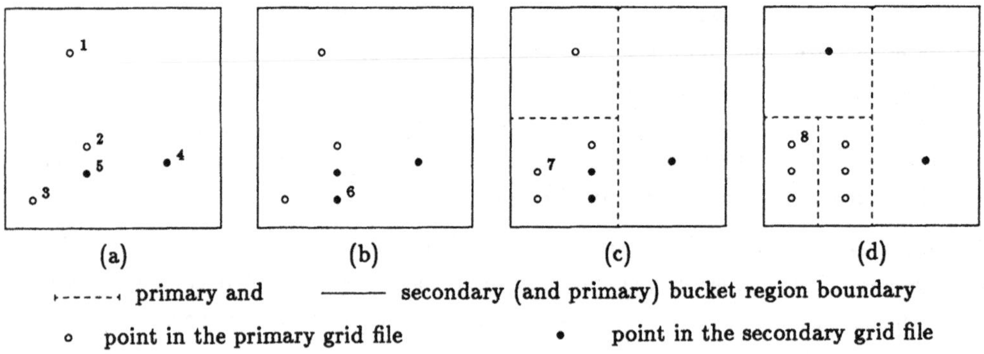

Figure 1.2: A twin grid file ($b = 3$)

In the example of Figure 1.1 (b), an extra bucket is needed to store point 6. In the twin grid file, no extra bucket is needed if point 6 is stored in S and not in P (see Figure 1.2 (b)). Since point 7 comes to lie in a bucket region of a full bucket in P and in S, the bucket in P is split. In our example, it needs to be split twice before it actually can be used to store point 7 (see Figure 1.2 (c)). Similarly, point 8 causes the split of a bucket in P. Unless further optimization takes place, one secondary and three primary buckets store our eight points. The twin grid file utilizes storage space better than that: points 5 and 6 are stored in a primary instead of a secondary bucket, and point 1 is stored in a secondary instead of a primary bucket, as shown in Figure 1.2 (d). Hence, only three instead of four buckets suffice to store the eight points. In addition, further insertions are more efficient in the twin grid file as against the standard grid file: any next point can be inserted into the twin grid file without the need for an extra bucket, whereas in the standard grid file an extra bucket may be necessary.

2 The optimal static twin grid file

Let us first define the split and merge strategy of the standard grid file G underlying the optimal (static) twin grid file O. A split line partitions a bucket region by cutting the longer of the edges into halves. If the bucket region is square, the split line is vertical (we do not use the freedom in choosing horizontal or vertical split lines). The merge operation is symmetric to the split operation, i.e., two buckets can only be merged if their bucket regions are two halves created by a split. Hence, for each bucket there is at most one buddy with which it can be merged. Of course, a bucket and its buddy are merged only if all points in both buckets fit into one bucket. As a consequence, by splitting and merging we only arrive at situations that can also be created by only splitting the initially unpartitioned data space.

Let us now consider the problem to achieve the highest possible — the optimal — space utilization for a set of points in a twin grid file. Each point in a set of x points has to be stored in one of the two grid files. We divide the points among the grid files such that the total number of buckets is minimal. Depending on the number x of points being stored, we distinguish 4 cases.

If no point is stored ($x = 0$), the twin grid file is empty, and therefore it is optimal.

If $0 < x \leq b$, one bucket in either of the two grid files is optimal. The other grid file is empty.

If $b < x \leq 2b$, two buckets are necessary and sufficient. In general, one bucket for each grid file always realizes the optimum. It may sometimes be possible to store all points in two buckets belonging to one grid file, in which case the other grid file is empty.

If more than $2b$ points are stored ($x > 2b$), the data space of at least one grid file has to be split, because at least three buckets are needed to store the x points. We distinguish two cases, depending on whether the data space of the other grid file has to be split to achieve the optimal storage space utilization.

In the first case, the data space of the primary grid file is split, whereas the data space of the secondary grid file is not. For this situation to be optimal, the points have to be distributed among the two grid files in such a way that the number of buckets in P is minimal, using one bucket in S.

In the second case, the data space of both grid files is split into two subspaces. For this situation to be optimal, the points in each of the two subspaces have to be organized storage space optimally within their respective subspaces.

An algorithm that computes the subset of a given set of points to be stored in the secondary grid file S along the lines of the above characterization of storage space optimality has been presented in [5]. If the solution is not unique, a smallest subset of points to be stored in S is computed. The time complexity of this algorithm can be shown to be linear in the number of bucket regions needed in a standard grid file to store the whole set of given points.

From the constituent grid files and the twin principle it should be clear how to carry out the operations range query and exact match in an optimal static twin grid file.

3 The suboptimal dynamic twin grid file

The suboptimal dynamic twin grid file is useful to keep space utilization high in an initially optimal twin grid file, where insertions and deletions are allowed to occur. It may also be viewed as an independent, totally dynamic structure. In the dynamic twin grid file D, insertions and deletions trigger local restructuring operations; range queries and exact match operations do not initiate restructuring. After a point is inserted in D or deleted from D a restructuring operation takes place. Note that an insertion or deletion within P or S may already lead to some reorganization within P or S, as prescribed by the standard grid file mechanism.

The dynamic twin grid file D is restructured by transferring a point from P to S, a *shift down* operation, or from S to P, a *shift up* operation, and carrying out the necessary adjustments within P and S, as prescribed by the standard grid file mechanism. After a sequence of shift operations, a bucket in P or in S may become empty, or it may become possible to merge two buckets. As a result, the points can be stored in one less bucket; we call this a *saving* in P or in S, respectively. If a sequence of shift operations necessitates a bucket split resulting in two non-empty buckets, or an empty bucket region becomes non-empty, an additional bucket in P or in S is needed. We call this a *loss* in P or in S, respectively. If the number of buckets in P or in S remains unchanged after a sequence of shift operations, we call this *neutral* in P or in S, respectively.

To locally minimize the number of buckets, we merge two buddies, if a saving in P or in S is possible without any shift operation. If there exists a sequence of shift down operations leading to i savings in P and less than i losses in S, for some integer $i > 0$, then we perform these shift operations. Note that this restructuring step for $i \geq 2$ is essential for achieving any savings at all when inserting points into the initially empty dynamic twin grid file.

To support future optimization efforts, we try to keep secondary regions large and secondary buckets rather empty. To this end, we perform a sequence of shift up operations if it leads to i

savings in S and at most i losses in P, for $i > 0$. For the same reason, we perform any shift up operation neutral in P and in S. As a consequence, all points stored in S lie in empty or full bucket regions of P.

4 Implementation and performance evaluation

We implemented all three kinds of grid files on an IBM-AT in Modula-2, the standard grid file G, the optimal static twin grid file O, and the dynamic twin grid file D, for two-dimensional points. Both twin grid files are based on two standard grid files with the same bucket capacity. Each standard grid file has a two-level directory and buckets with fixed capacity.

The optimization algorithm for O requires that first the points are stored in the primary grid file P. Then we compute the points to be stored in the secondary grid file S, to achieve the optimal storage space utilization. Last, we insert these points into S and delete them from P; all other points remain in P.

For D, in addition to the bucket addresses, the directory also contains the number of points stored in each bucket. This information is used to reduce the number of bucket accesses during restructuring operations in D. For our experiments, we used a set of 40000 uniformly distributed two-dimensional points. For G, O and D, we inserted the points in random order in 4 sequences of 10000 points each into the initially empty structure. We measured the storage space utilizations, the numbers of external accesses, and the numbers of points stored in S (for O and D). We carried out 300 range queries with square ranges of three different sizes at uniformly distributed positions, and we counted the average number of read operations to answer these range queries. We varied the bucket capacity b, measured in points per bucket, in steps of 4 from 8 to 40. To exhibit the effect of our experiments on the directories of the structures very clearly, we chose the fairly small directory page size of 512 bytes (equivalent to 128 bucket references per directory page).

In the following we deal with various aspects of storage access schemes. We first have a closer look at the storage space utilization of both, the buckets and the directory pages. Then we evaluate the performance of the operations insert, range query, and successful and unsuccessful exact match. Finally, we make some observations on the behavior of our point insertion algorithm for D.

The most important aspect is the storage space utilization of the buckets. Averaged over all bucket capacities, the storage space utilization turned out to be 69.5% in G. In O it was exactly 90.0%. That makes a difference of 20.5 percentage points, or 29.5% more buckets in G as against O. The differences in storage space utilizations for the same point set, but various bucket capacities, range from 12 to 28 percentage points. Surprisingly, for bad utilizations in G (around 65%) the optimization in O achieves values above 90%. For good utilizations in G (over 72%) O utilizes space at less than 90%. In general, the amount of improvement in O against G does not depend on the bucket capacity.

In D, the average storage space utilization was 89.6%. Storage space utilization in D at times was as good as in O; in the worst case, it was 2 percentage points less. The maximum deviation of the storage space utilization in G from its average was $\pm 5\%$, twice as big as the deviation in D. This deviation for G and D increases with increasing bucket capacity.

Next, we have a look at the space utilization of the directory pages. In G the average utilization was roughly 72% (directory page size 512 bytes or 128 entries). In O, in the directory pages of P and S together, the number of required entries decreases by 20%, compared with G. If the utilization in G was high (around 80%) the improvement in O was large (20%), for low utilization in G (60% or less) the improvement in O was small (5%). In few cases, in O even more directory entries than in G have been necessary. For other directory page sizes, this dependency has not been observed in this clarity. In D the number of directory entries is almost identical to that in O.

Keep in mind that a directory entry in D is 1 byte larger (3 instead of 2 bytes) than an entry in O and in G, so that the entire directory for D is always larger than that for O and for G, respectively. As a conclusion of the above, note that the directory space requirements for both types of twin grid files remain practically as low as for the standard grid file. In all cases, the directory requires less than 1% of the entire storage space, for 20 points per bucket and 256 entries per page.

In the following, we look at the number of external read and write accesses to carry out grid file operations. First consider the insert operation. In Table 4.1 we compare the number of external accesses for G and O.

bucket capacity	storage space utilization in G	storage space utilization: improvement in O as against G (G = 100 %)	external accesses in G (average per insertion)	external accesses: additional in O as against G (G = 100 %)
8	69.0 %	30.0 %	3.31	11.9 %
12	69.9 %	27.5 %	3.16	8.1 %
16	68.2 %	32.0 %	3.08	6.8 %
20	69.3 %	31.2 %	3.01	5.1 %
24	72.2 %	22.3 %	2.96	3.9 %
28	69.3 %	29.1 %	2.92	4.1 %
32	65.2 %	40.8 %	2.89	4.0 %
36	64.7 %	42.7 %	2.86	3.5 %
40	69.6 %	32.8 %	2.83	2.6 %

Table 4.1: Accesses to optimize in O, compared to the accesses in G for inserting 40000 points

The additional accesses necessary to build an optimal twin grid file, starting with all points stored in a standard grid file, decrease linearly, with some variation, at the same rate as the bucket capacity grows. The basis of the percentage of additional accesses in O is the number of accesses in G; this number is tightly coupled to the bucket capacity, and therefore does not yield a reason for this variation. Instead, the variation depends mainly on the amount of improvement achieved by the optimization. In Table 4.2 we show the growth of necessary accesses to insert points in D as against in G for bucket capacity 20 (40), during the last 10000 of a sequence of 40000 insertions.

	read		write		read and write	
bucket	71 %	(40 %)	55 %	(30 %)	61 %	(33 %)
directory	50 %	(31 %)			131 %	(100 %)
bucket and directory	63 %	(36 %)	134 %	(100 %)	87 %	(57 %)

Table 4.2: Additional accesses in D as against G for inserting the last 10000 of 40000 points into buckets with capacity 20 (40)

Here, the number of additional accesses decreases when the bucket capacity grows, in the same manner as for O. The table entry for directory write operations is left empty, because here the values vary from 800% to 2000%, not depending on the bucket capacity, due to the small absolute number of directory write operations in G. If the number of bucket splits is high, i.e. when the storage space utilization decreases, the number of directory writes is high and the ratio for D is low, and vice versa. But it should be mentioned that the absolute number of directory writes in D shrinks with growing bucket capacity, like all the other external accesses, too.

Second, we look at the most important operation, the range query. The efficiency of range queries in O and D as against G varies with the amount of improvement in storage space utilization in the twin grid files, compared with the standard grid file. For the average improvement of 20 percentage points, range queries with ranges intersecting the regions of roughly 22 blocks (buckets and directory pages) or more can be answered more efficiently in O and D than in G. For much higher improvement of storage utilization in O and D as against G (more than 25 percentage points), the break-even point is lower (18 blocks), and vice versa (24 blocks for 15 percentage points). In G range queries are at most 2 accesses better than in O and in D; this is the case for very small query ranges. The bigger the ranges, i.e. the more blocks are affected, the better perform the twin grid files. For answering range queries with ranges affecting 250 blocks in G, the twin grid files need 40 accesses less than G. In O and D the range query performances are almost equal.

Third, let us consider the exact match operation. For exact match, we distinguish between successful and unsuccessful exact match. In G we need two accesses (one to a directory and one to a bucket) for both cases. In the worst case in the twin grid files, we need twice as many accesses as in G, since we have to search two grid files. But for successful exact match the worst case only occurs if the search point is stored in the secondary grid file S. The amount of points stored in S varies linearly with the bucket capacity b from 14% of all points ($b = 8$) to 10% ($b = 40$) on the average for the optimal static twin grid file O. So we need 1.14 to 1.10 times as many block accesses in O as against G, on the average.

In D as compared with O, roughly 2.5% to 3% more of all points are stored in S, independent of the bucket capacity. Hence, the successful exact match operation is slightly less efficient in D. The variation of the amount of points in S from its mean value is larger for bigger bucket capacities.

Unsuccessful exact match also costs twice as many accesses in O and D as against G in the worst case. But, due to the invariant that points in the secondary grid file S only lie in full or empty bucket regions of P, we don't have to search both grid files under all circumstances. If the search point falls into a region of a non-full and non-empty bucket in P, we don't have to access S for further search. In O, 45% of all buckets are non-full and non-empty, so that unsuccessful exact match here requires 1.55 times as many accesses as in G, under the assumption that each bucket region in P is affected by a search with the same probability.

In D this average block access factor is 1.58, since in D more buckets are full than in O. These factors vary slightly, independent of the bucket capacity b.

Finally, for the dynamic twin grid file, we look at the restructuring operations, triggered by point insertions. Table 4.3 shows the average number of restructuring operations for inserting 10000 points into D for several bucket capacities.

bucket capacity b	8	12	16	20	24	28	32	36	40
average number of restructuring operations	2936	2026	1560	1268	1070	920	807	716	635
restructuring operations per bucket	2.35	2.43	2.50	2.54	2.57	2.58	2.58	2.58	2.54

Table 4.3: Average number of restructuring operations for inserting 10000 points

A restructuring operation in D is carried out, if the insertion point would require an additional bucket if inserted into P as well as if inserted into S. Then the point is inserted into P, followed by a restructuring operation. In a sequence of b insertions of points, roughly 2.5 restructuring operations occur. For fairly small bucket capacities ($b \leq 16$), the number of restructuring operations is slightly less. A restructuring operation consists of a sequence of the four restructuring steps (S1) to (S4) local to the region of the bucket b_S in S within which the insertion point lies:

(S1) Try to merge b_S with its buddy in S. This achieves a saving if both merge partners are not empty.

(S2) Try to empty bucket b_S in S using one additional bucket in P. This action achieves no saving. If b_S is empty, retry (S1).

(S3) If b_S in S is not full, fill it, if this achieves at least one saving in P, or at least two savings if bucket b_S was empty.

(S4) If bucket b_S in S is not empty, split it, if this achieves at least two savings in P.

A restructuring step is carried out by first testing its conditions; if they are fulfilled, the corresponding action is performed. Table 4.4 summarizes the frequency of the restructuring steps, relative to the number of restructuring operations, and the frequency of the restructuring actions, as well as the frequency of achieving savings, relative to the number of restructuring steps.

restructuring step	(S1)	(S2)	(S3)	(S4)
frequency of restructuring steps (in % of restructuring operations)	120 %	100 %	93 %	99 %
frequency of restructuring actions (in % of restructuring steps)	20 %	20 %	20 %	30 %
frequency of restructuring actions with bucket savings (in % of restructuring steps)	5 %	0 %	20 %	30 %

Table 4.4: Frequencies of restructuring steps, actions, and savings

These values are averaged over all bucket capacities. There are no significant differences in these values for different bucket capacities. Step (S4), together with (S3), is the most important and most successful one. Steps (S1) and (S2) contribute to keep secondary bucket regions rather big and rather empty, to support future optimization.

The dynamic twin grid file can be adapted to various situations and user requirements in numerous ways. For instance, for a sequence of insertions, restructuring actions at the beginning have a weaker effect on space utilization than at the end. If the total number of insertions is known beforehand, our experiments show that for achieving a space utilization of roughly 90%, it is sufficient to restructure only during the last fifth of all insertions, saving more than 20% of all block accesses as against D. Otherwise, block accesses can be saved by restructuring less frequently than in D, during the whole sequence of insertions, at the cost of decreasing space utilization. In a different attempt to save block accesses, we may restrict the restructuring actions applied in D. For example, 10% of all block accesses can be saved, by restructuring according to steps (S3) and (S4) only, with a resulting space utilization of 88%. Furthermore, applying step (S1) is slightly less costly than step (S2), but brings out better space utilization.

References

1. M. Freeston:
   The BANG file: a new kind of grid file, Proc. ACM SIGMOD International Conference on Management of Data, 1987, 260–269.
2. K.H. Hinrichs:
   The grid file system: implementation and case studies of applications, Doctoral Thesis No. 7734, ETH Zürich, 1985.
3. A. Hutflesz, H.-W. Six, P. Widmayer:
   Globally Order Preserving Multidimensional Linear Hashing, Proc. IEEE 4th International Conference on Data Engineering, 1988, 572–579.
4. A. Hutflesz, H.-W. Six, P. Widmayer:
   The Twin Grid File: A Nearly Space Optimal Index Structure, Proc. International Conference Extending Database Technology, 1988, 352–363.
5. A. Hutflesz, H.-W. Six, P. Widmayer:
   Twin Grid Files: Space Optimizing Access Schemes, Proc. ACM SIGMOD International Conference on Management of Data, 1988.
6. H.-P. Kriegel, B. Seeger:
   Multidimensional Order Perserving Linear Hashing with Partial Expansions, Proc. International Conference on Database Theory, 1986, 203–220.
7. R. Krishnamurthy, K.-Y. Whang:
   Multilevel Grid Files, IBM Research Report, Yorktown Heights, 1985.
8. D.B. Lomet:
   Partial Expansions for File Organizations with an Index, ACM Transactions on Database Systems, Vol. 12, 1, 1987, 65–84.
9. J. Nievergelt, H. Hinterberger, K.C. Sevcik:
   The Grid File: An Adaptable, Symmetric Multikey File Structure, ACM Transactions on Database Systems, Vol. 9, 1, 1984, 38–71.
10. E.J. Otoo:
    Balanced Multidimensional Extendible Hash Tree, Proc. 5th ACM SIGACT/SIGMOD Symposium on Principles of Database Systems, 1986, 100–113.
11. J.T. Robinson:
    The K-D-B-Tree: A Search Structure for Large Multidimensional Dynamic Indexes, Proc. ACM SIGMOD International Conference on Management of Data, 1981, 10–18.

ON SEPARABLE AND RECTANGULAR CLUSTERINGS

H. Heusinger, H. Noltemeier
Lehrstuhl fur Informatik I, University of Würzburg
Am Hubland, 8700 Würzburg, West Germany

ABSTRACT

For a finite set of points S in the Euclidean plane we introduce the so called C-s-clustering problem which can be stated as:
Partition S into C subsets S_i such that S_i is separable from $S \setminus S_i$ by a line and $|S_i| = k_i$, where k_i are given numbers. For a function f which maps C subsets S_i of S into $\mathbb{R}$ we present an algorithm which finds an, with respect to f, optimal C-s-clustering in $O(Cn^{3/2}\log^2 n + p_C(Cn^{3/2}U_f(n) + P_f(n)))$ steps. (where $P_f(n)$ resp. $U_f(n)$ are the time to calculate resp. to update f, if the arguments are slightly changed and p_C is the number of, for the algorithm distinct, orderings of the k_i. These orderings are also a part of the input of the algorithm.) Then an $O(n^{C-1}\log n)$ solution for the C-r-clustering problem of finding all sets of C ($C \leq 3$) axis-parallel rectangles R_i such that $|R_i \cap S| = k_i$ and $R_i \cap R_j \cap S = \emptyset$ for $i \neq j$ is given. If we assume in addition that $R_i \cap R_j = \emptyset$ for all $i \neq j$ we give an $O(n^{C-2})$ algorithm for $C > 2$ and an $O(n)$ algorithm for $C = 2$.

I. INTRODUCTION

Given a set S of points in the plane, a centralized C-clustering of S (as defined by Dehne, Noltemeier in [7]) is a partition of S into C disjoint subsets of S, such that every subset S_i can be covered by a

---

This work was supported by the DFG (Deutsche Forschungsgemeinschaft) under grant No 88/6-1

circle and there is no point of S in any intersection of two circles. If we assume in addition, that the centers of the circles are at infinity we see that every cluster S_i can be separated from $S \setminus S_i$ by a line; in other words S_i is a $|S_i|$-set of S (cf. [10, 11, 14]). Another problem which we considered in this paper is obtained from the above one when the word "circle" is substituted by "axis-parallel rectangle".

This paper is organized in the following way.

In section II.1 we give the basic notations and definitions for treating with separable clusterings. Section II.2 deals with the relationship between Voronoi diagrams and C-s-clusterings, and it contains the theorem that gives us the idea for the algorithm which will be presented in section II.3. This algorithm has as input the set S, C numbers k_i and a clustering measure $f : (S_1, S_2, \ldots , S_C) \longmapsto r \in \mathbb{R}$; it needs $P_f(n)$ (resp. $U_f(n)$) time to be computed (resp. updated) and the algorithm constructs in time $O(Cn^{3/2}\log^2 n + p_C(Cn^{3/2}U_f(n) + P_f(n)))$ an optimal C-s-clustering with respect to f, where p_C is the number of orderings of the k_i that can not be transformed by a cyclic permutation into one another. These orderings are also given as input to the algorithm. A sketch of the proof of the correctness and the analysis of the complexity of the algorithm is also given in this section. In section II.4 an example is presented which demonstrates that the number of C-s-clusterings can be exponential in n.

In section III we present the algorithm for constructing all C-rectangular clusterings of S ($C \leq 3$). Finally in section III.2 an algorithm for a variant of the C-rectangular clustering problem is given.

II SEPARABLE CLUSTERINGS

II.1 Notations and definitions

Let $S \subset \mathbb{R}^2$ be a finite set of at least three points in the plane in general position, this means that no three points are collinear and no four points are cocircular.

Notation 1:

Let g be a line and r a normal vector of g. Then

$$H(g,r) := \{x \in \mathbb{R}^2 \mid r^T x \geq r^T a \text{ for an } a \text{ on } g\}$$

is the halfplane which is bounded by g and contains all points of the side of g to which r points to.

Definition 1:

Let $\emptyset \neq A \subsetneq S$.

1. We say A is separable from S\A if and only if there exists a line g with normal vector r, such that

$$S \cap H(g,r) = A \text{ and } S \cap g = \emptyset.$$

2. A partition $\{S_1, S_2, \ldots, S_C\}$ of S is called a C-s-clustering (s stands for separable) if and only if S_i is separable from $S \setminus S_i$ for all $1 \leq i \leq C$.

For $1 \leq k < |S|$ we denote by k-VoD(S) the order k Voronoi diagram of S and we denote by V(A) ($A \subset S$, $|A| = k$) the Voronoi region of k-VoD(S) containing those points of S which have A as the set of their k nearest neighbours. A is called the label of V(A).

From [10] we know: $\emptyset \neq A \subsetneq S$ is separable from S\A, if and only if $V(A) \in |A|$-VoD(S) is unbounded.
In addition we know, if $A = S \cap H(g, r)$, then r is the direction of a ray which is completely contained in V(A). Thus we see immediately that a partition $\{S_1, S_2, \ldots, S_C\}$ of S is a C-s-clustering if and only if all $V(S_i) \in |S_i|$-VoD(S) are unbounded.

As $|S| \geq 3$ and as S is in general position no unbounded Voronoi region has two parallel Voronoi rays.

Notation 2:

Let V(A) be an unbounded Voronoi region and let R be a Voronoi ray of V(A) such that V(A) lies to the left side of the oriented line that contains R and has the same orientation as R. Then we say R is the right Voronoi ray of V(A).

Notation 3:

Let V(A) and V(B) be two unbounded Voronoi regions. We denote by $\sphericalangle(V(A), V(B))$ the angle between the right Voronoi rays of V(A) and V(B), measured in counterclockwise order, starting with the ray of V(A).

It is easy to see, that if A and B are disjoint labels of nonempty Voronoi regions V(A) and V(B), then also int(V(A)) and int(V(B)) are disjoint. The following observation is obvious.

Observation:

Let $\emptyset \neq A, B, C \subsetneq S$, such that $V(A)$, $V(B)$, $V(C)$ are unbounded Voronoi regions and $A \cap B = C \cap A = \emptyset$ and $|B| = |C|$ and $\sphericalangle(V(A), V(B)) < \sphericalangle(V(A), V(C))$. For every $X \in \{A, B, C\}$ select $p_X \in V(X)$ and $r_X \in \mathbb{R}^2$ not parallel to one of the Voronoi rays of $V(X)$ such that $p_X + \lambda r_X \in \text{int}(V(X))$ for all nonnegative λ. Then $\sphericalangle(r_A, r_B) < \sphericalangle(r_A, r_C)$ holds.

II.2 A property of C-s-clusterings

In this section we present a theorem giving us the basic idea for the algorithm which will be presented in the next section.

Theorem 1:

Let $\{S_1, S_2, \ldots, S_C\}$ be a C-s-clustering of S, such that the $V(S_i)$ are sorted with respect to their angles. (That is: $0 < \sphericalangle(V(S_1), V(S_i)) < \sphericalangle(V(S_1), V(S_{i+1})) < 2\pi$ for all $i = 2, 3, \ldots, C-1$.) Then $V(S_{i+1})$ has minimal angle to $V(S_i)$.

Proof:

To get a contradiction we assume:
There exists a subset $A \subset S$, $|A| = |S_{i+1}|$, $A \cap S_i = \emptyset$, $V(A)$ unbounded and $\sphericalangle(V(S_i), V(A)) < \sphericalangle(V(S_i), V(S_{i+1}))$.
Since $A \neq S_i, S_{i+1}$ there exists $j \in \{1, \ldots, C\} \setminus \{i, i+1\}$ such that $A \cap S_j \neq \emptyset$. Let $a \in A \cap S_j$. Because $V(A)$, $V(S_j)$, $V(S_i)$, $V(S_{i+1})$ are unbounded, there exist lines g_A, g_j, g_i, g_{i+1} with normal vectors r_A, r_j, r_i, r_{i+1}, such that
$A \subset H(g_A, r_A)$, $S \backslash A \subset H(g_A, -r_A)$, $S_j \subset H(g_j, r_j)$, $S \backslash S_j \subset H(g_j, -r_j)$,
$S_i \subset H(g_i, r_i)$, $S \backslash S_i \subset H(g_i, -r_i)$,
$S_{i+1} \subset H(g_{i+1}, r_{i+1})$, $S \backslash S_{i+1} \subset H(g_{i+1}, -r_{i+1})$.
W.l.o.g we can assume that $a = 0$, $r_A = -e_2 = -(0, 1)$ and that the above lines intersect each other. If we denote by g_A' the x-axis we have $H(g_A', r_A) \subset H(g_A, r_A)$.
From the assumption follows $\sphericalangle(r_i, r_A) < \sphericalangle(r_i, r_{i+1})$. As the clusters are sorted we can conclude:
$\sphericalangle(r_A, r_{i+1}) < \sphericalangle(r_A, r_j) < \sphericalangle(r_A, r_i)$.
1st case: g_j intersects the x-axis to the left of the origin. (see fig. 1).

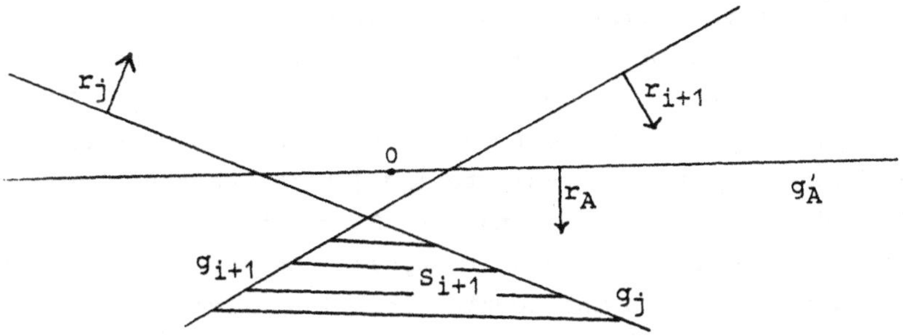

fig. 1

Since $0 \in S_j$ the angle $\sphericalangle(r_A, r_j)$ is smaller than π.
$0 \notin S_{i+1}$ and $\sphericalangle(r_A, r_{i+1}) < \pi$ imply that g_{i+1} intersects g_A' to the right of the origin and then we conclude that g_{i+1} intersects g_j in $H(g_A', r_A)$. From this follows:
$S_{i+1} \subset H(g_{i+1}, r_{i+1}) \cap H(g_j, -r_j) \cap S \subset H(g_A', r_A) \cap S \subset A$.
This leads to a contradiction.

2nd case: g_j intersects the x-axis to the right of the origin. (cf. fig. 2)

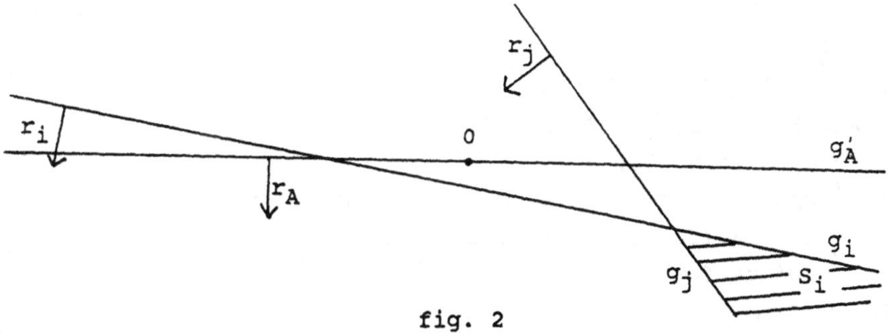

fig. 2

Because $0 \in S_j$ the angle $\sphericalangle(r_A, r_j)$ is larger than π. Now it follows $\pi < \sphericalangle(r_A, r_i)$. $0 \notin S_i$ implies that g_i intersects g_A' to the left of 0, thus g_i intersects g_j in $H(g_A', r_A)$. Then:
$S_i \subset H(g_i, r_i) \cap H(g_j, -r_j) \cap S \subset H(g_A', r_A) \cap S \subset A$,
in contradiction to $S_i \cap A = \emptyset$.

$\square$.

II.3 An algorithm for constructing sorted optimal C-s-clusterings

Theorem 1 gives us the basic idea for the following algorithm:
We calculate all unbounded order k_i Voronoi regions and we sort them

with respect to the angles of their Voronoi rays.
Now we take $V(S_1)$ from the k_1-VoD(S), then we determine $V(S_2)$ from k_2-VoD(S) with minimal angle to $V(S_1)$, then we determine $V(S_3)$ from k_3-VoD(S) with minimal angle to $V(S_2)$ etc. This is done as long as we have a chance to find a C-s-clustering (which means as long as $S_i \cap S_1 = \emptyset$). If we reach S_C and $S_C \cap S_1 = \emptyset$ then $\{S_1, \ldots, S_C\}$ is a C-s-clustering of S and we have to update the last optimal value of f. Then we update S_1 (goto the counterclockwise neighbour of S_1). The other S_i are updated by proceeding as above.

Algorithm:

1. Determine the unbounded order k_i Voronoi regions, sorted with respect to the angles of their Voronoi rays in counterclockwise order. The result are C cyclic lists L_i containing the labels of the above Voronoi regions. We denote by first(L_i) the Voronoi label of the Voronoi region which contains the positive end of the x-axis. (In the case when there are two such Voronoi regions, first(L_i) denotes the one in the upper halfplane.)
   Initialize the boolean array flag[i] := FALSE for all $2 \le i \le C$, and let S_i := first(L_i) for all $1 \le i \le C$.
   Determine $S_i \cap S_{i-1}$ and $S_i \cap S_1$ for all $2 \le i \le C$ and calculate f.
2. i := 2
3. IF flag[i] = FALSE THEN

```
        WHILE S_i ∩ S_{i-1} = Ø DO
          ⎧ S_i := next(S_i);
      (*) ⎨ update S_i ∩ S_{i-1}, S_i ∩ S_1, S_{i+1} ∩ S_i (if i < C)
          ⎪ setflag(flag, i); (* see later *)
          ⎩ update f
   WHILE  S_i ∩ S_{i-1} ≠ Ø DO
       (*)
   flag[i] := TRUE;
```

4. IF $S_1 \cap S_i \neq \emptyset$ THEN

```
        ⎧ S_1 := next(S_1);
        ⎪ setflag(flag, 1);
        ⎪ IF S_1 ≠ first(L_1) THEN
        ⎪    update S_1 ∩ S_i for all 2 ≤ i ≤ C;
  (**)  ⎨    update f
        ⎪    GOTO 2;
        ⎪ ELSE
        ⎩    STOP (* all posibilities are checked *)
   ELSE
```

```
      IF i = C THEN
(***)    update the optimal value of f;
         (**)
      ELSE
         i := i + 1;
         GOTO 3;
```

5. report the best C-s-clustering, if at least one C-s-clustering was found.

The procedure <u>setflag</u> has the following form:

```
setflag(flag, i)
      IF i < C THEN
         IF flag[i+1] = TRUE THEN
            flag[i+1] := (S_i ∩ S_{i+1}) = Ø;
```

To prove the correctness of the previous algorithm we have to show the following four facts.

1. Let $S \subset \mathbb{R}^2$ be a finite set of at least three points in the plane and let $\emptyset \neq S_1, S_2, \ldots, S_i \subsetneq S$, such that $V(S_j)$ is unbounded for $j = 1, 2, \ldots, i$ and $S_j \cap S_{j+1} = \emptyset = S_1 \cap S_{j+1}$ for $j = 1, 2, \ldots, i-1$ and $0 < \sphericalangle(V(S_1), V(S_j)) < \sphericalangle(V(S_1), V(S_{j+1})) < 2\pi$ for $2 \leq j \leq i - 1$. Then the S_j are pairwise disjoint.
2. Let $S \subset \mathbb{R}^2$ be a finite set of more than two points in the plane and let A, B, C be labels of unbounded Voronoi regions, such that $A \cap B = B \cap C = \emptyset$ and the angle $\sphericalangle(V(B), V(C))$ is minimal. Then either $A \cap C \neq \emptyset$ or $A \cap C = \emptyset$ and $\sphericalangle(V(B), V(C)) < \sphericalangle(V(B), V(A))$.
3. If flag[i] = TRUE then the current S_i and S_{i-1} fulfill $S_i \cap S_{i-1} = \emptyset$ and $V(S_i)$ has minimal angle to $V(S_{i-1})$.
4. If $\{S_1, \ldots, S_C\}$ is a sorted C-s-clustering of S then the algorithm will examine it.

Assume we have proved 1., 2. and 3. then at (***) the algorithm correctly recognizes that the current sets $S_1, \ldots, S_C$ form a C-s-clustering of S. This follows from the following consideration:

If the algorithm reaches (***) flag[i] = TRUE for all $2 \leq i \leq C$ and we have by fact 3 that $S_i \cap S_{i-1} = \emptyset$ and that $V(S_i)$ has minimal angle to $V(S_{i-1})$ for all $2 \leq i \leq C$. In step 4 of the algorithm we have $S_1 \cap S_i = \emptyset$ for all $2 \leq i \leq C$ hence by fact 2 one obtains $\sphericalangle(V(S_{i-1}), V(S_i)) < \sphericalangle(V(S_{i-1}), V(S_1))$ for all $3 \leq i \leq C$. This implies $0 < \sphericalangle(V(S_1), V(S_{i-1})) < \sphericalangle(V(S_1), V(S_i)) < 2\pi$ for all $3 \leq i \leq C$. From fact 1 it follows that the S_i are pairwise disjoint.

Fact 4 guarantees that the algorithm examines all C-s-clusterings.

Proof of fact 1:

We prove this fact by induction. The case $i = 2$ is obvious. Now we show the step from i to $i+1$. The inductive assumption says that the S_j are pairwise disjoint for $1 \leq j \leq i$.

We assume conversely: There exists $j \in \{2, \ldots, i-1\}$ with $S_{i+1} \cap S_j \neq \emptyset$. Let $a \in S_{i+1} \cap S_j$ (w.l.o.g.: $a = 0$), and for $k \in \{1, i, i+1, j\}$ let g_k be some line with normal vector r_k which separates S_k from $S \backslash S_k$. W.l.o.g. let g_k intersect each other and let $r_{i+1} = -e_2$. Furthermore denote the x-axis by g'_{i+1}. Since $0 \in S_{i+1}$, we have $H(g'_{i+1}, r_{i+1}) \subset H(g_{i+1}, r_{i+1})$.

1st case: $\sphericalangle(r_{i+1}, r_1) < \pi$

As $0 \notin S_1$ g_1 intersects the x-axis to the right of 0.

α) g_j intersects g'_{i+1} in the point $(s, 0)$ with $s < 0$. We now move g_1 to the point $(s, 0)$ and obtain a line g'_1 which is parallel to g_1 (cf. fig. 3). $0 \in S_j$ implies that $\sphericalangle(r_{i+1}, r_j) < \pi$ and as the $V(S_j)$ are sorted with respect to their angles one obtains $\sphericalangle(r_{i+1}, r_1) < \sphericalangle(r_{i+1}, r_j)$. We conclude $H(g'_1, r_1) \cap H(g'_{i+1}, -r_{i+1}) \subset H(g_j, r_j)$ and:

$S_1 \subset H(g_1, r_1) \cap S \cap H(g_{i+1}, -r_{i+1}) \subset$
$H(g'_1, r_1) \cap S \cap H(g'_{i+1}, -r_{i+1}) \subset H(g_j, r_j) \cap S = S_j$.

This is a contradiction to the assumption $S_1 \cap S_j = \emptyset$.

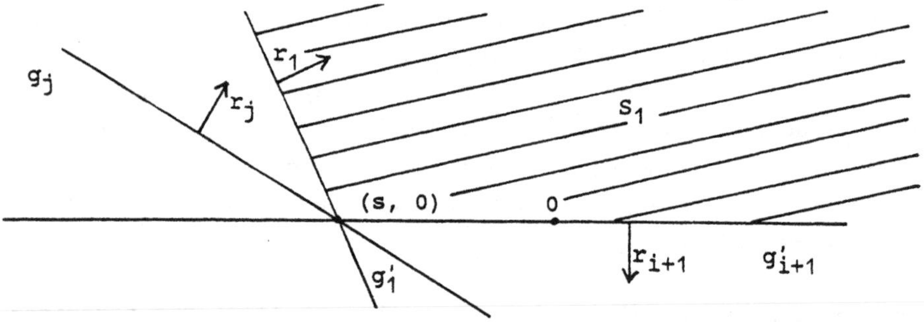

fig. 3

β) g_j intersects the x-axis in the point $(s, 0)$ with $s > 0$. (cf. fig. 4) Then we have the following inequalities:

$\sphericalangle(r_j, r_i) < \sphericalangle(r_j, r_{i+1})$ and $\sphericalangle(r_{i+1}, r_i) > \sphericalangle(r_{i+1}, r_j) > \pi$.

Since 0 is no member of S_i, g_i intersects the x-axis g'_{i+1} to the left of 0. Therefore g_i intersects g_j in $H(g'_{i+1}, r_{i+1})$ and because of the inductive assumption it follows that

$S_i \subset H(g_i, r_i) \cap S \cap H(g_j, -r_j) \subset H(g'_{i+1}, r_{i+1}) \cap S \subset S_{i+1}$, which is again a contradiction to $S_i \cap S_{i+1} = \emptyset$.

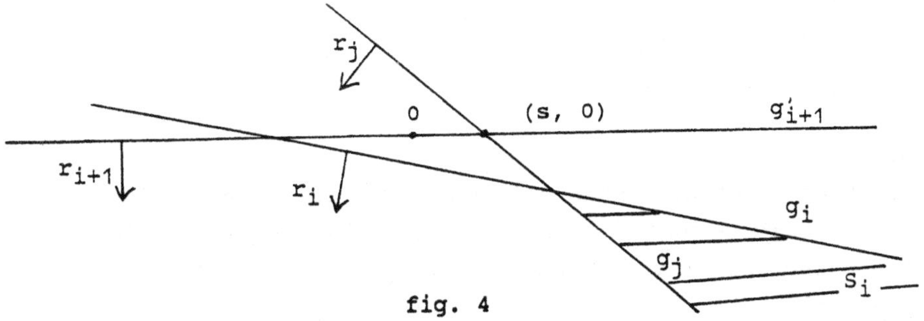

fig. 4

2nd case: $\sphericalangle(r_{i+1}, r_1) > \pi$.

Because of the order of the $V(S_k)$ we have $\sphericalangle(r_{i+1}, r_j) >$ $\sphericalangle(r_{i+1}, r_1) > \pi$ and therefore this is the same as the case 1ß.

□.

The proofs of the other three facts can be carried out in similar ways. But because of the forced shortness of the paper we must omit them here.

Now we want to estimate the worst case complexity of our algorithm. When we look carefully at the angles which are passed by the right Voronoi rays of the Voronoi regions $V(S_i)$ we see that this angles are at most 6π. It follows that every list L_i is passed at most three times.

The updates of the sets $S_i \cap S_{i-1}$ and $S_i \cap S_1$ can be done in constant time since S_i and $next(S_i)$ only differ in one element.

The unbounded Voronoi regions can be constructed in time

$$O\left(\sum_{i=1}^{C} nk_i^{1/2}\log^2 n\right) \subset O(Cn^{3/2}\log^2 n).\ \text{(cf. [6,lemma 5.4])}$$

The number of unbounded Voronoi regions is $O(nk_i^{1/2}) \subset O(n^{3/2})$ according to [10]. Thus the lists L_i can be constructed in time $O(Cn^{3/2}\log^2 n + Cn^{3/2}\log n) = O(Cn^{3/2}\log^2 n)$.

As every list has $O(n^{3/2})$ elements and every element is visited at most three times the total number of moves is $O(3Cn^{3/2})$. This leads to an overall complexity of $O(C(n^{3/2}\log^2 n + n^{3/2}U_f(n)) + P_f(n))$ time.

We can start the above algorithm with all possible orderings of $k_1, \ldots, k_C$ as additional input and maintain the best value of f. Let p_C be the number of orderings of the k_i such that two such orderings can not be transformed into one another by a cyclic permutation. (for

example (1 2 1 4) and (2 1 4 1) are the "same" in the above sense). For example in the case $k_i = \lfloor n/C \rfloor$ for $i < C$ and $k_C = n - (C-1)\lfloor n/C \rfloor$ we have $p_C = 1$.
This proves the following

Theorem 2:

Let S be a set of n points in the plane then there is an $O(Cn^{3/2}\log^2 n + p_C(Cn^{3/2}U_f(n) + P_f(n)))$ time algorithm for solving the C-s-clustering problem, if the p_C orderings of the k_i are available.

II.4 The C-s-clustering problem is not polynomial

We finally give an example, where the number of C-s-clusterings is exponential in n.
Choose $n = 3k$ points $p_1, \ldots, p_n$ on the plane, from which the points $p_1, \ldots, p_{2k}$ are on a circle, sorted in clockwise order. In the intersection of the triangles $\Delta(p_i, p_{i+1}, p_{i+2})$, $\Delta(p_{i+1}, p_{i+2}, p_{i+3})$ (for $i \equiv 1 \bmod 2$) we put another point. (cf. fig. 5)
The clusters with two points (,) (see fig. 5) can be chosen independently, so we have at least $2^k = (2^{1/3})^n$ 2k-s-clusterings of the type $(\underbrace{2, 2, \ldots, 2}_{k}, \underbrace{1, \ldots, 1}_{k})$.

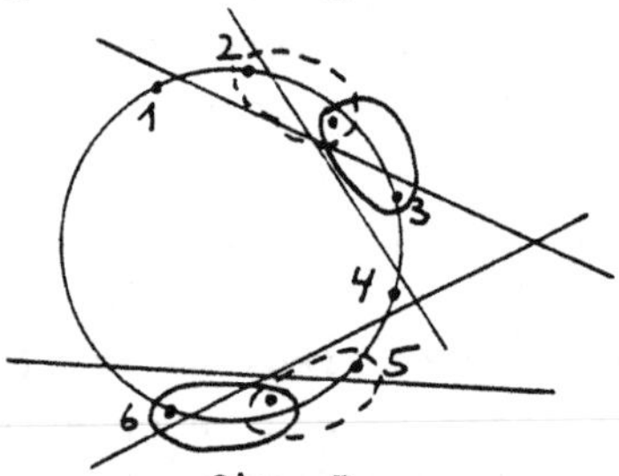

fig. 5

III. RECTANGULAR CLUSTERINGS

III.1 The general case

Dehne and Noltemeier considered in their paper [7] only the problem of finding centralized clusterings for a set S of points in the Euclidean plane. When we use instead of the L_2-metric the L_∞-metric we get:

$A \subset S$ is a centralized cluster, if there exists a square Q with edges parallel to the coordinate axis such that A is the subset of S which is covered by Q.

This leads to the following definitions:

A ⊂ S is a r(ectangular)-cluster of S if there exists an axis-parallel rectangle R such that A is the subset of S which is covered by R.

$\{S_1, S_2, \ldots, S_C\}$ ($C \in \mathbb{N}_+$) is a C-r-clustering if $\{S_1, S_2, \ldots, S_C\}$ is a partition of S into r-cluster S_i.

Now we can state the problem:
We have given a set S of n points in the plane and C positive integers k_i such that $\Sigma_{i=1}^{C} k_i = n$ and we want to determine all sets of C axis-parallel (minimal) rectangles R_i such that $|R_i \cap S| = k_i$ and $R_i \cap R_j \cap S = \emptyset$ for all $i \neq j$, $i, j \in \{1, \ldots, C\}$.

Clearly the sets $R_i \cap S$ build a C-r-clustering of S.

Let $a = (a_x, a_y)$, $b = (b_x, b_y)$, $c = (c_x, c_y)$, $d = (d_x, d_y)$ be points in the plane. If we set $l(R) = a$, $r(R) = b$, $b(R) = c$ and $t(R) = d$ we say that the above points define the rectangle $R = [a_x, b_x] \times [c_y, d_y]$. We say R is a correct rectangle if $\{a, b, c, d\} \subset R$.

For the following we shall assume that in S there are no two points with common x- or y-coordinate.

Prototype of a scene of C axis-parallel rectangles

Let $\{R_1, R_2, \ldots, R_C\}$ be a set of C axis-parallel rectangles such that no two vertical or horizontal edges are colinear. Then we can sort the x-coordinates of the vertical edges and the y-coordinates of the horizontal edges of the rectangles and we shall get unique rank numbers for the edges. For $m \in \{1, \ldots, C\}$ let l_m resp. r_m be the rank number of the left (resp. right) edge of R_m in the set of the vertical edges and let t_m (resp. b_m) be the rank numbers of the top (resp. bottom) edge of R_m in the set of the horizontal edges.
The prototype of our scene of rectangles is then the (2C-1) by (2C-1) matrix $P = (P_{ij})$ where each P_{ij} is the list of integers $m \in \{1, \ldots, C\}$ such that $l_m \leq j < r_m$ and $b_m \leq i < t_m$.
An example is given in figure 6.

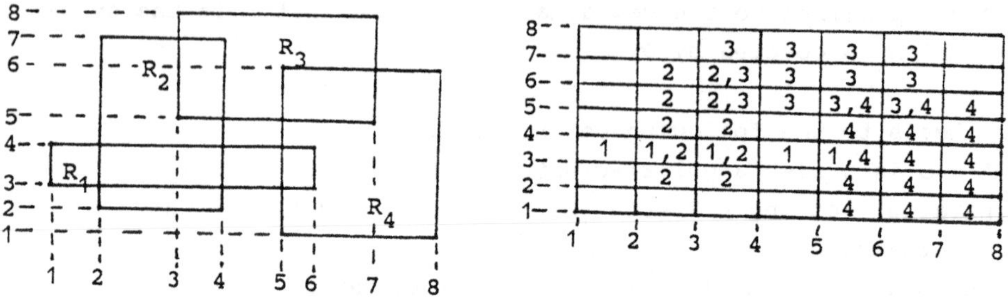

fig. 6

Now we consider the case C = 2. Then $k_1 = k$ and $k_2 = n - k$ and we are searching for two rectangles R_1 and R_2.
In linear time we can determine the smallest rectangle R with edges parallel to the axis which covers S.
We distinguish between the following cases according to their prototype.

1. $R_1 \cap R_2 = \emptyset$.
   It is easy to see that R_1 and R_2 can be separated by a horizontal or a vertical line. Therefore R_1 and R_2 are unique and they can be determined in linear time by searching for a k- and (k + 1)-smallest or largest element and three maxima and three minima by using an algorithm of [1].

2. $R_1 \cap R_2 \neq \emptyset$.
   Since no edge of one rectangle is contained in the other (every edge contains a point of S!) we know that two edges of R_1 and two edges of R_2 lie on the boundary of R.
   We have two possibilities (up to rotations with multiples of 90 degrees) which can be distinguished in a constant time with the aid of the prototype.
   a) or b)

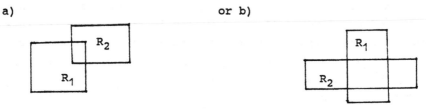

fig. 7

case a)

1. We sort the points of S according to their x- and y-coordinates in ascending order. This yields two arrays A_v and A_h.
   $l(R_1) := A_h[1]$, $b(R_1) := A_v[1]$, $t(R_2) := A_v[n]$, $r(R_2) := A_h[n]$.

2. Determine $r(R_1)$ and $t(R_1)$ such that R_1 contains exactly k points of S and R_1 is a correct rectangle and R_1 has maximal breadth under these rectangles.
3. $l(R_2) := \min\{x \in A_h \mid x \notin R_1\}$
   $b(R_2) := \min\{x \in A_v \mid x \notin R_1\}$
   Determine $D := R_1 \cap R_2 \cap S$
4. if ($l(R_2)$ is right of $r(R_1)$) or ($t(R_2) = t(R_1)$) or ($b(R_2) = b(R_1)$) then STOP
5. if $D = \emptyset$ and R_1 and R_2 form the right prototype then $R_1 \cap S$ and $R_2 \cap S$ are a 2-r-clustering. Report R_1 and R_2.
6. $r_1' := r(R_1)$
   $t(R_1)$:= next larger element in A_v which is to the left of $r(R_1)$; during this scan update $D = R_1 \cap R_2 \cap S$ after each transition
   $r(R_1)$:= next smaller elment in A_h which is below $t(R_1)$; and update of $D = R_1 \cap R_2 \cap S$.
   if $l(R_2) \in R_1$ then
      $l(R_2)$:= next larger element in A_h which is not in R_1; and update $D = R_1 \cap R_2 \cap S$.
   if r_1' is below $b(R_2)$ then
      update D by scanning A_v from $b(R_2)$ to r_1'
      $b(R_2) := r_1'$
   goto 4.

If one of the above searches can not be performed then the algorithm stops.

It is easy to see that after each iteration of steps 4 to 6 $S \subset R_1 \cup R_2$, $|R_1 \cap S| = k$ and $D = R_1 \cap R_2 \cap S$ holds. Therefore in step 5 we correctly recognize that $R_1 \cap S$ and $R_2 \cap S$ build a 2-r-clustering. Obviously the algorithm passes through all possible correct rectangles R_1 with decreasing breadth.
Step 1 needs O(nlogn) time and all the other steps together need O(n) time. This leads to an overall complexity of O(nlogn).

case b) can be handled in a similar way by moving a window which contains k points of S from left to right and updating the top and bottom edge of R_2 during this pass.

Now we turn over to the case C = 3. The three given positive integers are k_1, k_2 and k_3.
First of all we calculate in linear time the smallest axis-parallel rectangle R which covers S. Since four edges of the searched rectang-

les R_1, R_2 and R_3 have to lie on the boundary of R we can w.l.o.g. assume that two edges of R_1 are on the boundary of R. We get the kind of these two edges in a constant time with the aid of the prototype. When we choose one remaining edge of R_1 the last edge of R_1 is unique and it can be obtained in linear time. (searching for the k_1-largest or smallest x- or y-coordinate of the points in the corresponding halfstrip)

Now we have to distinguish between several cases. The first two cases are:

1. Two opposite edges of R_1 lie on the boundary of R.
2. Two edges of R_1 which meet in one corner of R_1 are on the boundary of R.

Every above case has then to be split into several subcases corresponding to whether the intersections $R_i \cap R_j$ are empty or not.

Here we only demonstrate the case when two opposite edges of R_1 are on the boundary of R and $R_i \cap R_j \neq \emptyset$ for all $i,j \in \{1, 2, 3\}$. The other cases can be treated in a similar way.

Assume we have rectangles R_1, R_2, R_3 which fulfill the above conditions. Let U be the smallest axis-parallel rectangle which covers $R_2 \cup R_3$. Then it is easy to see that $R_1 \cap (R_2 \cup R_3) = R_1 \cap U$. This leads to the following algorithm:

<u>Algorithm:</u>

for all possible correct rectangles R_1 do
  calculate the minimal rectangle U which covers $S \setminus R_1$
  if $R_1 \cap U \cap S = \emptyset$ then
    calculate all 2-r-clusterings of $S \setminus R_1$ with the previous algorithm
    if R_1, R_2, R_3 are of the current prototype then R_1, R_2, R_3 is a 3-r-clustering of the current prototype.

Since there are only n possibilities for R_1 and as U can be calculated in O(n) time the above algorithm needs $O(n(n + n + n\log n)) = O(n^2 \log n)$ time.

By a carefull examination of the other cases it can be seen that all the other cases do not need more time. So we have:

<u>Theorem 3:</u>

The C-r-clustering problem can be solved in $O(n^{C-1} \log n)$ time for $C = 2, 3$.

The following example demonstrates that $\Omega(n^{C-1})$ is a lower bound for the C-r-clustering problem.

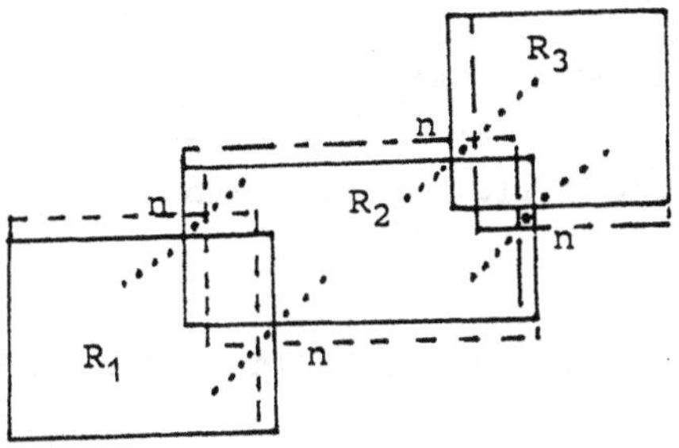

fig. 8

III.2 A variation

A C-r'-clustering of a set S of points in the plane is a C-r-clustering of S such that the (minimal) rectangles which constitute the C-r-clustering are pairwise disjoint.

Suppose we have given a set of C axis-parallel rectangles $R_1, R_2, \ldots, R_C$ which are pairwise disjoint. In addition we assume that no two edges of the rectangles are colinear. Then the following lemmata can be easily verified.

Lemma 1:

If $C \leq 3$ then there exists a horizontal or vertical line which separates one rectangle from the other(s).

Lemma 2:

If $C \geq 4$ then there exists a rectangle $R_{\hat{\imath}} \in \{R_1, R_2, \ldots, R_C\}$ such that the first quadrant of the coordinate system which has the left lower corner of $R_{\hat{\imath}}$ as the origin is not intersected by any other rectangle R_j, $j \neq \hat{\imath}$.

(cf. figure 9)

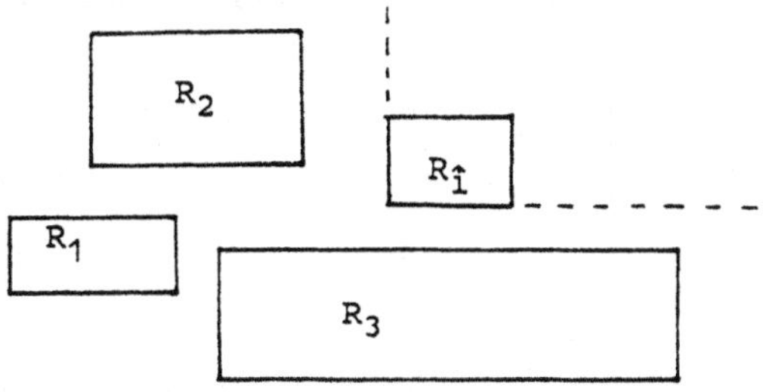

fig. 9

Lemma 1 shows that in the case $C \leq 3$ there is at most one C-r'-clustering of S for a given prototype. It follows that all C-r'-clusterings can be constructed in linear time by searching some minima, maxima and k_i-largest or smallest elements.

If $C \geq 4$ we can use the following algorithm:

Algorithm:

1. Determine $\hat{i}$ such that $R_{\hat{i}}$ fulfills lemma 2 by using the prototype.
2. Choose $l(R_{\hat{i}}) := p \in S$.
3. Set
   $t(R_{\hat{i}})$:= vertical maximum of the points of S to the right of $l(R_{\hat{i}})$
   $b(R_{\hat{i}})$:= $k_{\hat{i}}$-vertical-largest element of the points of S to the right of $l(R_{\hat{i}})$
   $r(R_{\hat{i}})$:= horizontal maximum of the points of S to the right of $l(R_{\hat{i}})$ and above $b(R_{\hat{i}})$.
4. if $R_{\hat{i}}$ is a correct rectangle then
   determine all (C - 1)-r'-clusterings of $S \backslash R_{\hat{i}}$ using the prototype which represents the rectangles $\{R_1, R_2, \ldots, R_C\} \backslash \{R_{\hat{i}}\}$ and check whether the resulting rectangles are from the correct prototype.
   goto 2.

Step 3 needs linear time and steps 3 and 4 are executed O(n) times. If C is looked as a constant the time needed by the algorithm is $O(n^{C-2})$. This leads to:

Theorem 4:

The C-r'-clustering problem can be solved in time O(n) if $C \leq 3$ and in time $O(n^{C-2})$ if $C \geq 4$ is regarded as a constant.

The example in figure 10 demonstrates that there are $\Omega(n)$ 4-r'-clusterings for a set of n points. This example can be generalized such that there are $\Omega(n^{(d-1)^2}$ d^2-r'-clusterings for a set S of O(n) points in the plane. The construction is very technical and so we omit it here.

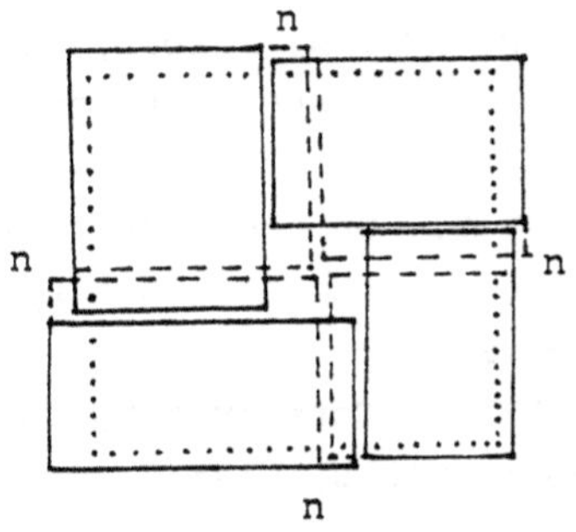

fig. 10

ACKNWOLEDGMENT:
The authors wish to thank M. B. Dillencourt and F. Jarre for a lot of helpful hints and improvements.

REFERENCES:

1. AHO, HOPCROFT, ULLMAN, The design and analysis of computer algorithms, Addison Wesley, 1974
2. A. ALON, U. ASCHER, Model and solution strategy for placement of rectangular blocks in the Euclidean plane, IEEE Transactions on Computer Aided Design, Vol. 7, No. 3, 1988, pp. 378 - 386
3. F. AURENHAMMER, Power diagrams: properties, algorithms and applications, SIAM J. Comp., Vol. 16, No. 1, 1987, pp. 78 - 96
4. R. COLE, M.SHARIR, C. YAP, On k-hulls and related problems, ACM SIGACT, Symp. on. Theory of Computing, 1984, pp. 154-166
5. F. DEHNE, An $O(n^4)$ algorithm to construct all Voronoi diagrams for K nearest neighbor searching in the Euclidean plane, Proceedings of the 10th International Colloquium on Automata, Languages and Programming (ICALP '83), Barcelona, Spain, Lecture Notes in Comp. Sci., No. 154, pp. 160 - 172
6. F. DEHNE, H. NOLTEMEIER, Clustering methods for geometric objects and applications to design problems, The Visual Computer, Vol. 2, Springer 1986, pp. 31 - 38
7. F. DEHNE, H. NOLTEMEIER, A computational geometry approach to clustering problems, Proc. 1st ACM Siggraph Symp. Comput. Geom., Baltimore, MD, USA, 1985
8. F. DEHNE, H. NOLTEMEIER, Clustering geometric objects and applications to layout problems, Proc. Comput. Graph., Springer Tokyo, 1985

9. H. EDELSBRUNNER, J. O'ROURKE, R. SEIDEL, Constructing arrangements of lines and hyperplanes with applications, SIAM J. Comput., Vol. 15, No. 2, 1986, pp. 341 - 363

10. H. EDELSBRUNNER, E. WELZL, On the number of line separations of a finite set in the plane, Journal of Combinatorial Theory, Vol. 38, No. 1, 1985, pp. 15 - 29

11. H. EDELSBRUNNER, E. WELZL, Constructing belts in two-dimensional arrangements with applications, SIAM J. Comput. Vol. 15, No. 1, 1986, pp. 271 - 284

12. D. T. LEE, On k-nearest neighbor Voronoi diagrams in the plane, IEEE Trans. on Comp., Vol. c-31, No. 6, 1982, pp. 478 - 487

13. F. P. PREPARATA, M. I. SHAMOS, Computational geometry, an introduction, Springer New York, 1985

14. E. WELZL, More on k-sets of finite sets in the plane, Discrete Comput. Geom., Vol. 1, 1986, pp. 95 - 100

15. F. F. YAO, A 3-space partition and its applications, Proc. 15th ACM Symp. on Theory of Comp., 1983, pp. 258 - 263

A sweep algorithm for the all-nearest-neighbors problem

Klaus Hinrichs*, Jurg Nievergelt, Peter Schorn
Department of Computer Science, University of North Carolina, Chapel Hill, NC 27599-3175, USA

Abstract

The 2-dimensional all-nearest-neighbors problem is solved directly in asymptotically optimal time O(n*log n) using a simple plane-sweep algorithm. We present the algorithm, its analysis, and a "foolproof" implementation which guarantees an exact result at the cost of using five-fold-precision rational arithmetic.

Keywords

Computational geometry, complexity, proximity problems, plane-sweep algorithms

1. Introduction

We consider the 2-dimensional *all-nearest-neighbors problem*: Given a set S of n points in the plane, find a nearest neighbor of each with respect to the Euclidean metric.

It is well known that Ω(n*log n) is a tight lower bound for this problem in the algebraic decision tree model of computation. Known algorithms with optimal worst case time complexity O(n*log n) are of two types:

1) Extract the solution (in linear time) from the answer to the more general problem of constructing the Voronoi diagram of the given points. The latter can be computed in time O(n*log n) both by a divide-and-conquer algorithm [SH 75], as well as by an intricate sweep based on transformed data [F 86].
2) Compute the desired result directly, without the additional information present in the Voronoi diagram. [V 86] describes an algorithm which works for any number of dimensions in any L_p-metric. It maintains a growing set of shrinking boxes; when a box has shrunk to a single point, its associated information determines a nearest neighbor.

---

* current address: Asea Brown Boveri, Corporate Research, Dept. CRBC, CH-5405 Baden, Switzerland

Algorithms of both types are complicated: their intricate logic and data structures are rarely specified in a sufficiently formal notation to allow the reader to assess the complexity of their implementation. We show how the 2-dimensional all-nearest-neighbors problem is solved effectively in asymptotically optimal time O(n*log n) using a plane-sweep algorithm [SH 76]. A vertical line (front, or cross section) sweeps the plane from left to right, stopping at every transition point (event) of a geometric configuration to update the cross section. All processing is done at this moving front, without any backtracking, with a look-ahead of only one point. Events to be processed are queued in an *x-queue*, the status of the cross section is maintained in a *y-table*. In the slice between two adjacent events, the relevant properties of the geometric configuration seen so far do not change; therefore the y-table needs to be updated only at transition points. Sweep algorithms have a simple structure typical of greedy algorithms:

```
initialize x-queue X;
initialize y-table Y;
while not emptyX do
   p := nextX;
   transition(p)
end
```

2. Left-right asymmetry

Sweep algorithms impose an arbitrary left-to-right order on the data to be processed, thus distinguishing a *known past* from a *future yet to be discovered.* Since the nearest neighbor of a point may lie "in the future" as yet unknown to the sweep algorithm, we study a simpler problem, *all-nearest-neighbors-to-the-left.* A solution to the original *all-nearest-neighbors* problem is trivially obtained in linear time O(n) from a solution to *all-nearest-neighbors-to-the-left* together with a solution to the analogous problem *all-nearest-neighbors-to-the-right.*

In the following figure an arrow points to the nearest neighbor of a point.

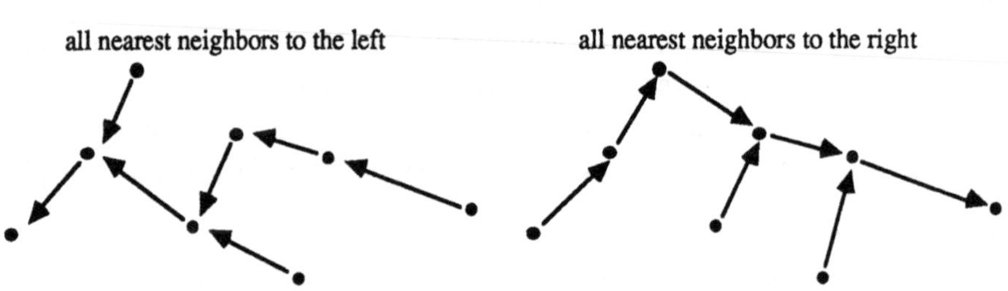

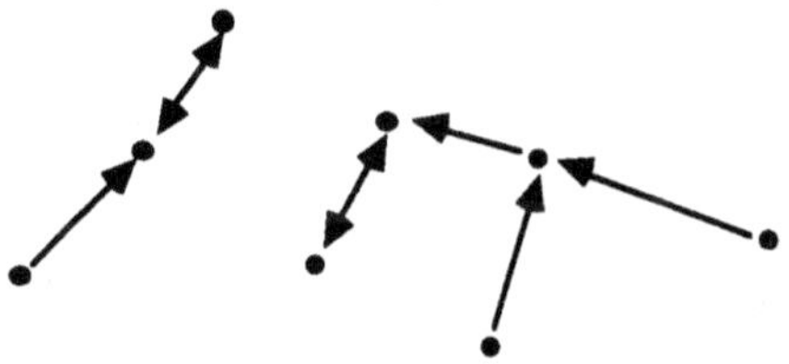

3. The sweep invariant

Let S_{left} denote the set of those points of S that lie to the left of the sweep line: initially empty, S_{left} grows one point at a time until at termination it equals S. During the sweep we maintain, as an invariant represented in the Y-table, the intersection $V \cap L$ of the Voronoi diagram V of S_{left} with the sweep line L. For $q \in S_{left}$, let V(q) denote the Voronoi polygon of q with respect to S_{left}. The points q whose polygon V(q) intersects the sweep line are called *active points*, and the edges of V that intersect the sweep line are called *active bisectors*. V partitions the sweep line into disjoint intervals I(q), one for each active point $q \in S_{left}$. All points in I(q) have q as their nearest neighbor to the left. To each active point there corresponds an *upper* and a *lower* active bisector. The intersections of the active bisectors with the sweep line define a total order on the active points and on their bisectors.

The intersection $V \cap L$ of the Voronoi diagram V of S_{left} with the sweep line must be updated in either of the following cases:

1) S_{left}, and hence V, remain unchanged, but there is a topological change in $V \cap L$ as L sweeps across a vertex of V. Type of transition: Intersection of two active bisectors. Action: An interval is deleted from $V \cap L$, its territory is distributed among its two neighbors.
2) S_{left} changes, and thus V changes, as L sweeps across a new data point. Type of transition: A data point $p \in S - S_{left}$. Action: A new interval is created at the expense of existing intervals, some of which may disappear completely, and is inserted into $V \cap L$.

We discuss both types of transitions.

3.1 Intersection point

An intersection of two active bisectors is processed by removing an interval I(p), then separating its two former adjacent intervals I(r) and I(s). As shown in the figure below, let r be the lower neighbor of p, s the upper neighbor of p. The two bisectors bs(r,p) and bs(p,s) are removed, the new bisector bs(r,s) is inserted into $V \cap L$.

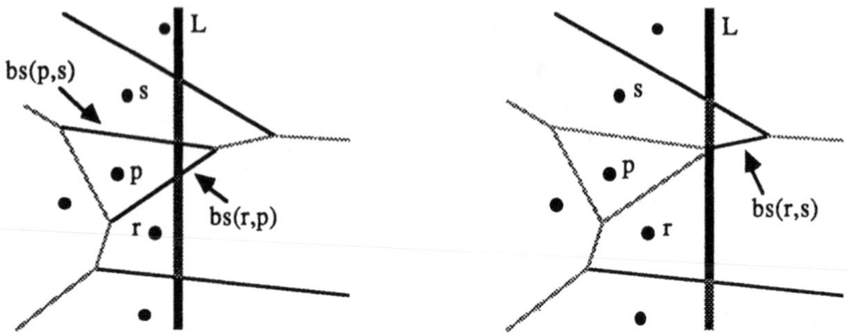

3.2 Data point

A new data point p generates a new interval I(p) created at the expense of intervals of currently active points. The interval I(nn) of p's nearest neighbor nn will always lose some territory to I(p). Some of its successor and predecessor intervals may also contribute to I(p). We find those by starting with nn and iterating upwards and downwards over active points. After computing I(p), and possibly deactivating some points, p becomes a new active point.

Updating the sweep invariant requires computing the new interval I(p) by determining its lower and upper bisectors. We begin by finding the interval I(nn) into which p falls and processing it. This involves removing from I(nn) the subinterval of those points closer to p than to nn, and initializing I(p) to this subinterval, as the following figure shows. The intersection of bisector bs(p,nn) of p and nn with L determines what needs to be done. If I(p) swallows up all of I(nn), nn is deactivated.

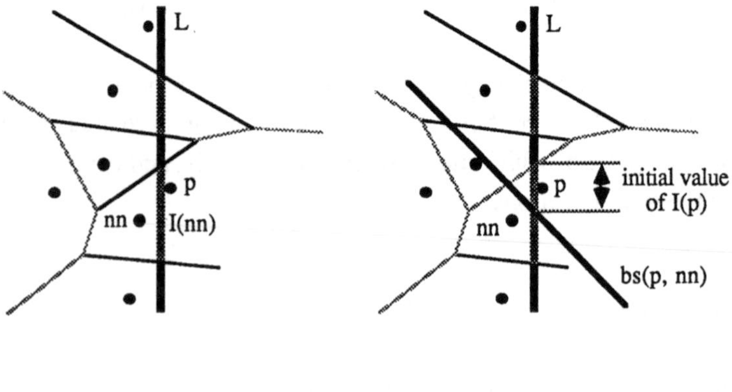

Having initialized I(p), we check whether I(p) needs to be extended upwards and/or downwards. We describe how the upper bisector of I(p) is computed; the lower is handled symmetrically. I(p) is extended upwards if nn has a successor, and bs(p,nn) intersects L outside the subinterval from p to the upper bisector of nn. The upward extension is the following loop:
Starting with q := nn, we immediately move upwards (q := successor(q)), and process each active point q in turn, until a termination condition T is satisfied. *Processing* q determines whether I(q) will lose some or all of its territory to I(p). The *termination condition* T is satisfied when T1: the bisector bs(p,q) of p and q intersects a test interval, namely the subinterval between p and the upper bisector of q, or T2: q has no successor.

The processing step depends on the termination test as follows:

P1: If q does not satisfy the termination test, q is deactivated, and I(q) is swallowed up by I(p).

P2: If q satisfies the termination test T1, I(q) loses some of its territory to I(p), namely everything below bs(p,q).

P3: If q does not satisfy the termination test T1, but satisfies T2, q is deactivated, and I(q), which extends to infinity, is swallowed up by I(p).

The next three pictures illustrate steps of this iteration. In the first, the termination test is not satisfied because bs(p, q) intersect L above the test interval that extends from p to the upper bisector of q: Step P1 deactivates q and extends I(p). In the second, termination test T1 is satisfied: I(q) loses some of its territory to I(p), a new bisector is inserted, and the final state is shown in the third figure.

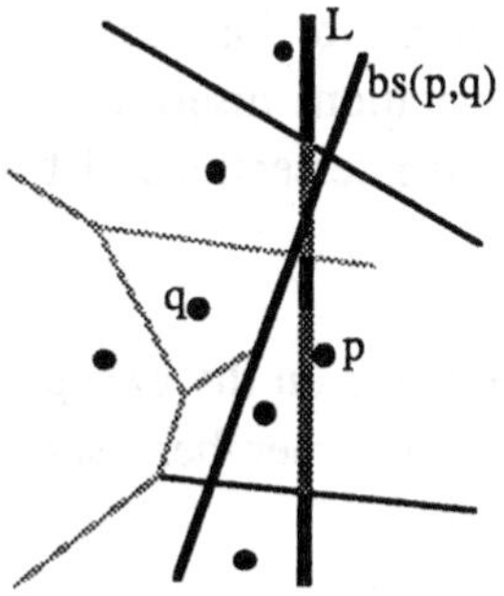

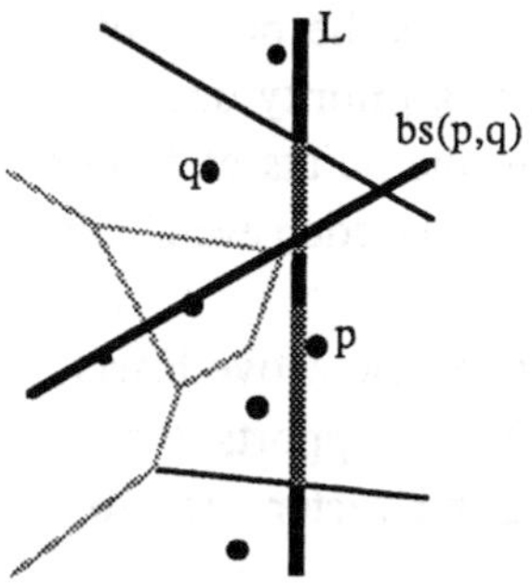

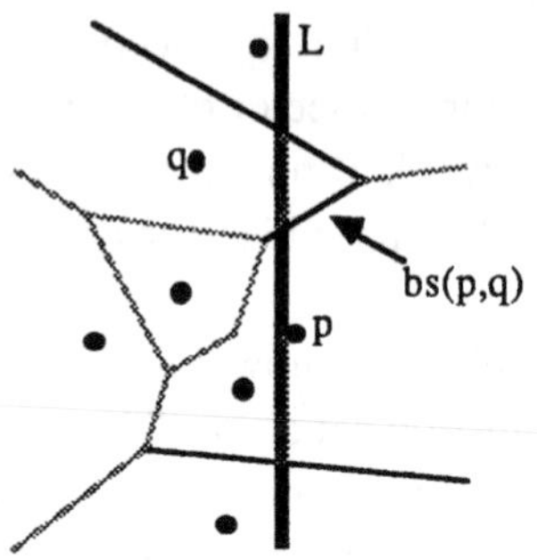

4. Embedding the invariant in an algorithm: data structures

The invariant developed in the previous section is embedded into a plane-sweep algorithm as the following figure suggests:

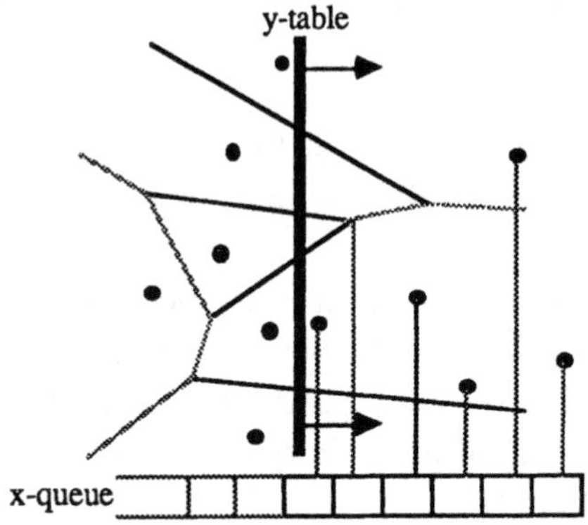

The x-queue stores the points of S and intersection points of bisectors sorted by x-coordinate. It is a priority queue which supports the operations insert and next, and is initialized with the points of S; during the sweep certain intersections of bisectors are inserted. Each intersection event points to its two bisectors.

The y-table stores the active bisectors sorted by their y-value on the sweep line. It is a dictionary which supports the operations insert, delete, member, successor and predecessor. Each bisector refers to the two points it bisects.

An intersection point in a Voronoi diagram is generated by two bisectors (its "parents") that belong to the same Voronoi polygon, and thus are adjacent. Thus a new active bisector is checked for intersection only against its successor and its predecessor. An intersection that lies to the left of the sweep line is ignored, one to the right (in the future) is inserted into the x-queue. Either or both of the parents may cease to be active (and thus be deleted from the y-table) before the intersection event they spawned is processed. Any intersection event in the x-queue that lacks two parents in the y-table is ignored.

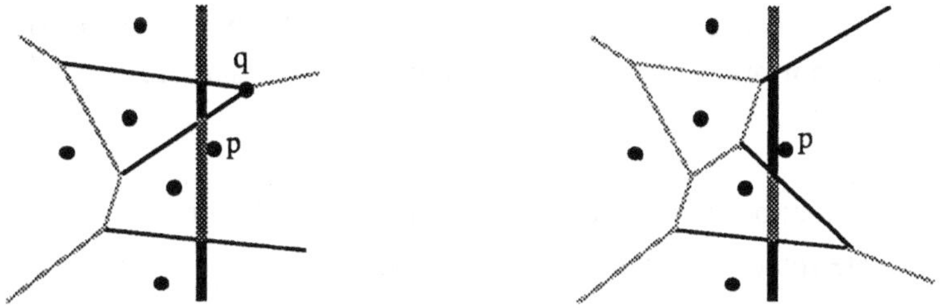

The figure above shows an example where an intersection q, although created, does not survive long enough to be processed as a transition when the sweep reaches q. Its parents are deactivated by the new point p. q could be deleted from the x-queue when p is processed; it is easier to leave it in the queue, and only process intersections whose parents are still active.

5. Program structure

We have implemented this all-nearest-neighbors algorithm as a MacPascal program. It lets the user enter points with the mouse and animates the algorithm's execution. The algorithm proper takes about 400 lines; additional code handles graphics and table management. The following procedures are called to process an event.

`findnnY(p, nn, lowerS, upperS)`
returns the nearest neighbor nn to the left of a given point p, and its lower and upper bisectors.

`inInterval(y, b1, b2)`
returns true if y lies in the interval bounded by b_1 and b_2, otherwise false.

If direction = $+\infty$, `nextP(bs, direction)`
returns the upper of the two points which determine bs, otherwise the lower one.

If direction = $+\infty$, `nextY(bs, direction)`
returns the successor of bs in the y-table, otherwise the predecessor.

`memberY(bs)` returns true if the bisector bs is contained in the y-table, otherwise false.

`bisector(pl, pu)` returns the bisector of two points p_l and p_u.

`deleteY(bs)` removes a bisector from the y-table.

`insertAndCheckY(bs)` inserts a bisector bs into the y-table and checks bs for intersection against its predecessor and successor. Any intersections found are inserted into the x-queue.

Two fictitious bisectors bound an interval extending toward $+\infty$ or $-\infty$. `infinity(bs)` determines whether bs is such a sentinel.

The procedure transition incorporates all the work to be done in processing a transition point.

```
procedure transition(p);
begin
  if p = newPoint then
```

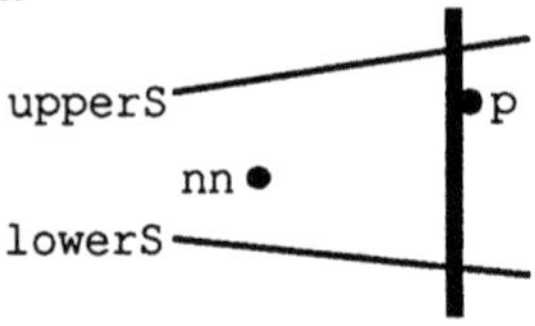


```
    findnnY(p, nn, lowerS, upperS);
    updateY(p, nn, upperS, +∞);
    updateY(p, nn, lowerS, -∞)
  elsif p = intersection then
```

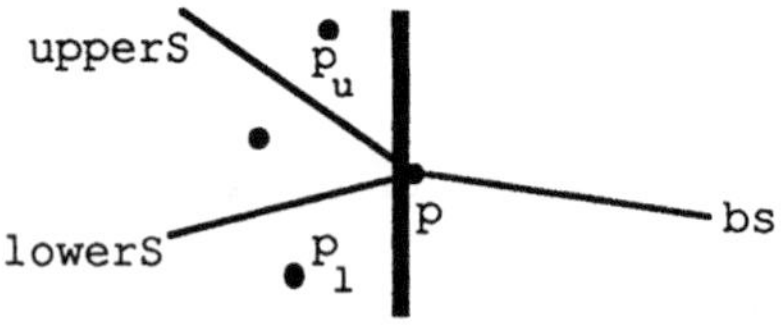


```
    if memberY(upperS) and memberY(lowerS) then
      bs := bisector(pl, pu);
      deleteY(upperS);
      deleteY(lowerS);
      insertAndCheckY(bs)
    end
  end
end transition;
```

If a new point p is encountered and lies in the interval I(nn) bounded by nextbs, the invariant in direction ($+\infty$ or $-\infty$) of nextbs is updated as explained in section 3:

```
procedure updateY(p, q, nextbs, direction);
begin
  bs := bisector(p, q);
  while not inInterval(bs_y, p_y, nextbs_y)
```

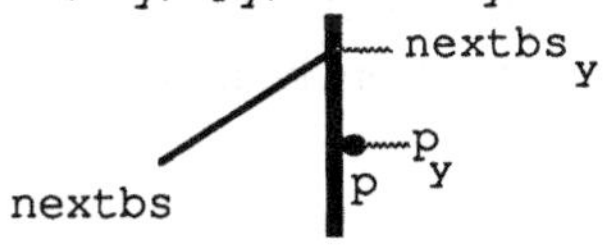


```
        and not infinity(nextbs) do
    q := nextP(nextbs, direction);
    n := nextY(nextbs, direction);
    deleteY(nextbs);
    nextbs := n;
    bs := bisector(p, q)
  end;
  if not infinity(nextbs) or
     inInterval(bs_y, p_y, direction) then
```

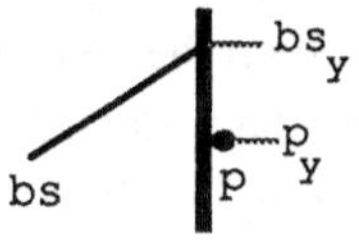


```
    insertAndCheckY(bs)
  else
    Ytable_direction := p
  end
end updateY;
```

6. Analysis

Each time a new point is encountered during the sweep at most two new bisectors are inserted into the y-table. The two bisectors may intersect each other, the upper bisector may intersect its upper neighbor in the y-table, and the lower bisector may intersect its lower neighbor. Therefore at most three intersection events are inserted into the x-queue. Each time an active intersection point is encountered a new bisector is inserted into the y-table. This bisector may intersect its upper and its lower neighbor in the y-table, and therefore at most two new intersection events may be generated. Since at each active intersection point an interval I(p) is removed, i.e. a point $p \in S$ is deactivated, there may be at most n active intersection points. Therefore at most 5n intersection events are inserted into the x-queue, and the x-queue contains no more than 6n events at any time.

If the x-queue is implemented as a heap the operations nextX and insertX can be performed in time O(log n). Initializing the x-queue with the points of S takes O(n) time.

The y-table is implemented by a balanced binary tree. The operations deleteY, memberY and nextY can be performed in time O(log n). The procedure insertAndCheckY inserts a new bisector into the y-table and at most two intersection events into the x-queue. Therefore it takes O(log n) time. The nearest neighbor nn of a given point p is determined by searching in the y-table for the bisectors bounding the interval I(nn); this can also be done in time O(log n).

The cost for updateY is O((k+1)*log n), where k is the number of points deactivated in a call of updateY. Since at most n points can be deactivated the total cost for all calls of updateY is O(n*log n). Hence the total cost for processing new points is O(n*log n). Since at most 5n intersection events are generated, the total cost for processing all intersection events is O(n*log n). Therefore the total cost of the plane-sweep algorithm for the all-nearest-neighbors problem is O(n*log n).

The constant factor in front of the (n*log n)-term of the cost can be obtained from the total number of O(log n)-operations performed on the heap and the balanced tree:

5n * InsertInHeap + 6n * NextFromHeap
+ 11n * SearchInTree + 3n * InsertInTree + 3n * DeleteFromTree

7. An exact implementation

We have implemented this algorithm on the Macintosh, using MacPascal, in such a way that it handles all degenerate configurations and achieves an exact result. The degeneracies to be dealt with are events of equal x-coordinate, in particular, intersection events that coincide. If input coordinates are integers, five-fold precision rational arithmetic guarantees exact results.

Degeneracies are handled by extending the order on events and bisectors to the case of equal x- and/or y-coordinates. Let (x, y, t), $t \in$ {intersection, data_point} denote an event.

$$(x_1, y_1, t_1) < (x_2, y_2, t_2) \Leftrightarrow \quad (x_1 < x_2)$$
$$\vee\ (x_1 = x_2) \wedge (y_1 < y_2)$$
$$\vee\ (x_1 = x_2) \wedge (y_1 = y_2) \wedge (t_1 = \text{intersection}) \wedge (t_2 = \text{data\_point})$$

Two distinct intersection events which coincide in the plane can be processed in any order - we need not break the tie in this case.

Let $a\,x + b\,y = c$ be the equation of a bisector ($b \geq 0$).

$$((a_1 x + b_1 y = c_1) < (a_2 x + b_2 y = c_2)) \text{ at } x_0 \Leftrightarrow$$
$$(b_1 > 0) \wedge (b_2 > 0) \wedge ((c_1 - a_1 x_0) / b_1 < (c_2 - a_2 x_0) / b_2)$$
$$\vee\ (b_1 > 0) \wedge (b_2 > 0) \wedge ((c_1 - a_1 x_0) / b_1 = (c_2 - a_2 x_0) / b_2) \wedge (-a_1 / b_1 < -a_2 / b_2)$$
$$\vee\ (b_2 = 0) \wedge (b_1 > 0)$$
$$\vee\ (b_2 = 0) \wedge (b_1 = 0) \wedge (a_1 > a_2)$$

We guarantee an exact result under the following assumption: all given points lie on a finite integer grid, i.e. there exists $M \in \mathbf{N}$ such that $|x| < M$ and $|y| < M$ for any given point $(x, y) \in \mathbf{Z} \times \mathbf{Z}$. We represent all numbers as a pair (numerator, denominator) and perform all operations with the precision necessary to avoid overflow. By examining all formulas used and expressions evaluated we bound the size of any intermediate integer generated in terms of the size of the input data. Four kinds of calculations exist:

1) Computing the coefficients in the equation of the bisector of two given points. For the equation of the bisector written as $a x + b y = c$ we find $|a| < 4 M$, $|b| < 4 M$ and $|c| < 2 M^2$.
2) Calculating the intersection point of two bisectors. The intersection point can be represented as $(x/d, y/d)$ with $|x| < 12 M^3$, $|y| < 12 M^3$ and $|d| < 12 M^2$.
3) Determining the order of two bisectors in the y-table. There are two cases depending on whether the bisectors have to be compared at a given point or at an intersection point. The latter is worse and requires handling integers whose absolute value is bounded by $288 M^5$.
4) Determining the order of two events in the x-queue. Here we have the three cases *given point against given point*, *given point against intersection* and *intersection against intersection*. The latter is worse and requires handling integers whose absolute value is bounded by $144 M^5$.

Conclusion: Let m be the number of bits used to represent the input data. Rational arithmetic using integers represented by $\lceil \log_2 288 \rceil + 5 m = 9 + 5 m$ bits will compute results exactly.

8. Conclusions

The algorithm presented is relatively simple both from the point of view of theory and implementation: it calls upon standard techniques in algorithm design (plane-sweep) and data structures (priority queues and dictionaries), and has been programmed to animate algorithm execution in just a few hundred lines of Pascal. It is plausible to assume that other optimal algorithms which first compute the Voronoi diagram of the given set S of points are less efficient, in practice, since the Voronoi diagram provides much more information about S than is needed to solve the problem.

Two issues need to be considered before a geometric program can be considered practical: 1) can it handle degenerate configurations, e.g. three lines that intersect in the same point? and 2) is it numerically robust, e.g. is it certain to avoid overflow, can errors be bounded? Geometric algorithms have rarely been studied from this point of view. Our implementation addresses these problems by assuming that the given points lie on an integer grid and using multi-precision rational arithmetic. Degenerate cases are classified into two categories: those where order is irrelevant, as in the case of coinciding intersections, and those that are ordered lexicographically.

Acknowledgement

This paper originated in a course on computational geometry held at the University of North Carolina. It was supported in part by the National Science Foundation under grant DCR 8518796.

References

[F 86] S. Fortune:
A Sweepline Algorithm for Voronoi Diagrams,
Proc. 2nd Ann. Symp. on Computational Geometry, ACM, 313-322, 1986.

[SH 75] M. Shamos, D. Hoey:
Closest-Point Problems,
16th Annual IEEE Symposium on Foundations of Computer Science,
151 - 162 (1975).

[SH 76] M. Shamos, D. Hoey:
Geometric intersection problems,
17th Annual IEEE Symposium on Foundations of Computer Science,
208 - 215 (1976).

[V 86] P. Vaidya:
An Optimal Algorithm for the All-Nearest-Neighbors Problem,
Proc. 27th IEEE Symp. Foundations of Computer Science, 117-122, 1986.

On Continuous Homotopic One Layer Routing

(Extended Abstract)

Shaodi Gao*, Mark Jerrum**, Michael Kaufmann*
Kurt Mehlhorn*, Wolfgang Rülling*, Christoph Storb*

* FB 10, Universität des Saarlandes, 6600 Saarbrücken, West Germany.
** Dep. of Comp. Science, University of Edinburgh, Edinburgh EH9 3JZ, Scotland.

Abstract: **We give an $O(n^3 \cdot \log n)$ time and $O(n^3)$ space algorithm for the continuous homotopic one layer routing problem. The main contribution is an extension of the sweep paradigm to a universal cover space of the plane.**

1. Problem Definition

An input of the *continuous homotopic one layer routing problem* (CHRP) consists of a set $W = \{w_1, ..., w_l\}$ of paths (also called wires) and a set $O \subseteq \mathbf{R}^2$ of obstacles. A path is a continuous curve. We assume that wires do not intersect and that they are disjoint from all obstacles.

A solution to a CHRP is a set $P = \{p_1, ..., p_l\}$ of paths which is obtained by a homotopic shift (precise definition below) from W such that $U(p_i) := \{x; dist(x, p_i) < 1/2\}$ is a simply connected region of the plane, $U(p_i) \cap U(p_j) = \emptyset$ for $i \neq j$ and $U(p_i) \cap U(O) = \emptyset\ \forall i$. Here *dist* denotes the Euclidian distance. An output to a CHRP is either a solution or the indication that no solution exists. Figure 1 gives an example of a CHRP.

Let us now give precise definitions.

A path is a continuous function $p : [0,1] \to \mathbf{R}^2$. A set $W = \{w_1, ..., w_l\}$ of paths is *collisionfree* if $w_i(t) \neq w_j(s)$ for $(i,t) \neq (j,s)$ and $w_i(t) \notin O$ for all i and t. So an input of a CHRP consists of a set O of obstacles and a collisionfree set W of paths.

We call two paths p and q ***homotopic*** ($p \sim q$) with respect to a set $F \subseteq \mathbf{R}^2$ of holes in the plane if there is a continuous function $h : [0,1] \times [0,1] \to \mathbf{R}^2$ such that

(1) $h(0,t) = p(t)$ and $h(1,t) = q(t)$, for $t, 0 \leq t \leq 1$
(2) $h(\lambda, 0) = p(0) = q(0)$ and $h(\lambda, 1) = p(1) = q(1)$ for $0 \leq \lambda \leq 1$
(3) $h(\lambda, t) \notin F$ for $0 \leq \lambda \leq 1, 0 < t < 1$

Let $T = \{w_i(0), w_i(1); 1 \leq i \leq l\}$ be the set of endpoints, also called terminals, of the paths in W and let $F = T \cup O$. We call F the set of features. A set $P = \{p_1, ..., p_l\}$ of paths is a ***homotopic shift*** of W if

i.) p_i is homotopic to w_i with respect to F, $1 \leq i \leq l$ and
ii.) P is collision-free.

Supported by DFG, SFB 124, TP B2.

For a set $S \subseteq \mathbf{R}^2$ let $U(S) = U_{1/2}(S) = \{x \in \mathbf{R}^2; dist(x,s) < 1/2 \text{ for some } s \in S\}$ be the open 1/2 - neighborhood of S. For a path p we frequently use p to denote the set $\{p(t); 0 \leq t \leq 1\}$ of points of p.

So a solution to a CHRP given by O and W consists of a homotopic shift P of W such that $U(p_i)$ is simply connected (simplicity condition), $U(p_i) \cap U(O) = \emptyset$ and $U(p_i) \cap U(p_j) = \emptyset$ $\forall i$ and $j, i \neq j$ (disjointness condition). We call P a λ - solution if the above conditions hold for $U_{1/2 \cdot \lambda}$ instead of $U_{1/2}$.

A ***cut*** is either a straight-line segment connecting two features or a semi-infinite ray starting in a feature. In either case a cut must not cross any other feature. The ***capacity*** of a cut is its Euclidian length.

For a cut C and a path w let $cr(C,w)$ denote the number of points which C and w have in common, i.e.,

$$cr(C,w) = |\{t; 0 < t < 1 \text{ and } w(t) \in C\}|$$

and let

$$mincr(C,w) = \min_{p \sim w} cr(C,p)$$

be the minimal number of crossings of any homotopic shift of w with C, cf. Figure 2. Note that crossings at the endpoints do not count. The ***density*** of a cut is given by

$$dens(C) = \sum_{w \in W} mincr(C,w).$$

We are now in a position to state our main theorem. For part b.) and c.) of the theorem we assume that the input paths are polygonal paths. We denote the number of bends in wire w_i by b_i, the total number of bends by $b = \sum_i b_i$, the number of features by m and the size of the input by $n = m + b$.

Theorem 1:

a.) A CHRP has a solution iff the cut condition holds, i.e., iff

$$dens(C) + 1 \leq cap(C)$$

for all cuts C.

b.) The cut condition can be checked in time $O(b \cdot m + m^2 \cdot \log bm) = O(n^2 \cdot \log n)$.

c.) A solution (if there is one) can be constructed in time $O(b \cdot m^2 \cdot \log bm) = O(n^3 \cdot \log n)$ and space $O(max_i b_i \cdot m^2) = O(n^3)$. Moreover, this solution minimizes the total path length.

The first results on homotopic one-layer routing are due to Cole/Siegel [CS] and Leiserson/Maley [LM]. They proved a result analoguous to theorem 1 for the L_∞-Norm, i.e., *dist* corresponds to the Manhattan distance. Maley [M] extends the result to arbitrary polygonal distance functions and as a limiting case also to the Euclidian metric. His running time for part b.) is the same as ours and it is $O(n^4 \cdot \log n)$ and space $O(n^4)$ for part c.). Moreover, he does not minimize the total path length. Our result was obtained independently of his (the papers [CS] and [LM] were the common basis) and the techniques used in parts a.) and c.) differ widely; part b.) is a simple generalization of [LM]. Maley [M] gives a very detailed proof for part a.) based on notions of combinatorial topology. His algorithm for part c.) is a slight generalization of [LM]. Our proof for part a.) is elementary and intuitive but may not fulfill the highest standards of rigour. Our main emphasis is on part c.) and we introduce a novel algorithmic idea there: sweeping the universal cover space.

Consider a fixed wire w_i and let $s = w_i(0), t = w_i(1)$. The ***universal cover space*** of the plane with origin s and with respect to the set F of holes is given by
$C = \{(x,p); x \in \mathbf{R}^2$ and p is the homotopy class of a path from s to $x\}$;
cf. Figure 3, 4 and 5 for an illustration.

The solution path p_i corresponding to w_i is a shortest path from (s,ϵ) to $(t,[w_i])$ in $C - \mathcal{F}$ where $\mathcal{F}$ is the forbidden region for wire w_i, cf. section 3 and Figure 5. We construct the solution path p_i for w_i by a sweep of the cover space. More precisely, we first construct the rubberband equivalent $rbe_0(w_i)$, i.e., a shortest polygonal path homotopic to w_i. In Figure 1 the w_i are already given by their rubberband equivalents (RBE). RBE's were introduced in [LM] and it was also shown how to construct them in [LM].

The path $rbe_0(w_i)$ is a path from (s,ϵ) to $(t,[w_i])$ in C. Our idea is now to sweep a line perpendicular to $rbe_0(w_i)$ from (s,ϵ) to $(t,[w_i])$ (at bends of $rbe_0(w_i)$ the line turns into a semi-infinite ray), and to construct the solution path p_i as we move along, using the ***funnel method***.
In the funnel method ([T]) we maintain a partition of the sweep line, where two points x and y belong to the same interval if the shortest paths to x and y are combinatorially the same and we maintain the current intersection of the sweep line with the forbidden region, cf. Figure 5.

The details of the algorithm can be found in section 3; section 2 contains a short discussion of part a.), and section 4 mentions some extensions.

It is natural to ask whether theorem 1 can be extended to vertex-disjoint routings in planar graphs. A partial answer is provided by theorem 2.

Let the problem ***shortest vertex-disjoint routings on a planar graph (with homotopies) (SRPGH)*** be given by
a planar graph embedding G, vertex pairs $(s_1,t_1),\ldots,(s_k,t_k)$, a 'sketch' of the routings from s_i to t_i and a bound b (integer).
Question: Do vertex-disjoint paths $s_1 \to t_1, \ldots, s_k \to t_k$ exist with the described homotopies of total length $\leq b$?

Theorem 2: SRPGH is NP - complete.

2. Routability and the Cut Condition

We briefly discuss the necessity and sufficiency of the cut condition. Let $P = \{p_1, ..., p_l\}$ be a solution to the CHRP and let C be any cut connecting two features, say f and g. Let $C^0 = C - U(f) - U(g)$. It suffices to show that the length of $C^0 \cap U(p_i)$ is at least $mincr(C, w_i)$. Let $s, s' \in \{f, g\} \cup \bigcup_j (C \cap p_j)$, $s \neq s'$. If $U(s) \cap U(s') = \emptyset$ for any such pair then we are done. So let us consider a pair s, s' with $U(s) \cap U(s') \neq \emptyset$. Note first that $s, s' \in p_i$ for some i. This follows from the disjointness condition. Observe next that the subpath p connecting s and s' must be homotopic to the straight-line segment $\overline{ss'}$ because of the simplicity condition. Thus we can remove the intersections s and s' (or one of them if the other is a terminal) by a homotopic shift, cf. Figure 6. Continuing in this way we obtain a path $p_i' \sim p_i$ such that $C \cap U(p_i') \subseteq C \cap U(p_i)$ and any two intersections of p' with C have distance at least one. Thus the length of $C^0 \cap U(p_i)$ is at least $mincr(C, w_i)$ and the necessity of the cut condition is established.

For the proof of sufficiency and the algorithms we need some additional definitions. Let p be a path and let C be a cut. A path p_1 is an ***initial segment*** of p if there is a continuous function $\alpha : [0,1] \rightarrow [0,1]$ such that $\alpha(t) \leq t$ and $p_1(t) = p(\alpha(t))$ for all t. A ***crossing*** of p and C is an initial segment p_1 of p with $p_1(1) \in C$. Let p and q be paths with $p(1) = q(0)$. Then the ***concatenation*** $p \circ q$ is defined by $(p \circ q)(t) = p(2 \cdot t)$ for $0 \leq t \leq 1/2$ and $(p \circ q)(t) = q(2 \cdot t - 1)$ for $1/2 \leq t \leq 1$.

Let C be a cut and let p and q be paths with $p(0) = q(0)$ and $p(1), q(1) \in C$. We call p and q C-equivalent if p is homotopic to $q \circ \overline{q(1)p(1)}$.

Let C be a cut and p_1 a crossing of p and C. Then the C-equivalence class of p_1 is called the ***type*** of the crossing p_1.

Let C be a cut incident to a feature f, let $W = \{w_1, ..., w_l\}$ be a given set of paths and let $h = \sum_i mincr(C, w_i)$.
We can shift W into a collision-free set $P = \{p_1, ..., p_l\}$ with $h = \sum_i cr(C, p_i)$ as follows. Let $p_i = w_i$ initially. If $mincr(C, w_i) = cr(C, p_i)$ for all i then we are done. Otherwise there must be some p_j such that two of the crossings of p_j with C can be removed by the operation shown in Figure 6. Proceeding in this fashion we obtain the desired set P of paths.

Let $s_1, ..., s_h$ be the crossings of the paths in P with the cut C in the order of increasing distances from f, i.e., $dist(f, s_j(1)) < dist(f, s_{j+1}(1))$ for $1 \leq j \leq h-1$. Then $(type(s_1), ..., type(s_h))$ is called the ***signature*** of C with respect to f.

Note that the set P defined above is not uniquely determined by W; nethertheless, the signature of C is well-defined. Note also that if $mincr(C, w_i) > 0$ then some of the s_j will be intersections of p_i and C. We call the types of these s_j the types of the required crossings of w_i and C. Again, these types are independent of the particular set P into which W was shifted.

Let λ be a positive real. A path p is a ***λ-representative*** of a path w if

- p is homotopic to w and if
- for all features f and cuts C incident to f:
  if p_1 is a crossing of p and C and $type(p_1) = t_i$, where $(t_1, t_2, \ldots)$ is the signature of C with respect to f, then $dist(f, p_1(1)) \geq \lambda \cdot i$.

A path p is the ***λ-realization*** of w if it is the λ-representative of shortest length. We denote the λ-realization by $rbe_\lambda(w)$.

Remark:
An alternative definition of λ-realization is as follows. Let w be a path and let $s = w(0)$. Then the ***forbidden region*** for the λ-realization of w is given by
$\mathcal{F}_\lambda = \{(y,p) \in C$; there is a feature f, a cut C incident to f with signature $(t_1, t_2, \ldots)$, $type(q) = t_i$ is the type of a required crossing of C and w and $dist(f, y) < \lambda \cdot i\}$.
Then the λ-realization of w is the shortest path from (s, ϵ) to $(t, [w])$ in $C - \mathcal{F}_\lambda$, cf. Figure 5c.

Lemma 1: If the cut condition holds then the paths $rbe_1(w_i), 1 \leq i \leq l$, exist and form a solution to the CHRP.
Proof (sketch): Replace each wire w_i by a rubberband of width λ and replace each feature by a disk of diameter λ. Start with $\lambda = 0$ and let λ grow. A rubberband runs straight except if it is forced to run otherwise. In other words any bend of a path $rbe_\lambda(w_i)$ is forced. Suppose now that we cannot enlarge λ all the way to $\lambda = 1$. Then two rubberbands of opposite curvature must collide and hence the cut condition does not hold, cf. Figure 7.

q.e.d.

3. Algorithms

Throughout this section we will assume that our CHRP has a solution. Note that the cut condition can be checked in time $O(n^2 \cdot \log n)$ by part b.) which is well below the target time for part c.).
The algorithm for part c.) is outlined in program 1.

```
1     Preprocessing (* time O(b · m · log bm) *)
2     for each w ∈ W do
          (* the loop body takes
                  time O(b_w · m^2 · log bm)*)
2.1       for each f ∈ F do
              construct the forbidden
                  region F(f) induced by f
          od;
2.2       construct F = ∪_{f∈F} F(f)
2.3       construct rbe_1(w)
      od;
```

Program 1

Let $w \in W, s = w(0), t = w(1)$. Let C be the cover space with respect to origin s and the set F of holes. Let $\mathcal{F} = \mathcal{F}_1$ be the forbidden region as defined in the remark preceding lemma 1. The solution path $rbe_1 = rbe_1(w)$ is the shortest path from (s, ϵ) to $(t, [w])$ in $C - \mathcal{F}$. We construct rbe_1 by application of the funnel method. The funnel method was introduced by Tompa [T] in the following situation, cf. Figure 8.

Given points s and t in the plane with the same y-coordinate and a set of obstacles, which are semi-infinite open vertical rays, construct a shortest path from s to t. In the absence of obstacles the shortest path would be the horizontal line segment $\overline{st}$.

Tompa proposed to sweep a vertical line (= a line perpendicular to the segment $\overline{st}$) from s to t and to record for each intermediate position of the sweep line the current funnel; i.e., a partition of the sweep line into maximal intervals such that two points x and y belong to the same block iff the shortest paths from s to x and from s to y are combinatorially equivalent, i.e., bend at the same obstacles, cf. Figure 8.

The funnel changes whenever an obstacle is hit by the sweep line. It is either augmented by one additional interval (time $O(1)$) or it is reduced by one or more intervals (time O(number of discarded intervals)). Thus the running time of his method is proportional to the number of obstacles.

We extend the funnel method to our more general situation. There are two main difficulties:

i.) The obstacles have more complex shape, cf. Figure 5.

ii.) We do not sweep the plane but a more complex topological space.

Define the rubberband equivalent rbe_0 of the path w to be the shortest path homotopic to w. (More precisely, rbe_0 is a shortest path in the closure of the set of paths homotopic to w. Note that rbe_0 goes through features and hence is not 'really' homotopic to w.) It is a polygonal path whose vertices are features. The rubberband equivalents of all wires can be constructed in time $O(n^2 \cdot \log n)$, cf. [LM]. This constitutes the preprocessing phase.

For steps 2.1, 2.2 and 2.3 we use a *sweep of the cover space* which we now define. A connected subset $\mathcal{L} \subseteq C$ is a straight-line (ray) if its projection $proj(\mathcal{L})$ on the first coordinate is. A line $\mathcal{L}$ is perpendicular to rbe_0 at footpoint $(x,p) \in rbe_0 \cap \mathcal{L}$ if x lies in the interior of one of the straight-line segments constituting rbe_0 and if the projections are perpendicular. A ray $\mathcal{L}$ is perpendicular to rbe_0 at footpoint (x,p) if (x,p) is a vertex of rbe_0, (x,p) is the start point of ray $\mathcal{L}$ and if $\mathcal{L}$ is contained in the cone defined by the two rays starting at (x,p) and being perpendicular to the line segments of rbe_0 incident to $x,p)$; cf. Figure 9.
The sweep is now given by the continuous motion of the footpoint of a line (ray) perpendicular to rbe_0 from (s,ϵ) to $(t,[w])$, more precisely, if the footpoint is on one of the straight-line segments then the footpoint moves continuously and if the footpoint is a vertex then the ray turns continuously. Thus the sweep alternates between straight-line and angular motion.

Next we will explain the three steps 2.1, 2.2 and 2.3 in more detail.

Step 2.1: For a feature f let
$\mathcal{F}(f) = \{(y,q)$; there is a cut C incident to f, C has signature $(t_1, t_2, ...)$, $t_i = type(q)$, t_i is the type of a required crossing of w with C and $dist(y,f) < i\}$
be the forbidden region for rbe_1 induced by feature f. We show how to construct $\mathcal{F}(f)$ in time $O(b_w \cdot m \cdot \log bm)$ for each feature f.

A **sector** of rbe_0 with respect to f is a maximal subpath of rbe_0 which is monotonic with respect to f and completely visible from f, i.e., rays starting in f intersect the subpath only once and do not pass through a feature before they intersect the subpath. The sectors of rbe_0 correspond in a natural way to angular sectors with tip f, cf. Figure 5 and 10. Within each angular sector the boundary of $\mathcal{F}(f)$ is a circular arc of constant radius.
Next we show how to compute the sectors and circular arcs mentioned above. We first determine for each feature f the sorted order of the other features around f and then define $2 \cdot m - 2$ rays for each f, one for each feature g and one for each interval between two features. Next we compute for each wire w and feature f the intersections between the rays starting in f and $rbe_0(w)$ by plane sweep and sort the intersections for each ray and each $rbe_0(w)$. At this point we have determined the rank of the intersection on the ray for each intersection of a ray and a rubberband, cf. Figure 11. This constitutes a global preprocessing step for step 2.1 and takes time $O(b \cdot m^2 \cdot \log bm)$.

From now on we consider again a fixed w and f. The sectors of $rbe_0 = rbe_0(w)$ with respect to f are now easily computed. We move along rbe_0 and stop at each vertex and at each intersection with a ray starting in f. At each stop one decides in time $O(1)$ whether the current sector ends. Also, the radius of the circular arc is given by the rank of any intersection within the sector. The entire process takes time $O(b_w \cdot m)$ and yields the circular arcs bounding $\mathcal{F}(f)$. For each circular arc we know the radius and the sector of rbe_0 which it constrains.

Let us call a circular arc ***visible*** at a certain position of the sweep if the sweep line intersects the arc, cf. Figure 12. The positions where an arc a is visible form an interval which we will to compute in the following. Also note that if an arc is visible at all then one of its endpoints is visible.
Let r be the starting ray of the sector of arc a and let (x,p) be the intersection of r with rbe_0; the symmetric procedure is applied to the terminating ray. We position ourselves in point (x,p) and move towards t if

- $dist(x,f) \geq i$ and the angle between the oriented line segment $\overline{xf}$ and the rbe_0 is $\geq \frac{\pi}{2}$ or if
- $dist(x,f) < i$ and the angle is $< \frac{\pi}{2}$

and towards s otherwise; this is a rudimentary form of the cover space sweep. Let us assume that we move towards t and that the second case occurs; the other cases are similar.

During our walk along rbe_0 we maintain the intersection of the sweep line with the ray or arc. We stop whenever the sweep line passes through a feature or when it passes through the endpoint of the arc. In the former case we check whether the feature is between the sweep line and the ray or arc and if so we stop the sweep because from now on the ray or arc will not be visible from the footpoint of the sweep line. In the latter case we replace the ray by the arc at the starting point of the arc and stop the

sweep at the endpoint of the arc, cf. Figure 12. The time used for this process is proportional to the number of features passed. This assumes that we sort the features with respect to rbe_0 in time $O(b_w \cdot m \cdot \log b_w m)$ beforehand. The crucial observation is now that for any fixed intersection of a ray r and rbe_0 we will pass through any feature at most once and hence the time spent for each of the $b_w \cdot m$ intersections is $O(m)$ or $O(b_w \cdot m^2)$ in total.
We have now computed for each arc its visibility interval and step 2.1 is completed.

Step 2.2: The Forbidden Region $\mathcal{F}$

Step 2.1 provides us with at most $b_w \cdot m^2$ arcs and their intervals of visibility. The goal of step 2.2 is to find for each position of the sweep line the most constraining visible arc, i.e., the intersection of the boundary of $\mathcal{F}$ with the sweep line. In general there will be two intersections, one constraining rbe_1 from the left, the other from the right. We solve this problem by a sweep along rbe_0 from (s, ϵ) to $(t, [w])$. The sweep line datastructure DS contains the intersections of the sweep line with some of the currently visible arcs. We maintain the following
Invariant:
If a currently visible arc is not contained in DS then it will never be the most constraining arc at a later position of the sweep line.
The invariant implies that the most constraining arcs are always contained in DS. Note also that DS contains at most $O(m^2)$ arcs.
We stop the sweep at four kinds of events

i.) a starting point of a visibility interval
ii.) a terminating point of a visibility interval which is currently in DS
iii.) a change of motion, from straight-line to angular or vice-versa
iv.) an intersection of two arcs in DS.

In order to find these events we maintain two queues. The queue Q_A contains all events of kind i.), ii.) and iii.) and is easily precomputed in time $O(b_w \cdot m^2 \cdot \log bm)$ using the output of step 2.1. The queue Q_B contains all positions which lie on the current segment of the sweep (i.e., either on the same straight-line or the same angular segment as the current footpoint) at which an intersection of two arcs adjacent in DS is met by the sweep line.
We now describe the actions performed at the four kinds of events. The actions for events of kind i.) or ii.) are obvious. For kind i.) we insert an arc into DS and update Q_B (one deletion and two insertions); for kind ii.) we delete an arc from DS and update Q_B (two deletions and one insertion). Thus an action of type i.) or ii.) takes time $O(\log m)$ and there are $O(b_w \cdot m^2)$ of them.
For events of kind iii.) we scan through DS, compute for each adjacent pair of arcs their intersection and check whether this intersection is met during the next segment of the sweep. If so, the appropriate position of the sweep line is added to Q_B. This takes time $O(\log m)$ per pair, there are $O(m^2)$ pairs and there are $O(b_w)$ events of kind iii.).
For events of kind iv.) the sweep line meets the intersection of two arcs, say a and b, where a is the more constraining arc immediately after the intersection. We delete the arc b from DS und update Q_B. This takes time $O(\log m)$ per event; also, there are only $O(b_w \cdot m^2)$ events of kind iv.) since each such event eliminates an arc.

Lemma 2 (Looser Lemma):
The action needed for an event of kind iv.) maintains the invariant.
Proof (sketch): Let b be the arc deleted in an event of kind iv.), let f be the center of the arc and let S be the corresponding sector. Let r be the terminating ray of the sector and let H be the closed half-plane bounded by the line supporting r and not containing S, cf. Figure 13. The arc a intersects b and is more constraining than b immediately after the intersection. Since the sector S does neither contain any feature between f and b nor between f and rbe_0 we conclude that the center g of arc a lies in H. If the visibility interval of a does not end before the visibility interval of b then we are done. Assume otherwise. The visibility interval of a can end for two possible reasons. Either we sweep across a feature between rbe_0 and a or the arc a ends. In the former case the visibility interval of b would also end and hence only the latter case can apply. Let A be the endpoint of a. The arc a ends in A for either of two reasons. Either a monotonous sector of rbe_0 ends at the ray $\vec{gA}$ or there is a feature h on the line segment $\overline{gA}$. Again the former case cannot arise because the visibility interval of b has not ended yet. In the latter case consider the constraint generated by h.
Lemma 3 (Strength Lemma):
Let g and h be features, let r be a ray starting in g and let h lie between g and $x = r \cap rbe_0$ on r. Assume also that there is no other feature on the line segment $\overline{hx}$. Let i be the radius of the arc generated by g immediately to the left of h and let j be the radius of the arc generated by h in the direction r, cf. Figure 14. Then $j + dist(h, g) \geq i$.
Proof (sketch): Consider ray r' immediately to the left of r. Then rbe_0 is the i-th path crossing r' if we count intersections starting in g. Only $dist(g, h)$ of these paths can cross the ray between g and h or end at h by the cut condition. Hence rbe_0 is at least the $(i - dist(g, h))$-th path when counting starts in h. This proves $j \geq i - dist(g, h)$.

q.e.d.

The constraint generated by h is at least as strong on the ray $\vec{gh}$ as the constraint generated by g. Also, $h \in H$. Consider a ray immediately to the right of $\vec{gh}$. On this ray either the constraint generated by h or the one generated by g is stronger. In the former case we continue the argument with h instead of g and in the latter case we continue the argument with g.

q.e.d.

Step 2.3: The construction of rbe_1

As an output of step 2.2 we get two sorted lists containing the most constraining arcs bounding the wire to be routed from both sides. We compute the shortest path through the corresponding routing region by a combination of the funnel method and the sweep algorithm used in step 2.2. Like in [T] the sweep line is organized as a double ended queue of tangents to the constraining arcs seperating the intervals on the sweep line. We update this queue each time we reach the next arc in the two lists. The fact that we can give an unambiguous course for each wire rests on the
Lemma 4 (Monotonicity Lemma):
The path rbe_1 intersects the sweep line exactly once for each position of the sweep line.

Proof (sketch): Note first that rbe_0 intersects the sweep line exactly once for each position of the sweep (recall that we are in cover space). Thus, rbe_1 intersects at least once since rbe_1 and rbe_0 are homotopic. So we still have to show that there is no more than one intersection.
Assume otherwise and consider again the process of growing up rbe_λ from rbe_0 to rbe_1 and let λ be maximal such that rbe_λ intersects no sweep line twice. Then the sweep line is tangent to two arcs of rbe_λ that have different curvature, cf. Figure 15. So the cut between the features who caused these arcs is oversaturated and λ cannot be enlarged to $\lambda = 1$.

q.e.d.

4. Extensions

The algorithm can also deal with the following two generalizations.

i.) Obstacles are line segments instead of points.

ii.) The nets are multi-terminal nets. In this case each wire w_i is a tree, the leaves being the terminals. The trees w_i are pairwise disjoint, cf. figure 16. The output trees must be homotopic to the input trees and must satisfy the simplicity and the disjointness condition. Our algorithm solves this problem, but, of course, it can no longer construct a minimum length solution.

5. Conclusion

This paper gives another illustration of the versatility of the sweep line paradigm. In particular, we show that it can be applied to more complex spaces than Euclidean spaces.

6. References

[CS] R. Cole and A. Siegel, "River Routing every which way, but loose," 24th Annual Symposium on Foundations of Computer Science (November 1983), pp. 112-121.

[LM] C. E. Leiserson and F. M. Maley, "Algorithms for Routing and Testing Routability of Planar VLSI Layouts", 17th Annual ACM Symposium on Theory of Computing (May 1985), pp. 69-78.

[M] F. M. Maley, "Single-Layer Wire Routing", Ph. D. Thesis, Massachusetts Institute of Technology, August 1987.

[ST] H. Seifert and W. Threlfall, "Lehrbuch der Topologie", Chelsea Publishing Company, 1947.

[T] M. Tompa, "An Optimal Solution to a Wire-Routing Problem", Journal of Computer and System Sciences 23, pp. 127-150.

7. Figures

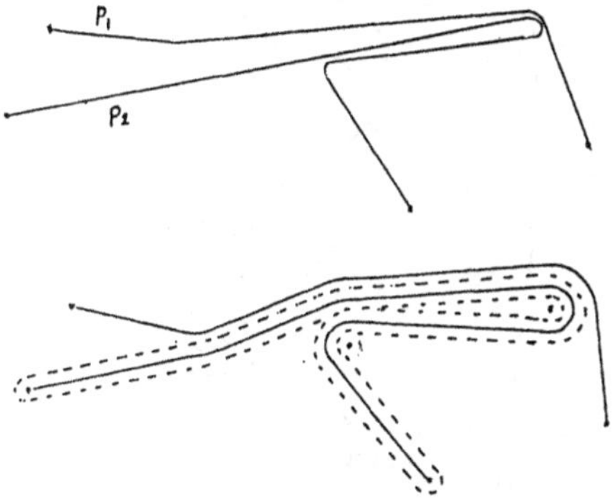

Figure 1: A CHRP and its solution. Obstacles and wire endpoints are shown as dots. $U(p_2)$ is indicated by a dashed line.

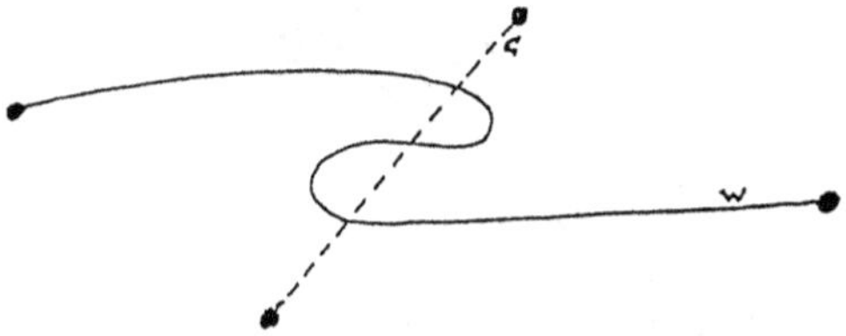

Figure 2: A cut C is shown as a dashed line. The path w has one necessary crossing with the cut, so $cr(C, w) = 3$ and $mincr(C, w) = 1$.

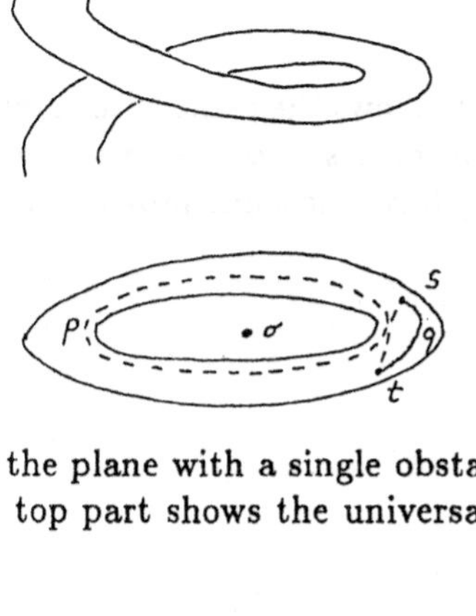

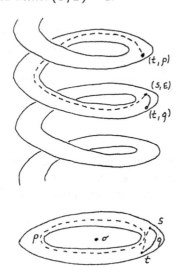

Figure 3: The bottom part shows the plane with a single obstacle o and two paths p and q with terminals s and t. The top part shows the universal cover space and the two paths in the cover space.

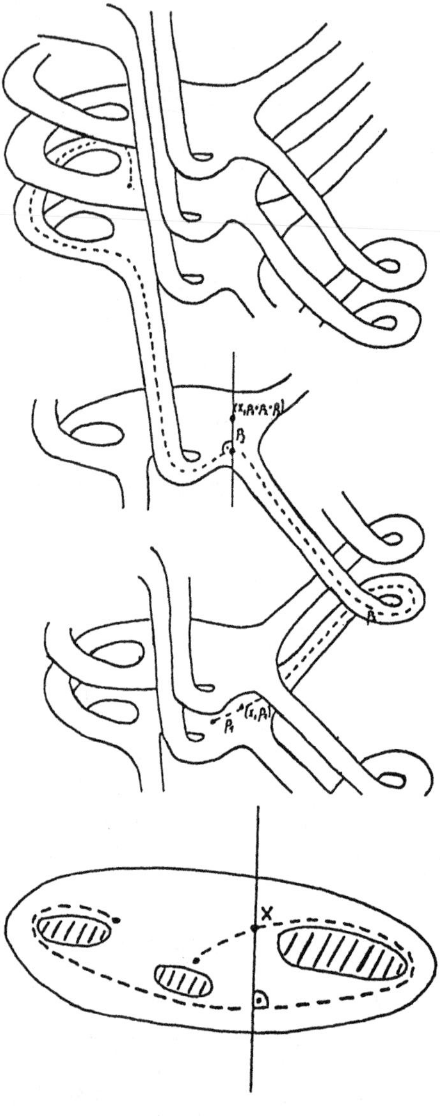

Figure 4: The bottom part shows the plane with three obstacles, a path and a position of the sweep-line. The top part shows the situation in the cover space. The points (x, p_1) and $(x, p_1 \circ p_2 \circ p_3)$ have the same projection into the plane.

5 a.)

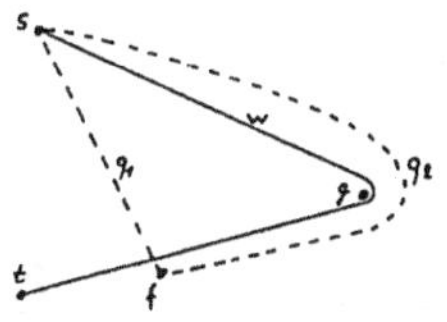

5 b.)

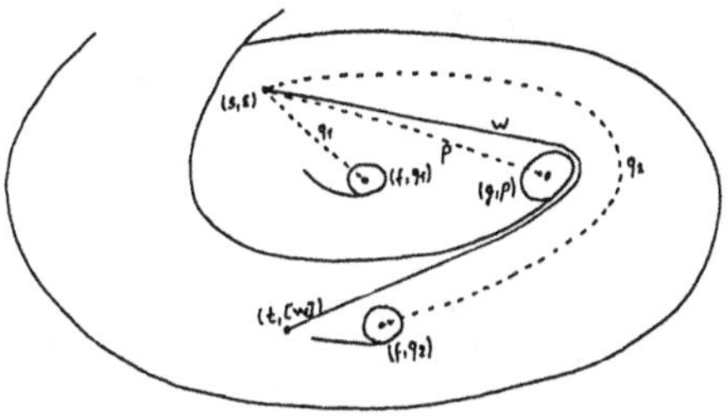

5 c.)

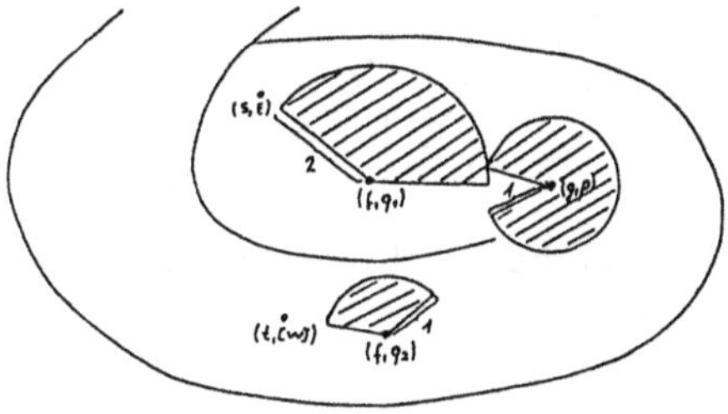

5 d.)

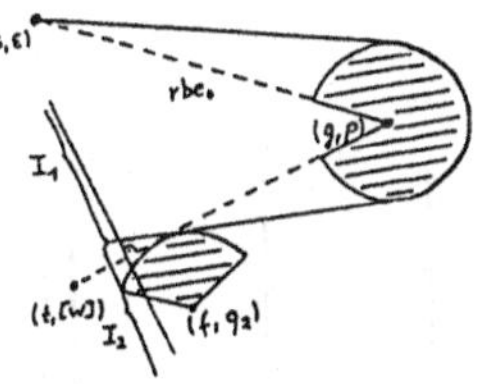

Figure 5: Part a.) shows a path w, two obstacles f and g and two paths q_1 and q_2 from s to f. Part b.) shows the relevant part of the cover space. Observe the points (f, q_1) and (f, q_2) in the cover space. Part c.) shows the forbidden region $\mathcal{F}$. The forbidden region is the union of the three sectors, one with center (f, q_1), radius 2 and rays $\vec{fs}$ and $\vec{fg}$, one with center (g, p), radius 1 and rays $\vec{gs}$ and $\vec{gt}$, and one with center (f, q_2), radius 1 and rays $\vec{fg}$ and $\vec{ft}$. Part d.) shows a position of the sweep-line. The RBE rbe_0 is shown as a dashed line. The sweep-line intersects the boundary of the sector with center (f, q_2). The partition of the sweep-line consists of two intervals I_1 and I_2. For $x \in I_1$, the last segment on the path is a tangent to the obstacle with center (g, p) and for $x \in I_2$ the last segment is a tangent to the obstacle with center (f, q_2).

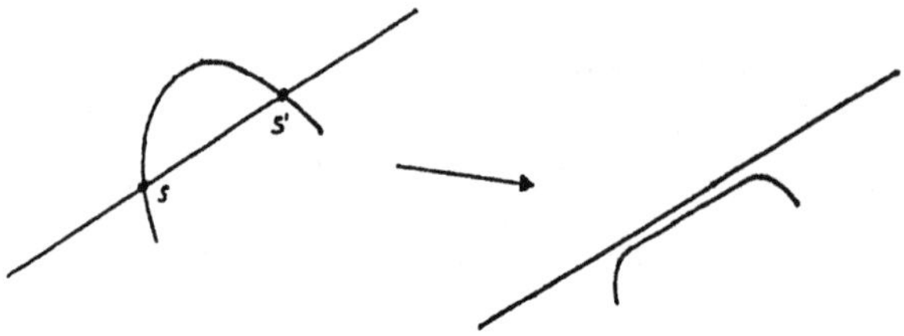

Figure 6: Removal of the two crossings s and s'.

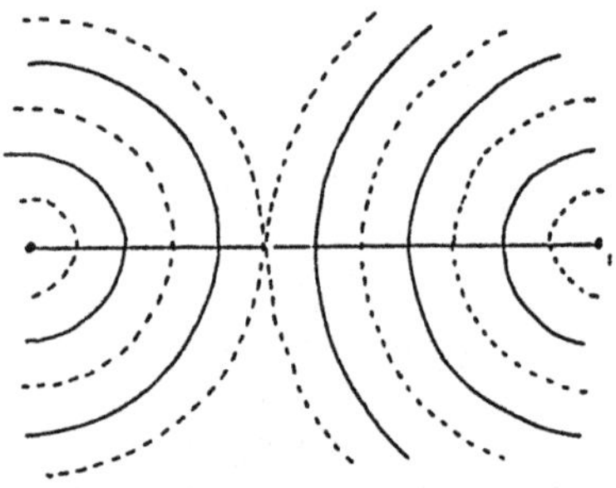

Figure 7: A collision of two rubberbands of opposite curvature gives rise to a cut whose density exceeds its capacity.

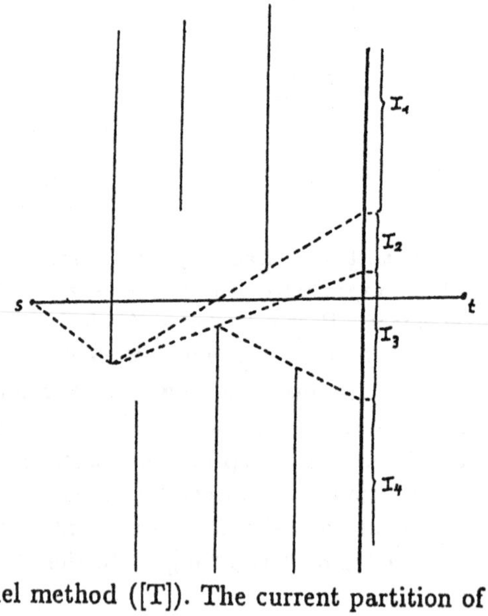

Figure 8: The funnel method ([T]). The current partition of the sweep-line consists of 4 intervals.

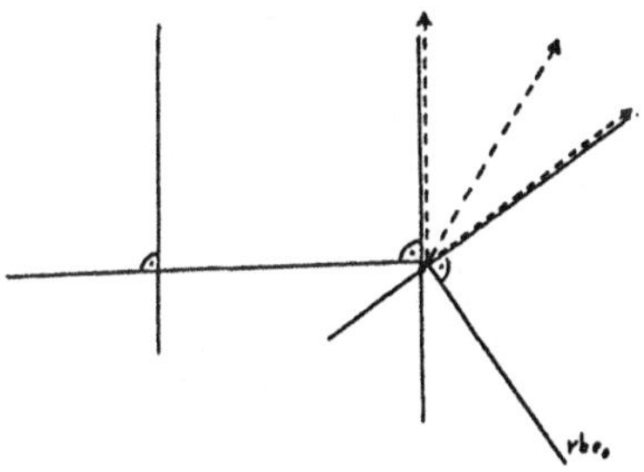

Figure 9: Lines (solid) and rays (dashed) perpendicular to rbe_0.

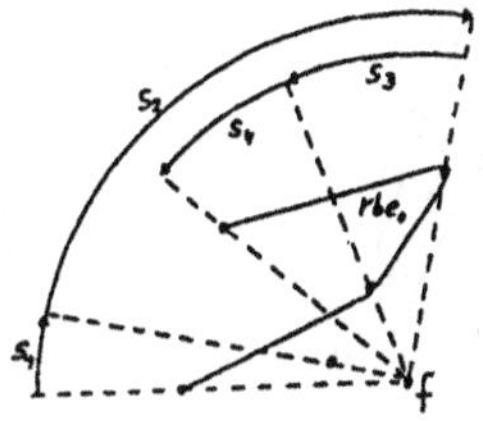

Figure 10: The sectors of an rbe_0 with respect to a feature f.

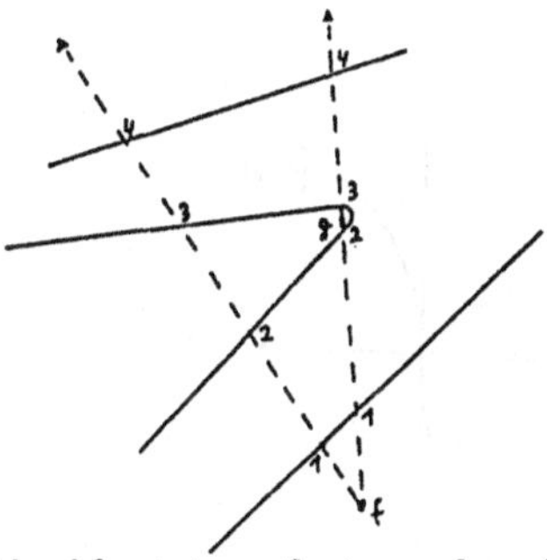

Figure 11: Two rays incident to a feature f and the ranks of intersections between the rays and rubberbands.

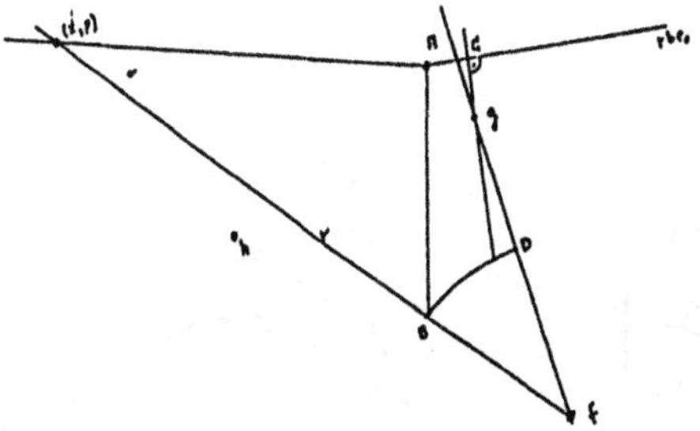

Figure 12: The arc BD becomes visible when the sweep-line has footpoint A and direction $\vec{AB}$ and stays visible until the footpoint is C. The intersection between the ray r and rbe_0 is (x,p). The ray stays visible when we sweep through h, the arc becomes invisible when we sweep through g.

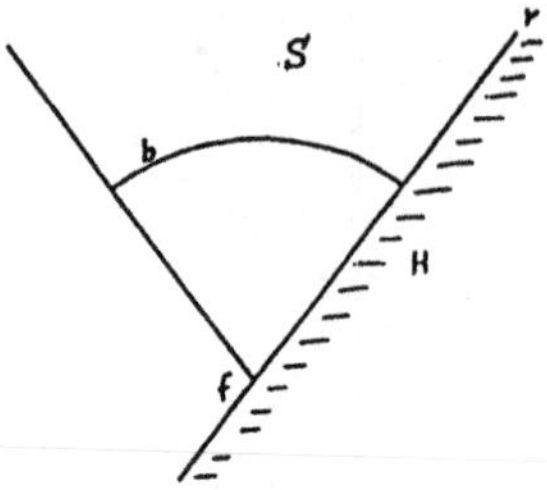

Figure 13

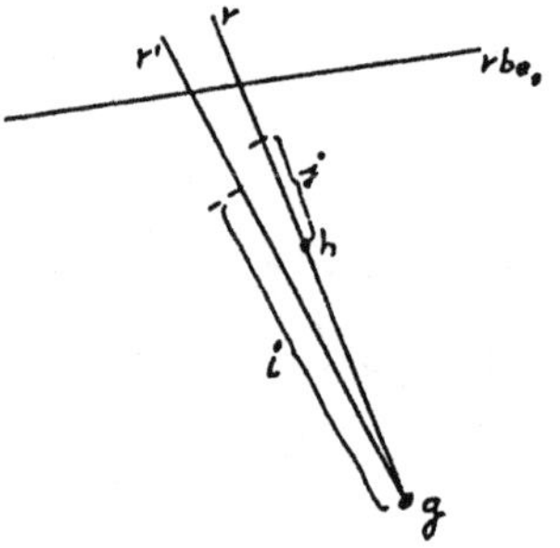

Figure 14

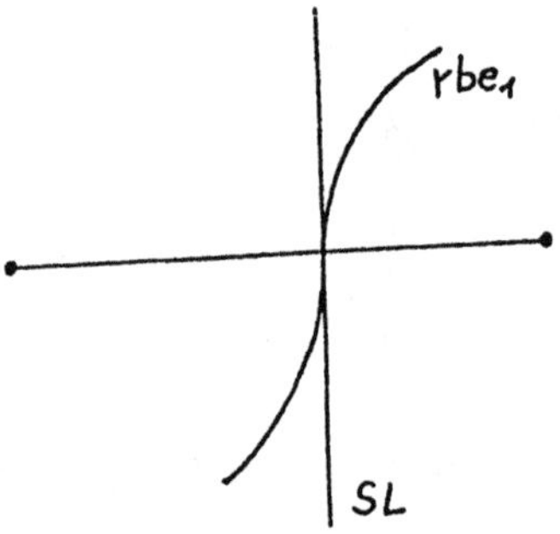

Figure 15

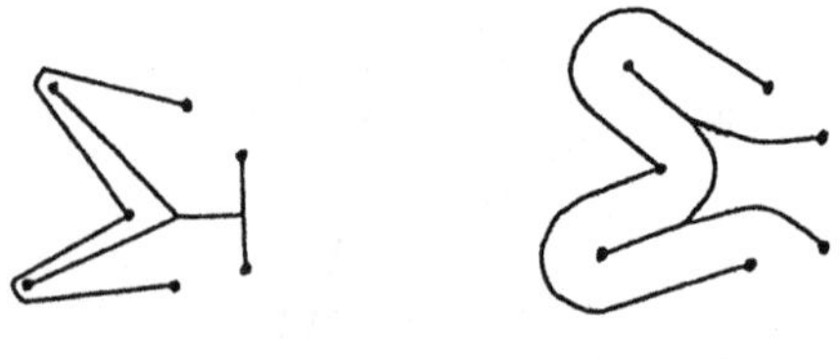

Figure 16: **A multiterminal net problem and a solution for it.**

Point Location in Arrangements

Extended Abstract

Stefan Meiser

FB10, Informatik
Universität des Saarlandes
D-6600 Saarbrücken

Abstract: **We present a data structure and an algorithm for the point location problem in arrangements of hyperplanes in E^d with running time $O(d^5 \log n)$ and space $O(n^{d+\kappa})$ for arbitrary $\kappa > 0$, where n is the number of hyperplanes. The main result is the d^5 factor in the asymptotic expression for the running time, whereas all previously known algorithms depend exponentially on d.**

Introduction

Point location in general subdivisions of higher dimensional spaces is a well-known problem in computational geometry, for example point location in planar graphs and in Voronoi regions. In the plane we know optimal algorithms for most of these problems. But unfortunately, the solutions found do not apply to higher dimensions. We therefore concentrate on a more special subdivision defined by hyperplanes.

A set of hyperplanes partitions the d-dimensional Euclidean space E^d into regions of various dimensions. The regions differ in their positions relative to the hyperplanes. The point location problem in arrangements of hyperplanes determines for a given point p, the region in which p lies.

Dobkin and Lipton [DL76] were the first to present an $O(f(d) \log n)$ algorithm. However, their algorithm needs space $O(n^{2^{d-2}})$. Clarkson [Cla86] describes an $O(f(d) \log n)$ algorithm with space $O(n^{d+\kappa})$ for arbitrary $\kappa > 0$. The running time of both algorithms is logarithmic in n but depends exponentially on d, i.e. $f(d)$ grows exponentially in d. Meyer auf der Heide [Mey84] solved the problem by partitioning E^d into small cubes in which the intersecting hyperplanes have nonzero common intersection. The running time of his algorithm is $O(d^4 \log C)$, where C is the greatest coefficient of all hyperplanes (cf. the conclusion.) Our work also improves the result of Meyer auf der Heide because the running time of our algorithm is independent of C. In the terms of complexity theory this means that our algorithm is strongly polynomial whereas the algorithm of Meyer auf der Heide is only polynomial.

Following Clarkson, our approach is by divide and conquer. We divide the search problem into smaller problems consisting only of a constant fraction of the hyperplanes actually considered and thus achieve recursion depth $\log n$. We distinguish the subproblems by considering only a constant number of hyperplanes. Here *constant* means independent of n, but depending on d. The existence of such a constant number of hyperplanes is shown by the use of the concept of ε-nets introduced by Haussler/Welzl [HW86]. In order to identify the subproblem we need to search a triangulation of the arrangement defined by selected hyperplanes. Our main contribution is the insight that we can exploit the way of triangulation in determinig the relevant subproblem, such that we need only a polynomial (in d) number of steps instead of an exponential number of steps.

Arrangements of Hyperplanes

We start with the fundamental definitions and terminology relying on [Edl87].

A **hyperplane** in E^d, the d-dimensional Euclidean space, is the affine hull of d affinely independent points. For every non-vertical hyperplane h, there are unique real numbers $\eta_1, \ldots, \eta_d$ such that h consists of all points $x = (x_1, \ldots, x_d)$ that satisfy $x_d = \eta_1 x_1 + \eta_2 x_2 + \ldots + \eta_{d-1} x_{d-1} + \eta_d$. We say that a point $p = (p_1, \ldots, p_d)$ is **above, on,** or **below** h if p_d is greater than, equal to, or less than $\eta_1 p_1 + \eta_2 p_2 + \ldots + \eta_{d-1} p_{d-1} + \eta_d$, respectively, and we write h^+ for the set of points above h and h^- for the set of points below h. Both, h^+ and h^- are open half-spaces.

The distinction between the half-space above a non-vertical hyperplane and the half-space below it allows us to specify the location of a point relative to the hyperplanes of a set H. Thus let $H = \{h_1, \ldots, h_n\}$ be a set of non-vertical hyperplanes and for a point p define

$$pv_i(p) = \begin{cases} + & \text{if } p \in h_i^+ \\ 0 & \text{if } p \in h_i \\ - & \text{if } p \in h_i^- \end{cases} \qquad , \ 1 \leq i \leq n$$

The vector $pv(p) := (pv_1(p), \ldots, pv_n(p))$ is called the **position vector** of p relative to H.

Note that these definitions apply only to non-vertical hyperplanes. A simple redefinition of the coordinate system is always sufficient to make all hyperplanes of a given finite set non-vertical.

For $0 \leq k \leq d$ a **k-flat** in E^d is defined as the affine hull of $k+1$ affinely independent points. A 0-flat is a point, a 1-flat is a line, a 2-flat a plane and a $(d-1)$-flat a hyperplane. There is only one d-flat in E^d, namely E^d itself.

A finite set H of hyperplanes in E^d defines a dissection of E^d into connected regions of various dimensions. We call this dissection the **arrangement** $\mathcal{A}(H)$ of H. if $pv(p) = pv(q)$ for two points p and q, then we say that p and q are equivalent. The equivalence classes so defined are called the **faces** of the arrangement $\mathcal{A}(H)$. Thus a face f consists of all points having the same position vector. Analogously, we define the position vector of a face f: $pv(f) := pv(p)$ for a point $p \in f$.

A face f is called k-face, if its (affine) dimension equals k. In particular 0-faces are also called **vertices**, 1-faces **edges** and d-faces **cells**. A face f is a **subface** or a **facet** of another face g, if the dimension of f is 1 less than the dimension of g, and f is contained in the boundary of g. Vice versa g is called **superface** of f. It follows that $pv_i(f) = 0$ for all $1 \leq i \leq n$ with $pv_i(f) \neq pv_i(g)$. If f is a subface of g, then we say that f and g are **incident**.

An arrangement $\mathcal{A}(H)$ of $n \geq d$ hyperplanes in E^d is called **simple**, if any d hyperplanes of H have a unique point in common and if any $d+1$ hyperplanes have no point in common. If the number n of hyperplanes does not exceed $d-1$, then we call $\mathcal{A}(H)$ simple if the common intersection of the n hyperplanes is a $(d-n)$-flat. It follows from this definition that in a simple arrangement the common intersection of $k \leq d$ hyperplanes of H has dimension $d-k$.

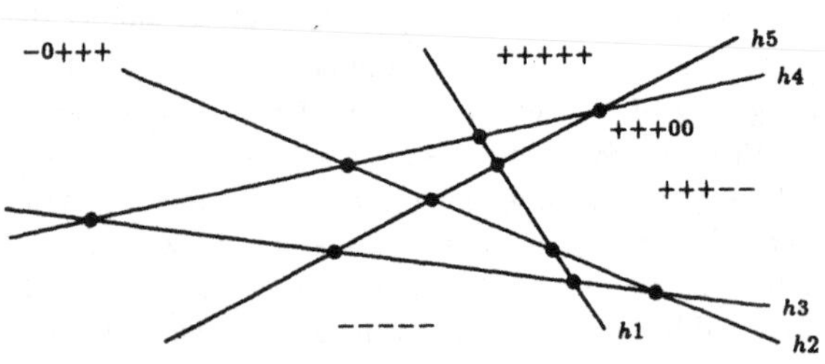

Fig. 1. A simple arrangement of 5 hyperplanes

Observation:
A face in an arrangement of n hyperplanes in E^d can only have n facets. A facet of a face f is built by intersecting f with one or more hyperplanes. Intersections with only one hyperplane can only yield n facets. On the other hand, an intersection of f with more than one hyperplane either yields a face with smaller dimension and thus not a facet or is equal to the intersection of f with one (arbitrary) of the hyperplanes. □

How do we represent arrangements in storage? In addition to the regular (proper) faces of $\mathcal{A}(H)$ we introduce two improper faces: the (-1)-face $\emptyset$ and the $(d+1)$-face $\mathcal{A}(H)$. For convenience we define the (-1)-face as being incident to all vertices of $\mathcal{A}(H)$, and the $(d+1)$-face as being incident to all cells of $\mathcal{A}(H)$. The **incidence graph** $\mathcal{I}(H)$ of $\mathcal{A}(H)$ can now be defined as follows:

> For each proper and improper face f of $\mathcal{A}(H)$, $\mathcal{I}(H)$ contains a node that represents f. An edge connects two nodes whose faces are incident.

Complexity of Arrangements

As we will see below, the complexity of the algorithms and data structures presented in this paper depends on the number of faces and incidences of the given arrangement. Let H be a set of hyperplanes in E^d. Then $f_k(H)$ denotes the number of k-faces in $\mathcal{A}(H)$, $0 \leq k \leq d$, and $i_k(H)$ denotes the number of incidences between k- and $(k+1)$-faces, $0 \leq k \leq d-1$.

Lemma 1.
Let H be a set of n hyperplanes in E^d and $\mathcal{A}(H)$ the arrangement of these hyperplanes. Then

$$f_k(H) \leq \sum_{i=0}^{k} \binom{d-i}{k-i}\binom{n}{d-i}, \quad \text{for } 0 \leq k \leq d,$$

$$i_k(H) \leq 2(d-k) f_k(H), \quad \text{for } 0 \leq k \leq d-1.$$

If $\mathcal{A}(H)$ is simple, we have equality.

Proof: [Edl87] □

Using these bounds on the number of k-faces and incidences we can compute a bound on the total number of faces. For a set H of hyperplanes in E^d let $|\mathcal{A}(H)| = \sum_{k=0}^{d} f_k(H)$ be the total number of faces in the arrangement $\mathcal{A}(H)$.

Lemma 2.
Let $\mathcal{A}(H)$ be the arrangement of n hyperplanes in E^d, $d \geq 1$. Then $|\mathcal{A}(H)| = O(n^d)$.

Proof: follows directly from lemma 1 □

Point Location in Arrangements

The point location problem in arrangements of hyperplanes is given as follows:

> Given is an arrangement $\mathcal{A}(H)$ of n hyperplanes. For a query point p find the face of $\mathcal{A}(H)$ that contains p.

In order to find the face f that contains p we have to make decisions concerning the position of p relative to a hyperplane. A simple algorithm for finding f considers all hyperplanes and thus computes the position vector of p. If we store the faces in a data structure that supports access via the position vector, we can find f in time $O(n)$. Clarkson [Cl86] suggested the following approach. Let $R \subset H$ be a suitable (cf. theorem 1 below) subset of the hyperplanes, let $\mathcal{A}(R)$ the arrangement defined by R and let $\Delta\mathcal{A}(R)$ be the triangulation of E^d obtained by triangulating all faces of $\mathcal{A}(R)$. Clarkson's major insight was the fact that there is always a subset R of size polynomial in d such that any triangular region of $\Delta\mathcal{A}(R)$ is only intersected by a constant fraction ε, $0 \leq \varepsilon \leq 1$, of the hyperplanes in H. The idea is then to locate the query point p in $\Delta\mathcal{A}(R)$ and then to solve the subproblem of size εn. The basis for this approach is provided by theorem 1:

Theorem 1.
For any set H of n non-vertical hyperplanes in E^d, $d \geq 2$, and any ε, $0 < \varepsilon < 1$, there is a set $R \subseteq H$ of $r = O(\frac{d^2 \log^3 d}{\varepsilon})$ hyperplanes such that no cell of the triangulated arrangement $\Delta\mathcal{A}(R)$ is intersected by more than εn hyperplanes of H.

Note that r only depends on d and ε, not on n !
Our proof of the theorem is based on the concept of ε-nets by Haussler/Welzl [HW86]:

Range Spaces and Epsilon-Nets

Definition 1.
*A **range space** S is a pair (X, R), where X is a set and R a set of subsets of X. Members of X are called **elements** of S and members of R are called **ranges** of S.*
*Let $A \subseteq X$ be a finite set of elements of S. Then $\Pi_R(A)$ denotes the set of all subsets of A that can be obtained by intersecting A with a range of S, i.e. $\Pi_R(A) = \{A \cap r \mid r \in R\}$. If $\Pi_R(A) = 2^A$, then we say A is shattered by R. The **Vapnik-Chervonenkis-dimension** of S $VCdim(S)$ (or simply the dimension of S) is the largest integer d such that there exists a subset A of X of cardinality d that is shattered by R. If no such maximal d exists, we say the dimension of S is infinite.*

Example:
Consider the range space $S = (E^2, T)$, where T is the set of all triangles in the plane. We show that $VCdim(S) = 7$, i.e. there exists a set of 7 points that can be partitioned into all subsets by intersections with triangles, but no set of 8 or more points:
Let $A \subset E^2$ be a set of points in the plane. Consider the polygon $P = \text{conv}(A)$, where conv is the convex hull. If there is a point $q \in A$ that is not a vertex of P, then we cannot construct the subset of A that consists exactly of the vertices of P, since by definition of P every triangle that contains the vertices of P also contains q. This shows that we can shatter A only if all points of A are vertices of a convex polygon. Vice versa it is easy to see that not all sets of vertices of convex polygons can be shattered by triangles. More precise, sets of up to 7 points can be partitioned into all subsets while a set of at least 8 points cannot. This shows that $VCdim(S) = 7$.

Definition 2.
Let (X,R) be a range space, A a finite subset of X and $0 \leq \varepsilon \leq 1$. Then $R_{A,\varepsilon}$ denotes the set of all $r \in R$ that contain a fraction of the points in A of size greater than ε, i.e. such that $\frac{|A \cap r|}{|A|} > \varepsilon$. A subset N of A is an **ε-net of A (for R)** *if N contains a point in each $r \in R_{A,\varepsilon}$.*

Theorem 2.
For every range space (X,R) of VC-dimension $d \geq 1$ and any finite set $A \subseteq X$ there exists an ε-net, $0 < \varepsilon < 1$, of A for R of size $\leq (\frac{8d}{\varepsilon} \log \frac{8d}{\varepsilon})$.

Proof: [HW86] □

Example (continued):
Here we give an application of theorem 2:

> Let $A \subseteq E^2$, $|A| = n$. Construct a triangulation of the plane, such that no triangle contains more than εn points of A in its interior.

Let N be an ε-net of A for the set T of all triangles in the plane. The existence of an ε-net of constant size is guaranteed by theorem 2 and the discussion above. Triangulate the plane with respect to N, i.e. use the points from N as vertices of the triangles. Then no triangle can contain more than εn points in its interior, because no point of N lies inside a triangle. Furthermore, N has fixed size independent of n.

The theorem implies that all we have to do in order to prove theorem 1 is to construct a range space with finite VC-dimension that meets our demands. We can then continue as in the example above. Let H^d be the set of all non-vertical hyperplanes in E^d. Consider the range space (H^d, DS), where

$$DS = \{H_c \mid c \text{ is a cell in a triangulated arrangement}\},$$
$$H_c = \{h \mid h \text{ is a non-vertical hyperplane that intersects cell } c\}.$$

In [Mei88] we prove the following

Lemma 3.
(H^d, DS) has finite VC-dimension $d_0 = O(d^2 \log^2 d)$.

Theorem 1 then follows immediately, if we choose R to be an ε-net of H for DS. Since by definition no cell of $\Delta\mathcal{A}(R)$ is intersected by a hyperplane from R, no cell of $\Delta\mathcal{A}(R)$ can be intersected by more than εn hyperplanes from H. The existence of an ε-net follows from lemma 3 and theorem 2. This completes the proof of theorem 1.

Note that theorem 1 does not only hold for cells but also for k-faces for arbitrary k: Each hyperplane that cuts a k-face and does not contain it, also cuts any adjacent cell.

Up to now we know that we can always partition the search problem into smaller subproblems. However, there is no deterministic algorithm known for computing an ε-net for a set H. Thus the construction of our data structure will take probabilistic time, since we construct R by random sampling. Note that only the construction of the data structure is in a sense nondeterministic. The resulting data structure is always guaranteed to give correct solutions. This result corresponds to the following result of Clarkson [Cla86]:

> Let $R \subset H$ be a random sample of a collection of hyperplanes H in E^d. Then with probability at least $\frac{1}{2}$, every cell in $\Delta\mathcal{A}(R)$ is cut by $r \cdot O(\frac{\log r}{r})$ hyperplanes of H, where $r = |R|$, $n = |H|$.

Theorem 1 implies a data structure and an algorithm for solving the point location problem in time $O(\log n)$: The data structure has to reflect the subdivision of the problem and the algorithm's main task is to find the *correct* subproblem, i.e. the face of $\Delta\mathcal{A}(R)$ that contains the query point. The subproblems themselves are subdivided recursively until the size of the considered problem is smaller than or equal to r (r as in theorem 1). Since in every partition step the problem size is reduced by factor ε, $0 < \varepsilon < 1$, we achieve recursion depth $\log n$. The time needed for the computation of the *correct* subproblem determines the factor preceding $\log n$ in the asymptotic expression for the running time. Hence this step is the critical part of our analysis of the problem. A naive approach consists in visiting all faces of $\Delta\mathcal{A}(R)$ and deciding which face contains the query point. Clearly, this costs time proportional to the number of faces of $\Delta\mathcal{A}(R)$. Even the number of cells of $\mathcal{A}(R)$, however, is exponential in d as can be seen from lemma 1. In order to get time polynomial in d we have to examine the triangulation of an arrangement. We will see that we can exploit the way of triangulation.

Triangulation

We will need the following definitions:

A **polyhedral set** in E^d is the intersection of a finite number of closed halfspaces, and a **polytope** is a bounded polyhedral set. A **k-face of a polyhedral set** P is the intersection of P with $d-k$ (non-redundant) hyperplanes defining P. **Vertices, edges** and **facets** are 0-, 1- and $(d-1)$-faces. The set of **extreme points** (vertices) of a polyhedral set P will be denoted by **ext**P. The set of **extreme rays** of P **extr**P is the set of rays e *emanating from the origin* 0 such that there is some point $q \in P$ for which $(q+e)$ is an edge of P. A **k-simplex** S is a k-dimensional polytope with $|\text{ext}S| = k+1$.

A polyhedral set P is a **cone with apex** 0, iff $\lambda x \in P$ for each $x \in P$ and each $\lambda \geq 0$. A **pointed** cone has only one apex point. A polyhedral set P is a **cone with apex** a, iff $-a+P$ is a cone with apex 0. The **characteristic cone** $cc_x P$ of a polyhedral set P with respect to a point $x \in P$ is given by $\{y \mid x + \lambda y \in P,\ \forall \lambda \geq 0\}$. From ([Grü67] 2.5.2) $cc_x P = cc_z P$ for any $x, z \in P$, so the subscript may be omitted and ccP denotes the characteristic cone of P. Note that the characteristic cone of a polyhedral set always is a cone with apex 0.

relint P is the interior of P relative to its affine closure and **relbd** P is the boundary of P relative to its affine closure. The ray $(p, s) = \{x \mid x = p + \lambda(s-p),\ \lambda \geq 0\}$ is the ray that starts in p and passes through s. **conv**(P) denotes the convex hull and **clconv**(P) the convex closure of P. A set is called **line-free** if it contains no straight lines.

By definition a face f of an arrangement of hyperplanes in E^d is a set of points with the same position vector. f can be seen as the intersection of a finite number of *open* halfspaces and hyperplanes. Hence f is the interior of a polyhedral set P, namely $P = f \cup \text{relbd}(f)$. relbd$(f)$ consists of all facets of f, the subfaces of the facets, their subfaces and so on. It follows by the definition of a face in an arrangement that $P = \text{clconv}(f)$.

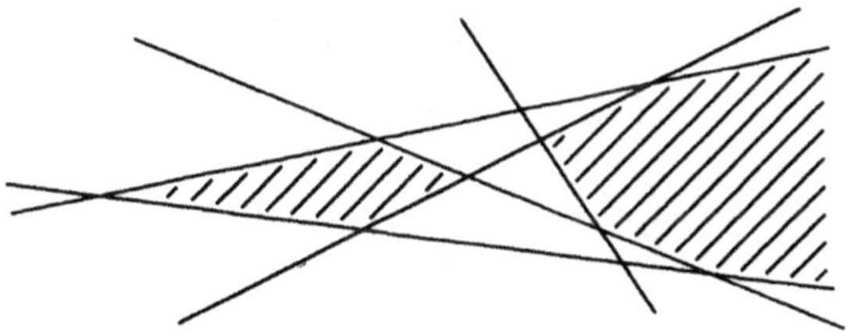

Fig. 2. Two faces are shown hatched

We therefore reduce the triangulation of a face to the triangulation of a polyhedral set. Since polyhedral sets are convex there are simple routines for triangulation. The following functions are taken from [Cla85]. In this paper the interested reader may also find the proof of correctness and further details.

Definition 3.
A triangulation of a k-dimensional line-free polyhedral set P is a collection T of k-dimensional polyhedral sets such that:

$$\bigcup_{A \in T} A = P$$

$$\bigcup_{A \in T} \mathrm{ext} A = \mathrm{ext} P$$

$$\bigcup_{A \in T} \mathrm{extr} A = \mathrm{extr} P$$

$$|\mathrm{ext} A| + |\mathrm{extr} A| = k + 1, \ \ A \in T$$

and any nonempty $A \cap B$ is a face of A and of B for any $A, B \in T$.

For simplicity we start with the triangulation of polytopes.

```
function triangulate_polytope (P: polytope): set T of regions;
begin
  if dim(P) ≤ 1
  then return {P}
  else T ← {};
       choose any vertex v ∈ ext P as reference point of the triangulation;
       for each facet F of P with v ∉ F
       do T' ← triangulate_polytope(F);
          for each S ∈ T'
          do add conv({v} ∪ S) to T; od;
       od;
  fi;
end;
```

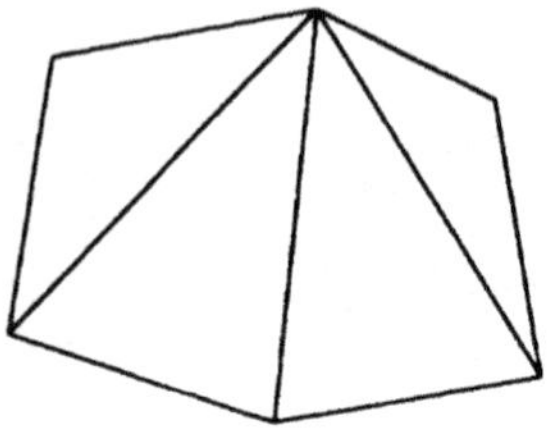

Fig. 3. A triangulated polytope

If we try to triangulate an unbounded polyhedral region, say U, with triangulate_polytope we will not get a complete partition of U into simplices. The very points in its characteristic cone are not contained in any constructed simplex. Hence we need a way of triangulating a cone. This construction results from a triangulation of a polytope, that is a slice of a cone. Clarkson shows that for any polyhedral pointed cone C there is always a hyperplane h such that the intersection of h and C is a polytope. Hence a polyhedral pointed cone may be triangulated by a procedure analogous to triangulate_polytope, in which the role of vertices is assumed by extreme rays. The following function triangulates arbitrary line-free polyhedral sets.

```
function triangulate (P: polyhedral set): set T of regions;
begin
  if dim(P) ≤ 1 then return {P}; fi;
  T ← {};
  choose any vertex v ∈ extP as reference point of the triangulation;
  if P is unbounded
  then (* P is not a polytope *)
         T' ← triangulate_cone(ccP);
         for each S ∈ T'
         do add v + S to T; od;
  fi;
  for each facet F of P with v ∉ F
  do T' ← triangulate(F);
      for each S ∈ T'
      do add conv({v} ∪ S) to T; od;
  od;
end;
```

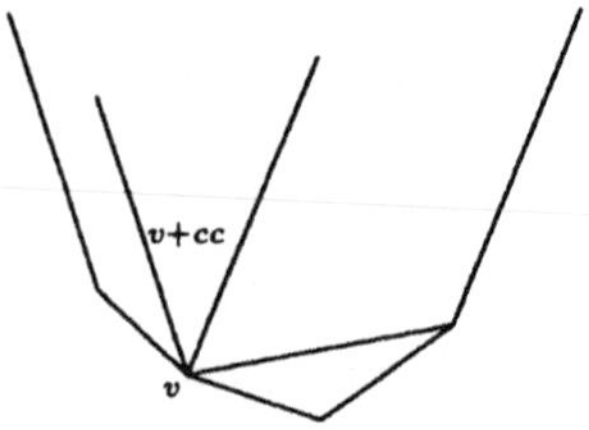

Fig. 4. A triangulated polyhedral set

The triangulation of a face f can be obtained from the triangulation of clconv(f) by removing the boundary of f from the constructed regions. The regions thus obtained are no longer simplices. They may be unbounded or open on some side. We therefore call them **degenerated simplices** or abbreviated **degsimplices**.

The routines for triangulation given above apply only to *line-free* polyhedral sets. If we add d hyperplanes, each spanned by $d-1$ axes of the coordinate system, then all faces of the resulting arrangement are line-free. The additional hyperplanes do not change the asymptotic bounds.

Determining the subproblem

We are now in a position to describe the crucial part of the algortihm, i.e. to determine the degsimplex in $\Delta\mathcal{A}(R)$ that contains the query point. We proceed in 2 steps: First we compute the face of $\mathcal{A}(R)$ that contains the query point: function compute_face. Secondly we look for the degsimplex within the face of step 1: function compute_degsimplex.

function compute_face (s: query point): face f from $\mathcal{A}(R)$;
begin
(1) compute the position vector of s with respect to the r hyperplanes from R;
(2) determine the face f from $\mathcal{A}(R)$ with the same position vector;
end;

In step (1) of compute_face we decide for every hyperplane from R whether s lies above, on or below it. For step (2) we have to maintain the faces in a datastructure such that they are accessable via their position vector. We use a **compressed TRIE** that contains all faces of the arrangement $\mathcal{A}(R)$ in its leaves, ordered by their position vector. Compressed TRIEs are explained for example in [Meh84], Vol. 1. A node of the TRIE has 3 sons, labelled +, 0 and −, according to the 3 possible positions of the query point relative to a hyperplane. As above, +, 0 and − denote the positive half-space, the hyperplane itself and the negative half-space. All nodes of the same depth represent the same hyperplane. In step (2) we then move downwards the tree from the root to a leaf according to the position vector computed in step (1). The correctness of compute_face follows from the definition of the position vector of a face and the fact that all faces are stored in the TRIE.

We can now turn to compute_degsimplex. As in the discussion of the triangulation we first concentrate on bounded faces. The degsimplices in a bounded face originate from degsimplices on the facets. The construction of a degsimplex can thus be pursued downwards via faces of smaller and smaller dimensions until we arrive at an edge. Hence we can to each k-degsimplex within a k-face f, $2 \le k \le d$, assign a sequence of faces $f^{k-1}, f^{k-2}, \ldots, f^1$ that participated in the construction of the degsimplex.

f^{k-1} is a $(k-1)$-face, a facet of f,
f^{k-2} is a $(k-2)$-face, a facet of f^{k-1},
⋮
f^1 is an edge, a facet of f^2.

Such a sequence of faces describes a unique degsimplex and allows us to determine the degsimplex in f, which contains the query point. We thus call the sequence the **face sequence** of the degsimplex.

For f we store the following r-ary tree for the degsimplices in f:

A node of depth i represents a $(k-i)$-face, the root represents f itself. A node has r sons, one son per facet of the corresponding face. The degsimplices of f are stored in the leaves of the tree such that a path from the root to a degsimplex corresponds to the face sequence $f^{k-1}, \ldots, f^1$ for the degsimplex as described above.

This tree distributes the degsimplices of f over the facets of f, then over their facets, and so on. Given a query point s we step downwards the degsimplex tree with the function compute_degsimplex:

function compute_degsimplex (s: query point, f: face of $\mathcal{A}(R)$): degsimplex DS of $\Delta\mathcal{A}(R)$;
(* $s \in f$, $\dim(f) = k$ *)
begin
(3) $B^k \leftarrow$ root of the degsimplex tree of f;
(4) $s^k \leftarrow s$;
(5) $f^k \leftarrow f$;
(6) **for** $i := k-1$ **downto** 1
do let p^{i+1} be the reference point of the triangulation for f^{i+1};
(7) $f^i \leftarrow$ facet of f^{i+1}, that is cut by (p^{i+1}, s^{i+1});
(8) $s^i \leftarrow \text{clconv}(f^i) \cap (p^{i+1}, s^{i+1})$;
(9) $B^i \leftarrow$ son of B^{i+1}, that corresponds to f^i;
od;
(10) $DS \leftarrow B^1$;
end;

The ray (p^{i+1}, s^{i+1}) in (7) must intersect the convex closure of a facet of f^{i+1}. The convex closures of the facets of f^{i+1} form a partition of the boundary of f^{i+1}. Moreover, no two facets have any point in common. It follows that the ray (p^{i+1}, s^{i+1}) either cuts a unique facet, or passes through the boundary of several facets. In the latter case we can take one of these facets arbitrarily. The query point then belongs to several degsimplices at one time.

Before we go on with a detailed description we will give a proof of the correctness of the function compute_degsimplex. We prove the following assertion:

Assertion:
For any k-face $f \in \mathcal{A}(R)$ and any point $s \in f$ let s^i and f^i, $1 \le i \le k$, be defined by a call compute_degsimplex(s, f). Then for all $1 \le i \le k$, there is a degsimplex $DS^i \subseteq f^i$ with face sequence $f^{i-1}, \dots, f^1$ and $s^i \in clconv(DS^i)$.

Proof: (*by induction on i*)
From step (8) we have $s^i \in \text{clconv}(f^i)$, $\forall i$.
The assertion is true for $i = 1$, since an edge consists of only one degsimplex, namely the edge itself. Then $DS^1 = f^1$ and $s^1 \in \text{clconv}(f^1) = \text{clconv}(DS^1)$. The face sequence of an edge is empty.
So assume that the assertion holds for all dimensions $\le i$. By induction hypothesis there is a degsimplex $DS^i \subseteq f^i$ with face sequence $f^{i-1}, \dots, f^1$, such that $s^i \in \text{clconv}(DS^i)$. Since s^{i+1} lies on the line segment $\overline{p^{i+1}s^i}$, the point s^{i+1} is contained in the convex closure of $\text{clconv}(DS^i)$ and p^{i+1}. Taking $\text{clconv}(p^{i+1} \cup \text{clconv}(DS^i))$ is just the procedure we use to build the convex closure of the degsimplices in f^{i+1}. Hence there is a degsimplex $DS^{i+1} \subseteq f^{i+1}$ with face sequence $f^i, \dots, f^1$ and $s^{i+1} \in \text{clconv}(DS^{i+1})$. □

We can directly conclude from this assertion that the face sequence $f^{k-1}, \dots, f^1$ computed in a call compute_degsimplex(s, f) for a k-face f and a point $s \in f$ is the face sequence for a degsimplex $DS \subseteq f$ with $s \in DS$. This shows that we always choose the correct son in step (9).

We will now elaborate the details of the algorithm, starting with step (7). We can easily access the facets of f^{i+1} in the incidence graph of $\mathcal{A}(R)$. For each facet f'_j of f^{i+1} we consider the i-flat F_j, in which f'_j is contained, i.e. the affine hull of f'_j. We intersect all F_j with the straight line through p^{i+1} and s^{i+1} and get s'_j as the points of intersection, cf. Figure 5. All s'_j naturally lie on the straight line through p^{i+1} and s^{i+1}. At this point we exploit the convexity of the faces: The point s'_j, that is the nearest to s^{i+1} on the ray (p^{i+1}, s^{i+1}) gives us the facet we are looking for. If (p^{i+1}, s^{i+1}) cuts only one facet, then there is only one such nearest point. On the other hand, if (p^{i+1}, s^{i+1}) passes through the boundary of several facets, then several points of intersection collapse into the same point and we can take one of these facets arbitrarily as explained above.

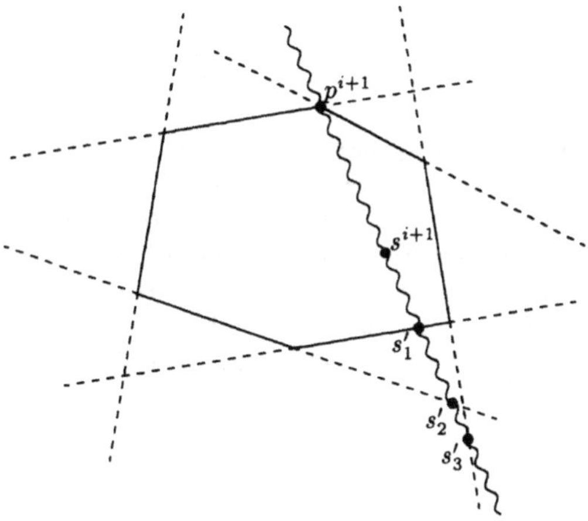

Fig. 5.

We compute the nearest point to s^{i+1} among the s'_j by comparing the length of the line segments $\overline{s^{i+1}s'_j}$ for each $s'_j \neq p^{i+1}$, that lies on the ray (p^{i+1}, s^{i+1}).

What data structures do we need in order to execute these operations? We have to visit the facets of f^{i+1}. Therefore we store the incidence graph $\mathcal{I}(R)$ of $\mathcal{A}(R)$. Moreover, we need for each j-face g in $\mathcal{A}(R)$ the j-flat that contains g, $1 \leq j \leq d-1$. We can describe a j-flat by a single point and j linearly independent vectors in E^d. We call this characterisation the **flat equation** of g.

Unbounded faces contain additional degsimplices in their characteristic cones. As we have seen above, the characteristic cone of a face f is triangulated by looking at the intersection S of the cone with a new hyperplane. Since S is a polytope, S consists only of bounded faces. Hence we can apply the same considerations as above to $cc\, f$: We need an incidence graph for S that reflects the incidences of the faces of S. Moreover, we store for each face of S its flat equation and degsimplex tree. The degsimplices of the characteristic cone originate from the degsimplices of S analogously to the degsimplices based on the facets. We can therefore simply treat S as an additional facet of f. Thus the degsimplex tree gets an $(r+1)$-ary tree. This discussion leads to the following final form of compute_degsimplex.

function compute_degsimplex (s: query point, f: face of $\mathcal{A}(R)$): degsimplex DS of $\Delta\mathcal{A}(R)$;
(* $s \in f$, $\dim(f) = k$ *)
begin
  $B^k \leftarrow$ root of the degsimplex tree of f;
  $s^k \leftarrow s$;
  $f^k \leftarrow f$;
  for $i := k-1$ **downto** 1
  do let p^{i+1} be the reference point of f^{i+1};
    if $s^{i+1} \in p^{i+1} + cc\, f^{i+1}$
    then $f^i \leftarrow$ intersection of $p^{i+1} + cc\, f^{i+1}$ with the hyperplane
        from the triangulation of $cc\, f^{i+1}$
    else for each facet f'_j of f^{i+1}

```
            do s'_j ← F_j ∩ (p^{i+1}, s^{i+1}); od;
            f^i ← a f'_j with s'_j ∈ (p^{i+1}, s^{i+1}), s'_j ≠ p^{i+1} and s^{i+1}s'_j minimal;
        fi;
        s^i ← clconv(f^i) ∩ (p^{i+1}, s^{i+1});
        B^i ← son of B^{i+1}, that corresponds to f^i;
    od;
    DS ← B^1;
end;
```

If a subproblem consists of at most r hyperplanes, the recursion terminates. In this case we do not have to subdivide and thus to triangulate any further. As above, let R be the set of hyperplanes currently considered. Then the data structure consists only of the compressed TRIE for the faces in $\mathcal{A}(R)$ and we use compute_face in order to find the face which contains the query point.

We have not yet discussed the relationship between the faces of $\mathcal{A}(R)$ and the leaves of our data structure. By means of triangulation a face of $\mathcal{A}(H)$ can be represented more than once in the data structure. We solve this problem by using pointers from the leaves of our data structure to the incidence graph of $\mathcal{A}(H)$.

Complexity

The time needed in a call compute_face is clearly $O(dr)$ since we visit r hyperplanes and step (1) costs $O(d)$ per hyperplane. Running downwards the tree also costs $O(r)$.

A call compute_degsimplex costs time $O(d^2 r)$: The outer for-loop inside compute_degsimplex is executed at most $d-1$ times. Now observe that any face in $\mathcal{A}(R)$ can have at most r facets and that the same is true for the characteristic cone of a face. Thus the test whether $s^{i+1} \in p^{i+1} + cc\, f^{i+1}$ costs $O(dr)$. Within the inner for-loop we compute at most r points of intersection. This can also be done in time $O(dr)$. The same holds for finding the minimum of the s'_j. The computation of the new s^i costs time $O(d)$. All other steps can be done in $O(1)$ time.

The space requirement for a subproblem consisting of r hyperplanes is $O\big((4d)^d r^{2d}\big)$: The incidence graph and the compressed TRIE for $\mathcal{A}(R)$ both need space $O(r^d)$. For each face we store the flat equation, the reference point of the triangulation, the degsimplex tree and a similar data structure for the characteristic cone. Since any degsimplex belongs to exactly one degsimplex tree and there are $O\big((4d)^d r^d\big)$ degsimplices in $\Delta\mathcal{A}(R)$ [Mei88], this term is also a bound on the total space needed for all degsimplex trees in $\Delta\mathcal{A}(R)$ outside the characteristic cones. The data structure in the characteristic cone deals only with bounded faces. Thus we can conclude from the discussion above that the space needed in a characteristic cone is $O\big((4d)^d r^d\big)$. This determines the space per face and yields the desired bound.

Now we can turn to the analysis of the overall data structure and the algorithm. We get the following recursive equations. Let $S(n)$ be the space of the data structure and $T(n)$ be the running time of the algorithm for a problem of size n. Then

$$S(n) = \begin{cases} O\big((4d)^d r^{2d}\big) + O\big((4d)^d r^d\big) S(\epsilon n) & \text{for } n > r \\ O(r^d) & \text{for } n \le r \end{cases}$$

and

$$T(n) = \begin{cases} O(d^2 r) + T(\epsilon n) & \text{for } n > r \\ O(dr) & \text{for } n \le r \end{cases}$$

From $r = O(\frac{d^2 \log^3 d}{\epsilon})$ we conclude the following theorem:

Theorem 3.
For a set H of n hyperplanes in E^d there is a data structure and an algorithm that solve the point location problem in $\mathcal{A}(H)$ in time $O(d^5 \log n)$ and $O(n^{d+\kappa})$ space, for $\kappa > 0$ arbitrary.

Preprocessing

The preprocessing steps are only of technical interest. The time we need for computing the triangulations and the degsimplex trees is proportional to the space of the data structure. However, the most expensive part of the preprocessing step is the establishment of the relationship between the leaves of the data structure and the faces of $\mathcal{A}(H)$. If we precompute an order on the faces of $\mathcal{A}(H)$ with respect to their position vectors similar to the TRIE in a subproblem then we need time $O(n)$ per leaf of the data structure. In total we get probabilistic time $O(n^{d+1+\kappa})$, for arbitrary $\kappa > 0$.

Conclusion

In a slightly modified form our data structure and algorithm also solve the following problem:

For a set $H = \{h_1, \dots, h_n\}$ of hyperplanes in E^d let $\mathcal{H} = \bigcup_{i=1}^{n} h_i$ be the set of all points on hyperplanes from H. For a given point $x \in E^d$ decide whether $x \in \mathcal{H}$.

Our algorithm decides this problem since x lies on a hyperplane iff x does not lie in a cell of $\mathcal{A}(H)$. The LSA (Linear Search Algorithm) introduced by Meyer auf der Heide in [Mey84] decides this problem in time $O(d^4 \log C)$, where C is the greatest coefficient of the hyperplanes from H. Hence the algorithm of Meyer auf der Heide is polynomial in the input length, if we take the input as binary encoded. If we change to arithmetic encoding, i.e. we take the input length as the number of input variables, then we cannot relate the running time of the algorithm to the input length, since C is not related to the input. Our algorithm however is independent of C and we have polynomial running time in the input length even if we take the unit cost measure. As similar problems have occurred concerning scaling algorithms the term *strongly polynomial* was introduced to characterize algorithms that are more than just polynomial.

An algorithm is strongly polynomial, if

- it runs on a RAM (Random Access Machine) with unit cost measure in polynomial time and
- the size of all computed numbers is bounded by a polynomial in the number of registers used by the input and the maximum size of an input number.

Since addition, subtraction, multiplication and comparisons take polynomial time on a RAM with logarithmic cost measure, a strongly polynomial algorithm also runs on a RAM with logarithmic cost measure in polynomial time. (Division is realized by taking rational numbers represented by nominator and denominator.) [Joh87]

The only numbers computed during an execution of our algorithm originate from the comparison of a point with a hyperplane or the intersection of a ray with a hyperplane. Hence the size of the numbers is always polynomial in the size of the coefficients of the hyperplanes and we conclude that our algorithm is strongly polynomial.

Acknowledgements

This work was done as a master thesis supervised by Prof. Kurt Mehlhorn. I wish to thank Prof. Mehlhorn and Otfried Fries for the numerous discussions and suggestions.

References

[Cla85] K.L. Clarkson
A Probabilistic Algorithm for the Post Office Problem
ACM, STOC (1985), pp. 175 – 184

[Cla86] K.L. Clarkson
Further Applications of Random Sampling to Computational Geometry
ACM, STOC (1986), pp. 414 – 423

[DL76] D. Dobkin, R.J. Lipton
Multidimensional Searching Problems
SIAM, J. Comp. Vol. 5 No. 2 (1976), pp. 181 – 186

[Edl87] H. Edelsbrunner
Algorithms in Combinatorial Geometry
Springer-Verlag (1987)

[Grü67] B. Grünbaum
Convex Polytopes
John Wiley & Sons (1967)

[HW86] D. Haussler, E. Welzl
Epsilon-Nets and Simplex Range Queries
ACM, Proc. of the 2. Symp. on Comp. Geometry (1986), pp. 61 – 71

[Joh87] D.S. Johnson
The NP-Completeness Column
Academic Press, J. of Algorithms 2 (1987), pp. 285 – 303

[Meh84] K. Mehlhorn
Data Structures and Algorithms, Vol. 1 – 3
Springer-Verlag (1984)

[Mei88] S. Meiser
Suche in einem Arrangement von Hyperebenen
Diplomarbeit, Fachbereich 10, Universität des Saarlandes, D-6600 Saarbrücken, (1988)

[Mey84] F. Meyer auf der Heide
A Polynomial Linear Search Algorithm for the n-Dimensional Knapsack Problem
J. ACM 31 (3), (1984), pp. 668 – 676

Internal and External Algorithms for the Points-in-Regions Problem - the INSIDE Join of Geo-Relational Algebra

Gabriele Blankenagel, Ralf Hartmut Güting

Fachbereich Informatik, Universität Dortmund, D-4600 Dortmund 50, West Germany

(Extended Abstract. The full paper is available as: Universität Dortmund, Fachbereich Informatik, Report 228, 1987, and is to appear in *Algorithmica*.)

Abstract: We consider the problem of collectively locating a set of points within a set of disjoint polygonal regions when neither for points nor for regions preprocessing is allowed. This problem arises in geometric database systems. More specifically it is equivalent to computing the *inside* join of geo-relational algebra, a conceptual model for geo-data management. We describe efficient algorithms for solving this problem based on plane-sweep and divide-and-conquer, requiring O(n (log n)+t) and O(log^2 n)+t) time, respectively, and O(n) space, where n is the total number of points and edges, and t the number of reported (point,region) pairs. Since the algorithms are meant to be practically useful we consider apart from the internal versions - running completely in main memory - also versions that run internally but use much less than linear space and versions that run externally, that is, require only a constant amount of internal memory regardless of the amount of data to be processed. Comparing plane-sweep and divide-and-conquer, it turns out that divide-and-conquer can be expected to perform much better in the external case even though it has a higher internal asymptotic worst-case complexity.

An interesting theoretical by-product is a new general technique for handling arbitrarily large sets of objects clustered on a single x-coordinate within a planar divide-and-conquer algorithm and a proof that the resulting "unbalanced" dividing does not lead to a more than logarithmic height of the tree of recursive calls.

1. Introduction

In Computational Geometry one usually assumes that all data and data structures fit into main memory. This assumption is not true in many practical cases (VLSI design, geometric database systems) which presents a major obstacle to utilizing the theoretical results. It is thus necessary to reconsider geometric algorithms and to systematically develop *external* variants, that is, algorithms that work with only a constant or at least sublinear amount of internal space with respect to the size of the input.

We consider the "points-in-regions" problem that occurs in the context of geometric database systems - it corresponds to the "inside"-join of "geo-relational algebra" [Gü2], a geometric database model. The *points-in-regions* problem is defined as follows:

Given a set of points POINTS and a set of disjoint polygonal regions REGIONS in the plane, determine all pairs (p,q) with $p \in$ POINTS and $q \in$ REGIONS such that p is inside q.

We develop efficient external solutions for the points-in-regions problem. Two standard set-processing strategies in computational geometry are *plane-sweep* and *divide-and-conquer*. Güting and Schilling [GüS] have shown that for the rectangle intersection problem efficient external solutions can be obtained with the divide-and-conquer technique. Therefore, although for the points-in-regions problem a simple and optimal internal plane-sweep algorithm is available, we also develop an internal divide-and-conquer algorithm to solve this problem. We then extend this algorithm to obtain an external solution.

2. The Points-In-Regions Problem: Internal Algorithms

2.1. A Plane-Sweep Solution

The idea of plane-sweep is conceptually to move a line through the plane and to "observe" the dynamically changing intersection of the sweep line with the set of geometric objects. The geometric objects are simple (hole-free with non-self-intersecting boundary) and closed regions, each initially described by its identifier and the ordered sequence of edges, and the set of points. In our case at any time the intersection of the sweep line with the set of regions is a (possibly empty) set of disjoint y-intervals. The set of regions currently intersecting the sweep line is given by a y-ordered sequence of edges. In the plane-sweep this sequence of edges is maintained while "moving" the sweep line. Whenever a query point is encountered, its y-coordinate is checked against the sequence of edges to determine whether it is inside some region. A suitable data structure to represent the sequence of edges and to support the update and query operations is an AVL-tree.

Theorem 2.1: For a set of k points and a set of disjoint polygonal regions with a total of m edges, the internal ps-algorithm solves the points-in-regions problem in $O(n (\log n) + t)$ time and $O(n)$ space, where $n=k+2m$ and t is the size of the output.

2.2. A Divide-and-Conquer Solution

We extend the techniques for planar divide-and-conquer (e.g. [GüW, Gü1]) to obtain a divide-and-conquer algorithm for the points-in-regions problem. The geometric objects are first transformed into their point set representation (that is the union of the set of points with the set of endpoints of edges where each edge is represented by its left and right endpoint) and sorted lexicographically by x- and y-coordinates. Divide-and-conquer is then applied to the ordered sequence of points. This results in a repeated splitting of this sequence into smaller and smaller subsequences so that groups of points with the same x-coordinate (called *x-groups*) are kept together. Hence the recursion stops when an x-group has been isolated (it may contain just a single point). For each group, we compute the following result sets of objects:

L, the set of edges extending from the group to the left,

R, the set of edges extending from the group to the right, and

Q, the set of query points within this group. A query point is "annotated" by information about the first edge above it within the group (if such an edge exists).

After the top-down splitting, the second phase of the algorithm is a bottom-to-top merging. We can associate with each x-group a vertical stripe of the plane with a non-zero x-extension (called an *x-region*) by using vertical splitting lines between the x-coordinates occurring in the set of points).

In general in each merge step two sets of points S_1 and S_2 within adjacent x-regions are merged into a single set S with one associated x-region. Each input set S_i has associated sets L_i, R_i, and Q_i (i = 1,2) and the merge step computes sets L, R, and Q for the complete set S. L, the set of edges extending from S to the left, is simply the union of L_1 with those edges from L_2 that do not end within S_1. R is formed similarly from R_1 and R_2. The set Q of query points within S is just the union of *updated* Q_1 and Q_2. So far we have mentioned only bookkeeping operations. The real work towards solving the points-in-regions problem performed in the merge step is the updating of the information attached to the points in Q_1 and Q_2. The assumption is that for a point q in Q_1, for instance, the closest edge above it *represented in S_1* has been determined before and is stored with q's representation. Clearly a change of q's "annotation" can only be caused by an edge that is represented in S_2, but not in S_1, in other words, by an edge that traverses S_1 completely. Since these edges do not intersect each other or anything else in S_1 (except, perhaps, a query point) they form a simple y-ordered sequence within the x-region of S_1. Hence during the merge step for each point q in Q_1 the first edge above it within this y-ordered sequence is determined. If it is closer to q than the closest edge stored with it then the stored edge is replaced by the new edge. The symmetric action is taken for Q_2. After the final merge step the algorithm returns a set Q whose points are annotated with the respective closest edges above them within the complete set of regions and the edge representation contains the corresponding region identifiers.

Theorem 2.2: For a set of k points and a set of disjoint polygonal regions with a total of m edges, the internal dac-algorithm solves the points-in-regions problem in $O(n (\log^2 n) + t)$ time and O(n) space, where n=k+2m and t is the size of the output (more precisely the time bound is $O(n(\log n) + k(\log n)*(\log m)+t)$ time).

3. Internal Algorithms with Sublinear Space Requirements

We define an instance of the points-in-regions problem (given sets of points and regions) to be *cross-section-restricted with parameter c* if there exists a constant c such that for any vertical line the total number of intersected edges and query points is at most c (c usually depends on n, e.g. $c=\sqrt{n}$). In this cross-section-restricted case the algorithms developed in the previous section can be modified to require less than linear (in the size of the input) internal space. For the analysis of "sublinear" and external algorithms four different measures of complexity are of interest, namely the internal time and space requirements and the external time and space requirements (the number of read/write operations of pages of secondary memory and the total number of pages used for storing the data, respectively). Let u denote the number of object representations (points, edges) that fit into one page.

Both the plane-sweep and the divide-and-conquer algorithm in a preprocessing step transform their input into an x-ordered sequence of points, a *p-sequence*, requiring internally O(n log n) time and O(1) space and externally O(n/u log n) time and O(n/u) space, or, as a shorthand, (O(n log n), O(n/u log n)) time and (O(1), O(n/u)) space.

For the cross-section-restricted case, the internal plane-sweep algorithm simply processes the p-sequence sequentially, maintaining a dynamic internal data structure of size O(c) requiring (O(n(logc)+t), n/u) time and (O(c), n/u) space.

The modified divide-and-conquer algorithm also processes the p-sequence sequentially, in a series of distinct *steps*, each step processing the next c points requiring O(n ($\log^2$ c) +t), n/u) time and (O(c), n/u) space.

4. External Algorithms

An external algorithm requires only a constant amount of internal space regardless of the size of the (externally stored) set of objects to be processed.

4.1. External Plane-Sweep

Basically the external plane-sweep algorithm must keep the set of edges intersecting the sweep line in an external file structure. The B*-tree is an appropriate structure since it allows to maintain the ordered sequence of edges under insertions, deletions, and searches with a query point during the sweep. It consists of leaf pages and index pages. Each leaf page contains a sequence of edges. An index page contains a pointer to a page followed by a sequence of entries where each entry consists of a line plus another pointer.

4.2. External Divide-and-Conquer

The external divide-and-conquer algorithm consists of three steps: In Step 1, the sequence of input points $p_1 \ldots p_n$ stored in a p-sequence is split into g groups of points $P_1 \ldots P_g$. Each group P_i is processed internally by the internal dac-algorithm which returns a triple of sets $M_i = (L_i, R_i, Q_i)$. These result sets are written back to secondary storage. - In Step 2, a series of external *merge phases* follows: Each phase processes a sequence of triples $M_1 \ldots M_h$ and returns a sequence $M_1' \ldots M_{\lceil h/2 \rceil}'$ by merging M_i with M_{i+1}, for odd i, into $M_{\lceil i/2 \rceil}'$ (if h happens to be odd, let $M_{\lceil h/2 \rceil}' := M_h$). The sequence $M_1' \ldots M_{\lceil h/2 \rceil}'$ forms the input for the next phase. Since each merge phase halves the number of triples, there are about ($\log_2$ g) such phases. Merging triples M_i and M_{i+1}, called a *merge step*, is done externally by reading the input sets from secondary storage and writing the result sets back to it. The actions performed in a merge step correspond to those of the merge step of the internal algorithm. Finally, in Step 3 the result set Q_1 of the last merge phase of Step 2 is scanned and the point enclosures defined by Q_1 are reported.

4.3. Comparison of Plane-Sweep and Divide-and-Conquer

Comparing plane-sweep and divide-and-conquer, plane-sweep appears to be much simpler. However, much complexity is hidden by relying on the powerful concept of balanced trees. Divide-and-conquer does not rely on such concepts which necessarily leads to a longer description but it is possible that the implementation effort is even less for divide-and-conquer than for plane-sweep. Plane-sweep is more efficient than divide-and-conquer with respect to asymptotic worst-case time complexity. This is due to the fact that divide-and- conquer cannot, as for other problems [Gü1, GüW], perform the merge step in linear time.

A comparison of the external algorithms indicates that divide-and-conquer is clearly preferable. Plane-sweep has a very bad performance in this case because basically a page access is required for each point processed. As a recommendation which algorithm to use under what circumstances we suggest to use one of the sublinear space algorithms if a large amount of internal memory is available. Which of the two should be preferred in practice remains to be seen. If very large collections of data have to be processed, or the internal space is very limited, the external divide-and-conquer algorithm is to be used.

The external dac-technique can more generally be used whenever the dac-algorithm uses mainly simple list structures which are merged by parallel scans (rectangle intersection problem [GüS] or the measure problem [Gü1]). The example shows that external dac-algorithms should be used even when there are internal, worst-case efficient ps-solutions or when dynamic external sweep-line structures are known.

References

[BlG] Blankenagel, G., and R.H. Güting, Internal and External Algorithms for the Points-In-Regions Problem - the INSIDE Join of Geo-Relational Algebra. Universität Dortmund, Abteilung Informatik, Forschungsbericht 228, 1987; to appear in *Algorithmica.*

[Gü1] Güting, R.H., Optimal Divide-and-Conquer to Compute Measure and Contour for a Set of Iso-Rectangles. *Acta Informatica 21 (1984)*, 271-291.

[Gü2] Güting, R.H., Geo-Relational Algebra: A Model and Query Language for Geometric Database Systems. In: J.W. Schmidt, S. Ceri, and M. Missikoff (eds.), Proc. of the Intl. Conf. on Extending Database Technology, Venice, March 1988, 506-527.

[GüS] Güting, R.H., and W. Schilling, A Practical Divide-and-Conquer Algorithm for the Rectangle Intersection Problem. *Information Sciences 42 (1987)*, 95-112.

[GüW] Güting, R.H., and D. Wood, Finding Rectangle Intersections by Divide-and-Conquer. *IEEE Transactions on Computers C-33 (1984)*, 671-675.

Geo-Relational Algebra: A Model and Query Language for Geometric Database Systems

Ralf Hartmut Güting

Fachbereich Informatik, Universität Dortmund, D-4600 Dortmund 50, West Germany

(Extended Abstract. The full paper appeared in: J.W. Schmidt, S. Ceri, and M. Missikoff (Eds.), Advances in Database Technology - EDBT '88. Proc. of the Intl. Conf. on Extending Database Technology, Venice, March 1988, 506-527.)

Abstract: *The user's conceptual model of a database system for geometric data should be*

simple and precise: easy to learn and understand, with clearly defined semantics,

expressive: allow to express with ease all desired query and data manipulation tasks,

efficiently implementable.

To achieve these goals we propose to extend relational database management systems by integrating geometry at all levels: At the conceptual level, relational algebra is extended to include geometric data types and operators. At the implementation level, the wealth of algorithms and data structures for geometric problems developed in the past decade in the field of Computational Geometry is exploited. - The paper starts from a view of relational algebra as a many-sorted algebra which allows to easily embed geometric data types and operators. A concrete algebra for two-dimensional applications is developed. It can be used as a highly expressive retrieval and data manipulation language for geometric as well as standard data. Finally, geo-relational database systems and their implementation strategy are discussed.

1. Introduction

In various application areas there is a need to store, retrieve and manipulate data with geometric components, for instance in geographic information systems, VLSI- and CAD-databases, pictorial databases etc. We use the general term *geometric database system* for a database system handling any kind of data with geometric properties. In this paper we address what we believe to be the two main problems in the development of a geometric database system:

1. Design a user model of data, including geometric data, and of data manipulation. Such a model should have the following characteristics:
   - It should be *simple*, so that a user can easily learn and understand it and the consequences of actions can be foreseen.
   - The semantics of the model should be *precisely defined* so that there is no ambiguity for the user or implementor of the model.

- It should be *highly expressive* so that it is easy to perform the desired retrieval and data manipulation tasks.

2. Develop an efficient implementation strategy for the user's conceptual model.

Our approach to achieve these goals consists of two parts:

1. Extend relational algebra by including geometric data types and operators. This is possible in a clean and simple way when relational algebra is viewed as a many-sorted algebra, as proposed in [GüZC87] (see the next section).
2. Implement the extended algebra by integrating geometric algorithms and data structures, as developed in Computational Geometry in the last few years, with the standard relational implementation technology.

A review of geometric data management and a comparison to other work is given in the full paper [Gü88a].

2. The Geo-Relational Algebra

Geo-relational algebra starts from a view of relational algebra as a *many-sorted* algebra, as (to our knowledge) first described in [GüZC87, GüZC88]. This means that the objects of the algebra are atomic values such as strings, numbers, or boolean values as well as relations. The operators of the algebra include arithmetic operators on numbers (e.g. addition) as well as the relational operators (e.g. selection, join).

Within such a formal framework it is easy to embed geometric data types and operators. The basic idea is the following:

- New data types are introduced for geometric objects, such as POINT, LINE, or REGION. In particular a relation may now have attributes of these new types. Furthermore specific operators are introduced for accessing objects of these data types (geometric objects). The representation of geometric objects is hidden from the user and they are in fact accessible only by the new geometric operators.
- A tuple of a relation describes an object as a combination of geometric and non-geometric attribute values.
- A relation as a whole describes a homogeneous collection of geometric objects, that is, a set of points, a set of lines, etc. (A homogeneous collection of geometric objects is the right kind of input for many geometric algorithms.)

Based on these principles one can design different algebras for specific applications. In the sequel we describe an algebra which contains rather fundamental data types and operators for the manipulation of two-dimensional geometric data. The choice of data types is perhaps somewhat biased towards geographic applications. Our specific algebra has the following object classes (or data types):

NUM	numbers	REL	relations
STR	strings		
BOOL	boolean values		

POINT	points in the plane, in a cartesian coordinate system
LINE	lines (a line is a sequence of line segments)
REG	regions in the plane (a region is a non-selfintersecting, hole-free polygon)

Actually there are two different types for regions called PGON and AREA, respectively. If an attribute of a relation is of type AREA then the polygons occurring as attribute values must be pairwise disjoint whereas for type PGON this is not required. Type AREA can be used to model planar subdivisions. The operators of the geo-relational algebra are listed in the following table:

1.	NUM × NUM	→ NUM	+, -, *, /
2.	BOOL × BOOL	→ BOOL	**and, or**
3.	BOOL	→ BOOL	**not**
4.	REL × REL	→ REL	∪, ∩, −, ×, |×|
5.	REL	→ REL	σ, π, λ
6.	NUM × NUM	→ BOOL	=, ≠, <, ≤, ≥, >
	STR × STR	→ BOOL	
	BOOL × BOOL	→ BOOL	
7.	REL	→ NUM	**count**
8.	NUM*	→ NUM	**sum, avg, min, max**
9.	REL	→ ATOM	**extract**

10.	POINT × POINT	→ BOOL	=, ≠
	LINE × LINE	→ BOOL	
	REG × REG	→ BOOL	
11.	GEO × REG	→ BOOL	**inside, outside**
12.	EXT × EXT	→ BOOL	**intersects**
13.	AREA × AREA	→ BOOL	**is_neighbour_of**

14.	LINE* × LINE*	→ POINT*	**intersection**
	LINE* × REG*	→ LINE*	
	PGON* × REG*	→ PGON*	
15.	AREA* × AREA*	→ AREA*	**overlay**
16.	EXT*	→ POINT*	**vertices**
17.	POINT* × REG	→ AREA*	**voronoi**
18.	POINT* × POINT	→ REL	**closest**

19.	POINT*	→ PGON	**convex_hull**
20.	POINT*	→ POINT	**center**
	EXT	→ POINT	

21.	POINT × POINT	→ NUM	**dist**
22.	GEO × GEO	→ NUM	**mindist, maxdist**
23.	POINT*	→ NUM	**diameter**
24.	LINE	→ NUM	**length**
25.	REG	→ NUM	**perimeter, area**

Table 2-1: Operators of the geo-relational algebra

The functionality of the operators is basically described in terms of the classes of objects mentioned above. Some additional notations are used to enhance clarity and readability. The notation ATTR* where ATTR represents one of the atomic object classes is used to refer to a relation of which an attribute of type ATTR is specified. Hence the ATTR* notation determines a column of a relation or, intuitively, a set of objects of class ATTR. The notation ATTR* used on the right hand side of an arrow means that the result relation will have a new attribute of type ATTR. REG, EXT, and GEO are generalizations of the basic geometric types, namely, REG = {PGON, AREA}, EXT = {LINE} ∪ REG, and GEO = {POINT} ∪ EXT. The meaning is that an occurrence of a set identifier in the listing may be replaced by each element of the set to obtain a valid functionality of the operator.

More formally, the algebra is a collection of sets and functions between these sets; it is described by an *S-sorted signature* Σ where S is a set of sorts (names for the sets) and Σ a family of sets $\Sigma_{w,s}$ of operator symbols (names for the functions), where $w \in S^*$ and $s \in S$ describe the functionality of operators in $\Sigma_{w,s}$ [GoTW78]. The set of sorts of geo-relational algebra is S = {REL, BOOL, NUM, STR, POINT, LINE, PGON, AREA} and the signature is basically given by Table 2-1. So the algebra consists of a set for each sort in S and a function for each element in Σ.

In the sequel we briefly discuss some of the operators and illustrate their use by example queries. For an explanation of the remaining operators see the full paper. A complete description and formal definition of the algebra is given in [Gü88b]. - For the example queries the following relations are used, describing cities, states, rivers, and highways in the United States.

cities	cname	center	cpop	states	sname	region	spop
	STR	POINT	NUM		STR	AREA	NUM
rivers	rname	route		highways	hname	way	
	STR	LINE			STR	LINE	

"List all cities with more than 500000 inhabitants!"

cities σ [cpop > 500000] π [cname]

"What is the average population of these cities?"

cities σ [cpop > 500000] **avg** [cpop]

The λ operator extends a relation by new attributes whose values are determined through algebra expressions appearing as parameters.

"List for each city its population as a percentage of the maximal population number!"

cities **max** [cpop] {MAX};
cities λ [(cpop/MAX)*100 {percentage}]

This is a "multistep-query". In the first step the maximum population number is determined and named "MAX", in the second step it occurs as an atomic object of type NUM. The result relation has the schema:

cities	cname	center	cpop	percentage
	STR	POINT	NUM	NUM

"List all cities with their state!"

cities states |×| [center **inside** region] π [cname, sname]

"Which rivers are totally or partially in California?"

states σ [sname = 'California'] rivers |×| [route **intersects** region] π [rname]

Whereas the operators in classes 10. - 13. only compare geometric objects, the operators in the next group (classes 14. - 18.) take one or more relations (viewed as sets of geometric objects) as operands and produce a result relation. They also, except for the **closest** operator, construct new geometric objects and embed them into the result relation. For instance, the **intersection** operator can be applied to two sets of lines and it constructs all intersection points between lines of the first operand and lines of the second operand, as illustrated in Fig. 2-1.

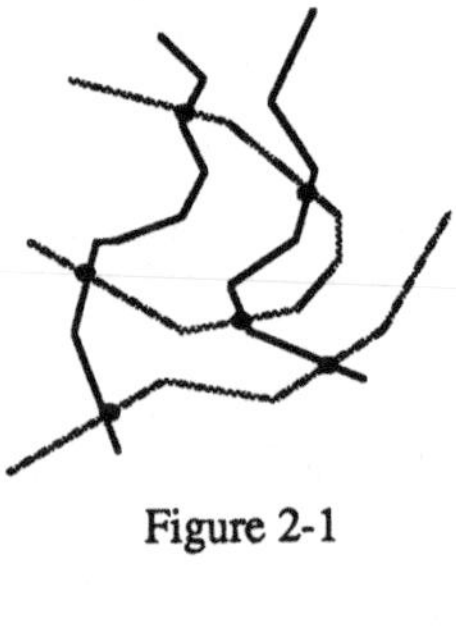

Figure 2-1

It is important to note that a single line of the first operand can intersect a single line of the second operand in several points. The result relation contains one tuple for each intersection point. This tuple consists of all attribute values of the first operand tuple, all attribute values of the second operand tuple and one additional attribute value of type POINT for the constructed intersection point. - The principle explained here for the

intersection operator applied to sets of lines is the same for the remaining functionalities of the **intersection** operator and the other operators in this group. So for instance the intersection of lines with regions will result in the construction of new lines and each result tuple will contain all information of the participating line and region plus the constructed intersection line.

"Which city with more than 100000 inhabitants is closest to the point where interstate highway 90 crosses the Mississippi river?"

(highways σ [hname = 'I90'] rivers σ [rname = 'Mississippi']
intersection [way, route, {crossing}] **extract** [**true**, crossing]) {crosspoint};

cities σ [cpop > 100000] crosspoint **closest** [center]

In a first step the intersection point of Interstate 90 with the Mississippi river is determined and called "crosspoint". This is done by selecting the right tuples from the highways and rivers relations and computing the intersections of their respective attributes of type LINE. The query assumes that we know there is only one intersection point (otherwise the verbal question would not make sense). The **intersection** operator will then produce a result relation with a single tuple containing the attribute values of the highway and the river, respectively, and additionally their intersection point in a new attribute called "crossing". The **extract** operator in this case just checks that its operand relation contains indeed only a single tuple and extracts the "crossing" attribute value from this tuple which is then called "crosspoint". - The second part of the query is then a simple application of the **closest** operator to the relation containing cities with more than 100 000 inhabitants.

"Find all cities within 100 miles from San Francisco!"

cities **extract** [cname = "San Francisco", center] {SF};
cities σ [**dist** (center, SF) < 100]

"Which US state has the highest population density?"

states λ [spop/**area**(region) {density}] {xstates} σ [density = (xstates **max** [density])]

For each state tuple its population density is computed and added as a new attribute, the result relation is named "xstates". From this relation the state tuple with maximal value of the density attribute is selected. Note that it is necessary to introduce the name "xstates" for the intermediate result to be able to refer to it within the selection condition.

In the full paper some remarks are made concerning the implementation of this data model. This topic is more thoroughly discussed in [Gü88b] where also a prototype implementation being carried out at the University of Dortmund is described.

3. Conclusions

In this paper we have described a conceptual model for a geometric database system. For a user the essential aspects of the model are:

(1) High-level geometric objects are available to model geometric aspects of real life entities and one can perform "calculations" with these objects in a manner similar to writing down simple numerical expressions. For instance, it is as easy to write down a test whether two polygons intersect as whether one number is greater than another. We believe a user does not care to see polygons represented by large collections of numbers stored in relations, nor does he/she want to program the intersection test in SQL, say, based on this representation. Our concept is to hide the internal representation of objects entirely; this representation is defined in the implementation language of the database system which is also used to implement operations efficiently based on this representation.

(2) Writing down a query is conceptually just the composition of functions initially applied to objects known to the system (such as stored relations). Since the domain and range types of functions fit together in many ways a very flexible and expressive query language results, as we hope to have demonstrated by the examples. The underlying mathematical concept is that of a many-sorted algebra.

References

[GoTW78] Goguen, J.A., J.W. Thatcher, and E.G. Wagner, An Initial Algebra Approach to the Specification, Correctness, and Implementation of Abstract Data Types. In: R. Yeh (ed.), Current Trends in Programming Methodology, Vol. IV, Prentice Hall 1978, 80-149.

[Gü88a] Güting, R.H., Geo-Relational Algebra: A Model and Query Language for Geometric Database Systems. In: J.W. Schmidt, S. Ceri, and M. Missikoff (Eds.), Advances in Database Technology - EDBT '88. Proc. of the Intl. Conf. on Extending Database Technology, Venice, March 1988, 506-527.

[Gü88b] Güting, R.H., Modelling Non-Standard Database Systems by Many-Sorted Algebras. Universität Dortmund, Fachbereich Informatik, Report 255, 1988.

[GüZC87] Güting, R.H., R. Zicari, and D. Choy, An Algebra for Structured Office Documents. IBM Almaden Research Center, San Jose, California, Report RJ 5559, 1987.

[GüZC88] Güting, R.H., R. Zicari, and D. Choy, An Algebra for Structured Office Documents. Revised Version November 1987, Universität Dortmund, Fachbereich Informatik, Report 254, 1988.

ELEMENTARY SET OPERATIONS WITH d-DIMENSIONAL POLYHEDRA

H. Bieri and W. Nef

Institut für Informatik und angewandte Mathematik
Universität Bern, Länggassstrasse 51, CH-3012 Bern

1. Introduction

When approximately 15 years ago we became interested in the computational geometry of polyhedra, we soon realized that certain basic notions (e.g. face of a polyhedron) were not properly defined, and that no appropriate theory was available. Most things therefore had to be done on a purely intuitive basis, leading into difficulties especially in certain singular cases (cf. [09]). Nevertheless a large number of publications in the field have since appeared, e.g. [08], [10], [13] to [18] and [21] to mention only some of the most significant.

Major problems have been caused by the fact that the family of polyhedra is not closed with respect to certain elementary operations. This situation can be got under control in two different ways:

- "regularization" of the operations ([21],[25]), or
- adaption of the definition of polyhedra.

Following the second way we established a system of definitions and theorems destined to serve as a basis for the whole field ([19]). The theory having proved to be appropriate for several applications (cf. [04],[05],[06]) we now present new algorithms (those in [19] being too complicated from today's standpoint) for the elementary algebraic and topological operations on polyhedra in $\mathbf{R}^d$, namely complement, intersection, union, difference, closure and interior. By their combination regularized operations too are realizable.

2. Fundamentals

<u>Definition 1</u> A **polyhedron** $P \subset \mathbf{R}^d$ is a set of points generated from a finite set of (open) halfspaces by forming complements and intersections (cf. [19], Definition 2;1).

Because of

$$(P_1 \cup P_2) = \mathrm{cpl}(\mathrm{cpl}\ P_1 \cap \mathrm{cpl}\ P_2)\ ,\quad P_1 \setminus P_2 = P_1 \cap \mathrm{cpl}\ P_2 \qquad (01)$$

the family of polyhedra is closed with respect to the operations "cpl", "$\cap$", "$\cup$", and "\". According to [19], Satz 3;20 the same is true for "clos" and "int".

Polyhedra in the usual sense are special polyhedra after definition 1. The latter need not be closed nor bounded nor connected.

An alternative definition equivalent to definition 1 has been given by Bieri [3]: A set $P \subset \mathbf{R}^d$ is a polyhedron if and only if P and cpl P both are unions of a finite number of relatively open sets.

A subset $Q \subset \mathbf{R}^d$ is called a **cone**, if a point $x \in \mathbf{R}^d$ exists such that

$$Q = x + \mathbf{R}^+ \cdot (Q - x) \quad (\text{with } \mathbf{R}^+ := \{\lambda \in \mathbf{R}: \lambda > 0\}). \tag{02}$$

x is called an **apex** of Q.

A cone may have more than one apex. The set N(Q) of all apices of a cone Q is a plane (i.e. a nonempty intersection of hyperplanes ([19], Sätze 1;2, 1;3.1). (03)

For every set $A \subset \mathbf{R}^d$, $Q := x + \mathbf{R}^+ \cdot (A - x)$ is a cone with x as an apex. (04)

Definition 2 A **pyramid** is a set $Q \subset \mathbf{R}^d$ which is a cone and a polyhedron. (05)

If P is a polyhedron and $x \in \mathbf{R}^d$ then there exists a neighbourhood U_o of x such that

$$P^x := x + \mathbf{R}^+ \cdot (U_o - x) = x + \mathbf{R}^+ \cdot (U - x) \tag{06}$$

for every neighbourhood $U \subset U_o$ of x ([19], Satz 3;8).

P^x is a pyramid with x as an apex ([19], Satz 3;3) and is called **locally adjoined to** P **in** x. (07)

The following are some rules for operations with adjoined pyramids:

$$x \in P^x \iff x \in P, \quad x \in \text{clos } P \iff P^x \neq \emptyset \tag{08}$$

$$x \in \text{int } P \iff P^x = \mathbf{R}^d, \quad x \in \text{ext } P \iff P^x = \emptyset$$

([19], Sätze 3;7, 3;19).

$$(P_1 \cap P_2)^x = P_1^x \cap P_2^x, \quad (P_1 \cup P_2)^x = P_1^x \cup P_2^x \tag{09}$$

$$(\text{cpl } P)^x = \text{cpl } P^x, \ (\text{clos } P)^x = \text{clos } P^x, \ (\text{int } P)^x = \text{int } P^x$$

([19], Sätze 3;13 to 3;15, 3;21).

Definition 3 The **faces** of a polyhedron are the equivalence classes of the relation $x \sim y \iff P^x = P^y$.

(This definition insofar differs slightly from [19], Definition 6;1, as ext P = int cpl P = $\{x \in \mathbf{R}^d: P^x = \emptyset\}$ now too becomes a face, provided it is $\neq \emptyset$. The properties of faces expressed by the following theorems are proved in [19] for faces $\neq$ ext P. They are, however, true for S = ext P ($\neq \emptyset$) too).

Faces according to definition 3 need not be bounded nor connected.

Among the faces of P are int P and ext P (provided they are $\neq \emptyset$). Their dimension (cf. theorem 2 below) is d. Faces of dimension 0 are called **vertices**. They contain just one point.

We denote with $\mathfrak{S}(P)$ the set of all faces of P. For a face S we introduce $P^S := P^x$ $(x \in S)$.

As an example we take a look at the closed (open) unit cubes in a orthogonal coordinate system in $\mathbf{R}^3$. Both cubes have the same 28 faces $S \in \mathfrak{S}(P)$, the difference being that all faces (except the exterior) are subsets of the closed cube, while (except the interior) they are disjoint to the open cube. The faces are:

number	type of S	P^S	$N(P^S)$	dim S (cf. th. 2)
8	vertices	closed (open) octants	points	0
12	edges	closed (open) quadrants	lines	1
6	facets	closed (open) half-spaces	planes	2
1	interior	$\mathbf{R}^3$	$\mathbf{R}^3$	3
1	exterior	$\emptyset$	$\mathbf{R}^3$	3

<u>Theorem 1</u> $\mathfrak{S}(P)$ is finite. All $S \in \mathfrak{S}(P)$ are polyhedra. ([19], Satz 6;1.1).

<u>Theorem 2</u> For all $S \in \mathfrak{S}(P)$: aff S = $N(P^S)$ (cf. (03)) and S is a relatively open subset of aff S. Therefore dim S = dim $N(P^S)$. ([19], Satz 6;2).

<u>Theorem 3</u> Every $S \in \mathfrak{S}(P)$ is a subset of or disjoint to P. ([19], Satz 6;4).

<u>Theorem 4</u> If S_o and S are faces of a polyhedron P, then $S_o \cap \text{clos } S = \emptyset$ or $S_o \subset \text{clos } S$. In the latter case, we say that **S is incident to** S_o. ([19], Satz 6.10).

We denote with $\mathfrak{S}_{S_o}(P)$ the set of all faces of P incident to S_o:

$$\mathfrak{S}_{S_o}(P) := \{S \in \mathfrak{S}(P): S_o \subset \text{clos } S\}. \tag{10}$$

(For a vertex {c} we shall write $\mathfrak{S}_c(P)$).

The following theorem has so far not been published:

<u>Theorem 5</u> For every $S_o \in \mathbb{S}(P)$ the following mapping is bijective:

$$S \in \mathbb{S}_{S_o}(P) \quad \langle\text{------}\rangle \quad \hat{S} = S^{S_o} \in \mathbb{S}(P^{S_o}).$$

Furthermore $P^S = (P^{S_o})^{\hat{S}}$.

For $S = S_o$ this means that aff $S_o = N(P^{S_o})$ (cf. theorem 2) is a face of P^{S_o}, and $S_o \subset P \iff N(P^{S_o}) \subset P^{S_o}$. (11)

The remaining theorems concern a polyhedron **in general position** with respect to a coordinate system. This means

$$N + H = \mathbf{R}^d \quad \text{or} \quad N_o \cap H_o = \{0\} \tag{12}$$

for all $N =$ aff S ($S \in \mathbb{S}(P)$) and all coordinate planes H of any dimension (N_o and H_o are the linear subspaces parallel to N, H).

The available space does not allow a detailed description of how to construct a coordinate system satisfying this condition for a given P. So we briefly mention that there are basically two ways:

1. Find an appropriate transformation numerically (cf. [22]).
2. Apply a formal linear transformation differing little from identity, the coefficients being considered as variables subject to certain order relations.

A face $S \in \mathbb{S}(P)$ will be called **unbounded to the left (right),** if it contains points with arbitrarily small (big) ξ_1-coordinate.

We denote with $\mathbb{S}_{-\infty}(P)$ ($\mathbb{S}_{+\infty}(P)$) the set of faces of P unbounded to the left (right). (13)

<u>Theorem 6</u> Let $P \subset \mathbf{R}^d$ be in general position. Then every face $S \in \mathbb{S}(P)$ is unbounded to the left or incident to a vertex {c} of P:

$$\mathbb{S}(P) = \mathbb{S}_{-\infty}(P) \cup \bigcup_c \mathbb{S}_c(P). \tag{14}$$

This theorem has so far not been published.

For $\theta \in \mathbf{R}$ we denote with $H(-\theta)$ the hyperplane $\xi_1^{-1}(-\theta)$, i.e. the set of all points with ξ_1- coordinate $-\theta$.

<u>Theorem 7</u> (not yet published) Let $P \subset \mathbf{R}^d$ be in general position. Then there exists a $\theta_o \in \mathbf{R}$, such that for all $\theta > \theta_o$:

1. A face $S \in \mathbb{S}(P)$ is unbounded to the left if and only if $S \cap H(-\theta) \neq \emptyset$.

2. The following mapping is bijective:

$$S \in \mathbb{S}_{-\infty}(P) \quad \langle\text{------}\rangle \quad S^* = S \cap H(-\theta) \in \mathbb{S}(P \cap H(-\theta)).$$

3. $(P \cap H(-\theta))^{S^*} = P^S \cap H(-\theta)$.

Instead of assigning a numerical value to θ_o we better consider θ as a variable and grant for $\theta > \theta_o$ by applying the following rule:

Everyone of the (finitely many) expressions $\alpha + \beta\cdot\theta$ with $\beta \neq 0$ occuring during the execution of an algorithm has the same sign as β.

<u>Theorem 8</u> Let $P \subset \mathbf{R}^d$ be in general position and $\{c\}$ a vertex of P. Then every face of P^c (except $\{c\}$ itself) is unbounded to the left or to the right:

$$\mathbb{S}(P^c) = \mathbb{S}_{-\infty}(P^c) \cup \mathbb{S}_{+\infty}(P^c) \cup \{\{c\}\}. \tag{15}$$

3. Information structure

In every implementation polyhedra will have to be described by a certain data structure. In our case, however, it seems to be recommendable to introduce a purely geometric intermediate level, which we shall call **information structure.**

<u>Definition 4</u> The **information structure** $\mathbb{Q}(P)$ of a polyhedron P is the set of pyramids P^S adjoined to P:

$$\mathbb{Q}(P) := \{Q = P^S\colon S \in \mathbb{S}(P)\}. \tag{16}$$

By analogy to (10),(13) we introduce

$$\mathbb{Q}_{\pm\infty}(P) := \{Q = P^S\colon S \in \mathbb{S}_{\pm\infty}(P)\} \subset \mathbb{Q}(P), \tag{17}$$

$$\mathbb{Q}_c(P) := \{Q = P^S\colon S \in \mathbb{S}_c(P)\} \subset \mathbb{Q}(P).$$

We ask any implementation to allow the following eight primitive operations with pyramids:

P1: given Q, find cpl Q (18)

P2: given Q_1, Q_2 with $N(Q_1) \cap N(Q_2) \neq \emptyset$, find $Q_1 \cap Q_2$

P3: given Q, find N(Q)

P4: given Q, find $Q \cap H(-\theta)$ with θ as a variable *)

P5: given $Q \cap H(-\theta)$ with θ as a variable, find Q *)

P6: given Q and $x \in \mathbf{R}^d$, find Q^x

P7: given Q, find all (finitely many) Q^x $(x \in \mathbf{R}^d)$

P8: given Q, find clos Q

*) in P4 and P5 we assume Q to be in general position, $\dim N(Q) \geq 1$ and $\dim(N(Q) \cap H(-\theta)) = \dim N(Q) - 1$.

With these primitive operations, pyramids represent an abstract data type (cf. sect. 6). The same is true for $\mathbb{Q}(P)$ as a set of pyramids.

The following four problems A1-A4 can be reduced to P1-P8 (A2-A4 without proof):

A1: Applying P7 to $Q_o = P^{S_o}$ we find (theorem 5) all $Q = P^S$ with $S \in \mathbb{S}(P)$, $S_o \subset \text{clos } S$. Given $\mathbb{Q}(P)$ it is therefore possible to find the **incidence relation** on $\mathbb{S}(P)$ resp. $\mathbb{Q}(P)$, containing all pairs (S_o,S) resp. (Q_o,Q) with $S_o \subset \text{clos } S$.

A2: Given $\mathbb{Q}(P)$ and $S \in \mathbb{S}(P)$ (represented by $Q = P^S \in \mathbb{Q}(P)$), decide wether $S \subset P$.

A3: Given $\mathbb{Q}(P)$, $S \in \mathbb{S}(P)$ (represented by P^S) and $x \in \mathbf{R}^d$, decide wether $x \in S$.

A4: Given $\mathbb{Q}(P)$, find $\mathbb{Q}_{\pm\infty}(P)$.

From A2, A3 follows that $\mathbb{Q}(P)$ uniquely determines $P = \bigcup_{S \subset P} S$ (theorem 3).

It is now our intention to establish algorithms determining $\mathbb{Q}(\text{cpl } P)$ and $\mathbb{Q}(\text{clos } P)$ from $\mathbb{Q}(P)$, and $\mathbb{Q}(P_1 \cap P_2)$ from $\mathbb{Q}(P_1)$, $\mathbb{Q}(P_2)$. Because of (01) and int P = cpl clos cpl P we will then be able to find $\mathbb{Q}(P_1 \cup P_2)$, $\mathbb{Q}(P_1 \setminus P_2)$ and $\mathbb{Q}(\text{int } P)$, and finally, by combination, to execute "regularized" operations (cf. [7], page 450).

4. Computing $\mathbb{Q}$(cpl P) and $\mathbb{Q}$(clos P)

Given $\mathbb{Q}(P)$, find $\mathbb{Q}(\text{cpl } P)$ and $\mathbb{Q}(\text{clos } P)$.

<u>Theorem 9</u> P and cpl P have the same 0, ... ,(d-1)-dim faces. ([19], Satz 6;14).

With (09) and ext P = int cpl P, int P = ext cpl P we get

$$\mathbb{Q}(\text{cpl } P) = \{\text{cpl } Q\colon Q \in \mathbb{Q}(P)\}. \tag{19}$$

So we find $\mathbb{Q}(\text{cpl } P)$ with the aid of the primitive operation P1.

<u>Theorem 10</u> Every face of clos P is a union of faces of P. ([19], Satz 6;26).

With (09) we therefore find

$$\mathbb{Q}(\text{clos } P) = \{\text{clos } Q\colon Q \in \mathbb{Q}(P)\} \tag{20}$$

and thus reduce the construction of $\mathbb{Q}(\text{clos } P)$ to P8.

5. Computing $\mathfrak{Q}(P_1 \cap P_2)$

Given $\mathfrak{Q}(P_1)$ and $\mathfrak{Q}(P_2)$, find $\mathfrak{Q}(P_1 \cap P_2)$.

The algorithm "IS" (for InterSection) we are going to describe is recursive with respect to the dimension d of $\mathbf{R}^d$. (In case of dimension 0 the problem is of course trivial). For the sake of simplicity we will just show how to reduce the d-dimensional case to the (d-1)-dimensional, assuming an algorithm to be known for the latter.

We assume P_1, P_2 and $P := P_1 \cap P_2$ to be in general position (cf.(12)). No restriction, however, is asked concerning the relative position of P_1, P_2.

The following is an immediate consequence of (14):

$$\mathfrak{Q}(P) = \mathfrak{Q}_{-\infty}(P) \cup \bigcup_{c} \mathfrak{Q}_c(P). \tag{21}$$

So we may write an algorithm IS as follows:

```
IS   IS1  find Q_{-∞}(P)
     IS2  find all vertices {c} of P
     IS3  for all vertices {c} of P
               do IS3*: find  Q_c(P)
               od
     IS4  find Q(P) according to (21). *)
```

*) Faces thereby are in most cases detected more than one time. This does not necessarily mean the CPU-time required to be a bad investement. For the algorithm easily provides as a supplement the incidences between vertices and other faces, and by recursion those between all pairs of faces.

Let us now take a closer look at IS1 to IS3!

<u>Subalgorithm IS1</u>

Given $\mathfrak{Q}(P_1)$, $\mathfrak{Q}(P_2)$, find $\mathfrak{Q}_{-\infty}(P)$.

Based on theorem 7 we find the following refinement of IS1:

IS1	IS1.1	find $\mathbb{Q}_{-\infty}(P_i)$ $(i=1,2)$	1)
	IS1.2	$\mathbb{Q}(P_i \cap H(-\theta)) := \{Q \cap H(-\theta) : Q \in \mathbb{Q}_{-\infty}(P_i)\}$ $(i=1,2)$	2)
	IS1.3	$\mathbb{Q}(P \cap H(-\theta)) := \mathbb{Q}((P_1 \cap H(-\theta)) \cap (P_2 \cap H(-\theta)))$	3)
	IS1.4	$\mathbb{Q}_{-\infty}(P) := \{Q : Q \cap H(-\theta) \in \mathbb{Q}(P \cap H(-\theta))\}$	4)

1) A4 2) P4 3) (d-1)-dim algorithm 4) P5

<u>Subalgorithm IS2</u>

Given $\mathbb{Q}(P_1)$, $\mathbb{Q}(P_2)$, find all vertices $\{c\}$ of $P = P_1 \cap P_2$.

<u>Theorem 11</u> Every vertex $\{c\}$ of $P = P_1 \cap P_2$ is a 0-dimensional intersection of two faces $S_i \in \mathbb{S}(P_i)$ $(i=1,2)$: $\{c\} = S_1 \cap S_2$. ([19], Satz 6;16).

(The reverse is not true!)

So we get

```
IS2  for { all / selected } pairs (Q1,Q2) ∈ (Q(P1), Q(P2))          5)
                    (Qi = P^Si (i=1,2))
     do IS2*: decide wether S1 ∩ S2 is a vertex of P
     od
```

Where: for $\left\{\begin{matrix}\text{all}\\ \text{selected}\end{matrix}\right\}$ pairs $(Q_1,Q_2) \in (\mathbb{Q}(P_1), \mathbb{Q}(P_2))$ $(Q_i = P^{S_i}$ $(i=1,2))$, do IS2*: decide wether $S_1 \cap S_2$ is a vertex of P.

5) Taking into consideration all pairs (Q_1,Q_2) is the easiest but least efficient way. For a selection two methods may be thought of:

1. Enclose every face S_i into a rectangular box $B(S_i)$ and consider (Q_1,Q_2) only if $B(S_1) \cap B(S_2) \neq \emptyset$. (cf. [24]; [20], page 364).

2. For every line L parallel to the coordinate vector e_d find (by a sweep method) the ξ_d-ordered set of intersections of L with the faces of P_1 and P_2. Consider (Q_1,Q_2) only if S_1,S_2 are neighbours in one of these ordered sets. This of course is a generalisation of the well known algorithms in [2], [23].

IS2* may be realized as follows:

IS2* if $\dim(N(Q_1) \cap N(Q_2)) = 0$ (with $\{c\} := N(Q_1) \cap N(Q_2)$) 6)
then if $c \in S_1 \cap S_2$ (i.e. $S_1 \cap S_2 = \{c\}$) 7)
then if $\dim N(Q_1) + \dim N(Q_2) = d$
then $\{c\}$ is a vertex 8)
else do $Q := P^c = P_1^{S_1} \cap P_2^{S_2}$ 9)
if $N(Q) = \{c\}$ 10)
then $\{c\}$ is a vertex
else $\{c\}$ is not a vertex
od
else $S_1 \cap S_2$ is not a vertex
else $S_1 \cap S_2$ is not a vertex

6) P3. (Since S_1, S_2 are relatively open: $S_1 \cap S_2 = \emptyset$ or $\dim(S_1 \cap S_2) = \dim(N(Q_1) \cap N(Q_2))$ (cf. theorem 2).

7) A3.

8) regular intersection, cf. [19], Satz 6;17

9) P2. (cf. (09)).

10) P3. (N(Q) is a plane containing c (theorem 2) but may be $\neq \{c\}$).

<u>Subalgorithm IS3*</u>

Given $\mathbf{Q}(P_1)$, $\mathbf{Q}(P_2)$,

$Q_i = P_i^{S_i} \in \mathbf{Q}(P_i)$ $(i=1,2)$ with $\{c\} = S_1 \cap S_2$ a vertex of P,

find $\mathbf{Q}_c(P)$.

Based on theorem 5 we get

$$\mathbf{Q}(P_i^{S_i}) = \mathbf{Q}_{S_i}(P) \;,\; \mathbf{Q}_c(P) = \mathbf{Q}(P^c). \qquad (22)$$

Furthermore the following is an immediate consequence of (15):

$$\mathbf{Q}(P^c) = \mathbf{Q}_{-\infty}(P^c) \cup \mathbf{Q}_{+\infty}(P^c) \cup \{P^c\}. \qquad (23)$$

So we find the following refinement for IS3*:

IS3* IS3.1 $\mathbf{Q}(P_i^{S_i}) := \mathbf{Q}_{S_i}(P_i)$ $(i=1,2)$ 11)
IS3.2 find $\mathbf{Q}_{\pm\infty}(P^c)$ 12)
IS3.3 find $\mathbf{Q}_c(P)$ 13)

11) (22) and A1

12) Apply IS1 to $P_1^{S_1}$, $P_2^{S_2}$ (IS1 determines $\mathbf{Q}_{-\infty}(P_1 \cap P_2)$ from $\mathbf{Q}(P_1)$, $\mathbf{Q}(P_2)$. So, with $\mathbf{Q}(P_1^{S_1})$, $\mathbf{Q}(P_2^{S_2})$ as input it will determine $\mathbf{Q}_{-\infty}(P_1^{S_1} \cap P_2^{S_2}) = \mathbf{Q}_{-\infty}(P^c)$ (cf. (09)). $\mathbf{Q}_{+\infty}(P^c)$ is found by analogy.

13) With (22),(23) and (with P2) $P^c = P_1^{S_1} \cap P_2^{S_2}$ (cf. (09)).

6. An implementation

The design of our algorithms for set operations on polyhedra has largely been based on the notion **locally adjoint pyramid** as well as on **primitive operations** with such pyramids. It seems natural, therefore, to implement these algorithms by a two-stage procedure, as follows:

- The first stage - which has essentially been described in the foregoing - implements the algorithms with the aid of an **information structure** (see section 3). This information structure can be represented in a straightforward way by making use of an **abstract data type pyramid**.
- The second stage implements the abstract data type pyramid, as usual, with the aid of a **data structure.** The implementation we will propose has been chosen by simplicity and generality as much as by efficiency. It will be possible to replace it, if required, without changing the principal structure of our algorithms - an obvious advantage of such a two-stage implementation.

In section 3 we introduced a second abstract data type, **set of pyramids**, representing $\mathbf{Q}(P)$. As our actual implementation of this ADT will use standard techniques it will not be discussed here any more.

6.1 The abstract data type pyramid and its data structure

We define an abstract data type **pyramid** according to common practice (cf. [1],p. 11) as a pair (OB,OP) of two sets: OB represents the set of **objects** of our ADT which are the pyramids $\subset \mathbf{R}^d$, and OP the set of **primitive operations** which are necessary to construct our algorithms. OP is defined by the list (18). It is important to notice, and easy to see, that by applying the primitive operations of (18), we do not leave the realm of objects of the ADT pyramid.

In [5], [6] and [19], we specified polyhedra mainly by means of Boolean expressions in (open) halfspaces which we implemented by ordinary characterstrings. Especially when dealing with pyramids, it seems an improvement, however, to work with **cell complexes** (or **arrangements**) $\mathbf{C}(\mathbf{H})$, where $\mathbf{H}$ denotes a finite family of hyperplanes $\subset \mathbf{R}^d$: The hyperplanes in $\mathbf{H}$ partition $\mathbf{R}^d$ into a finite number of **cells** (or **faces**), i.e. disjoint relatively open convex polyhedra of dimensions $0,\ldots,d$ (cf. [11], [12], [19]), and $\mathbf{C}(\mathbf{H})$ is the set of all these cells. Every polyhedron $P \subset \mathbf{R}^d$ can be represented as the union of certain cells of an appropriate cell complex (cf. [19], Satz 2;6). We call every cell complex $\mathbf{C}(\mathbf{H})$ of this kind **compatible** with P and, moreover, denote by $\mathbf{C}(\mathbf{H},P)$ the subset $\{C \in \mathbf{C}(\mathbf{H}): C \subset P\}$ of $\mathbf{C}(\mathbf{H})$. In the special case of a pyramid Q, there always exists a compatible cell complex $\mathbf{C}(\mathbf{H})$ with the additional property that the intersection of all hyperplanes $\in \mathbf{H}$ equals the set of apices N(Q) of Q (cf. (3) and [19], Sätze 2;8 and 2;9). In most practical cases such a $\mathbf{C}(\mathbf{H})$ occurs where $\mathbf{H}$ consists of at most d hyperplanes. This fact seems advantageous especially with respect of efficiency, and it is the main reason for our decision to choose this new kind of implementation.

In view of the above, our implementation of the ADT pyramid will principally consist in a way to implement (compatible) cell complexes. Let a cell complex $\mathbf{C}(\mathbf{H})$ in $\mathbf{R}^d$ be generated by the set $\mathbf{H} = \{H_1,\ldots,H_n\}$ of hyperplanes. It is convenient to define each H_i by an affine function f_i on $\mathbf{R}^d$; this also allows to identify the two open halfspaces belonging to H_i as positive resp. negative. f_i itself can be defined by its coefficients with respect to a given coordinatesystem of $\mathbf{R}^d$. Denoting the two open halfspaces belonging to a hyperplane H by H^+ and H^-, and H itself by H^o, we can then represent each cell C belonging to $\mathbf{C}(\mathbf{H})$ as an intersection

$$C = \bigcap_{i=1}^{n} H^{\sigma_i} \neq \emptyset , \tag{24}$$

where $\sigma_i \in \{+,o,-\}$ $(i=1,\ldots,n)$. This representation associates with each $C \in \mathbf{C}(\mathbf{H})$ a unique **sign tuple** $t = \{(i, \sigma_i): i=1,\ldots,n\}$ (i.e. $t(i) = \sigma_i$ $(i=1,\ldots,n)$) which can be used to implement C. Consequently, $\mathbf{C}(\mathbf{H})$ itself is associated with a set $\mathbf{T}(\mathbf{H})$ of such sign tuples, and for every pyramid $Q \subset \mathbf{R}^d$ its set of cells $\mathbf{C}(\mathbf{H},Q)$ with a corresponding subset $\mathbf{T}(\mathbf{H},Q)$. We arrive at the following implementation:

- A finite set $\mathbf{H}$ of hyperplanes in $\mathbf{R}^d$ (as well as the family of affine functions which defines it) will be represented by a set $F(H)$ of $(d+1)$-tuples of coefficients.

- The corresponding cell complex $C(H)$ will be represented by a set $T(H)$ of n-tuples of signs (n = card H) (and by $F(H)$).
- Any pyramid $Q \subset R^d$ with which $C(H)$ is compatible will be represented by a suhset $T(H,Q)$ of $T(H)$ (and by $F(H)$).

The two special cases $Q = \emptyset$ and $Q = R^d$ are included in this implementation: In both cases, $C(H)$ is compatible with Q for every finite family H of hyperplanes in R^d, including $H = \emptyset$; that is, we have $C(H,\emptyset) = \emptyset$ and $C(H,R^d) = C(H)$ for any choice of H. Choosing $H = \emptyset$ we get the representations $T(\emptyset,\emptyset) = \emptyset$ and $T(\emptyset,R^d) = \{\emptyset\}$ (i.e. $T(\emptyset,R^d)$ only contains the "empty tuple").

(Of course, the sets $F(H)$, $T(H)$, and $T(H,Q)$ must be implemented as well. But since the choice of appropriate data structures is not specific to our purpose, it will not be discussed here.)

6.2 Implementation of the primitive operations

First, it will be fundamental to dispose of a procedure to construct the cell complex $C(H)$ belonging to a given family H of hyperplanes $\subset R^d$ or, respectively, the set $T(H)$ for a given set $F(H)$. One possibility consists in employing our former algorithm CELLS (cf. [4]) which already uses data structures corresponding to $F(H)$ and $T(H)$. We could also adjust the more efficient but also more complicated algorithm described in [11] and [12] to our needs. In the following, we will simply assume an algorithm for the construction of $T(H)$ to be available. In most practical cases with pyramids, $T(H)$ will turn out to be efficiently constructable anyway.

Next, we solve three elementary problems, E1 – E3, to which we subsequently will refer:

E1: Given $T(H_1,Q_1)$ and $T(H_2,Q_2)$, find a cell complex $C(H)$ which is compatible with Q_1 and Q_2.
Solution: Putting $H := H_1 \cup H_2$ and constructing $T(H)$ we get $T(H,Q_j) = \{t \in T(H)\colon t|H_j \in T(H_j,Q_j)\}$ $(j=1,2)$.

E2: Given $T(H_1,Q_1)$ and $T(H_2,Q_2)$, decide if $Q_1 = Q_2$.
Solution: Put $H := H_1 \cup H_2$ and construct $T(H,Q_j)$ $(j=1,2)$ (see E1). Then test if $T(H,Q_1) = T(H,Q_2)$.

E3: Given $T(H,Q)$ and $H_o \subset H$, decide if $C(H_o)$ is compatible with Q.
Solution: $T(H_o,Q_o) := \{t|H_o\colon t \in T(H,Q)\}$ defines a pyramid Q_o

which is the union of certain cells $\in C(H)$. Test if $Q_o = Q$ (see E2).

Now, we propose an implementation fo the primitive operations **P1 - P8** of (18). Every occuring pyramid Q will be assumed to be given by a set $T(H,Q)$ of sign tuples belonging to the cells $\subset Q$ with respect to a compatible cell complex $C(H)$.

P1: Find cpl Q

Input: $T(H,Q)$. **Output:** $T(H,\text{cpl } Q)$.

$T(H,\text{cpl } Q)$ consists of all sign tuples $t \in T(H)$ which do not belong to $T(H,Q)$.

P2: Find $Q_1 \cap Q_2$

Input: $T(H_1,Q_1)$, $T(H_2,Q_2)$. **Output:** $T(H,Q_1 \cap Q_2)$. **Assumption:** $N(Q_1) \cap N(Q_2) \neq \emptyset$ (to assure that $Q_1 \cap Q_2$ is a pyramid).

We construct a cell complex $C(H)$ compatible with Q_1, Q_2 (see E1). Then $T(H,Q_1 \cap Q_2)$ consists of all sign tuples $t \in T(H,Q_1) \cap T(H,Q_2)$.

P3: Find N(Q)

Input: $T(H,Q)$. **Output:** $N(Q)$, as a subset of H.

We determine a subset H_o of H such that $C(H_o)$ is compatible with Q and H_o is minimal (i.e., for every proper subset of H_o, its cell complex is not compatible with Q) (see E3). According to [19], Sätze 2;8 and 2;9, the intersection of all hyperplanes $\in H_o$ equals $N(Q)$.

P4: Find $Q \cap H(-\theta)$

Input: $T(H,Q)$. **Output:** $T(H \cap H(-\theta),Q \cap H(-\theta))$ for all $\theta \in R$.

Assumptions: See (18); H minimal (see P3).

For every $H_i \in H$, we replace in $f_i(x) = \alpha_{i1}\xi_1+\alpha_{i2}\xi_2+\ldots+\alpha_{id}\xi_d+\beta_i$ ξ_1 by $-\theta$ and get for each value of θ a representation of $H \cap H(-\theta)$. Since each cell $C \in C(H,Q)$ intersects $H(-\theta)$ there is a bijection between $T(H,Q)$ and $T(H \cap H(-\theta),Q \cap H(-\theta))$: To every tuple $t \in T(H,Q)$ corresponds a tuple $t_\theta \in T(H \cap H(-\theta),Q \cap H(-\theta))$ with the same value, i.e. the same signs. This allows to get $T(H \cap H(-\theta),Q \cap H(-\theta))$ at once.

P5: Find Q from $Q \cap H(-\theta)$

Input: $T(H \cap H(-\theta),Q \cap H(-\theta))$ for all $\theta \in R$. **Output:** $T(H,Q)$.

Assumptions: See (18).

Given $H \cap H(-\theta)$, we reconstruct H by replacing $-\theta$ by ξ_1 (cf. P4). In order to find $T(H,Q)$, we again only have to consider that every tuple $t_\theta \in T(H \cap H(-\theta),Q \cap H(-\theta))$ determines a tuple $t \in T(H,Q)$ with the same value, i.e. the same signs.

P6: Find Q^x

Input: $T(H,Q)$, $x \in \mathbf{R}^d$ (by its coordinates). **Output:** $T(H_x,Q^x)$.

Let $C \in \mathbf{C}(H,Q)$ be represented by (24), with $t \in T(H,Q)$ as its corresponding sign tuple, and let f_i be the affine function defining H_i $(i=1,\ldots,n)$. For every $H_i^{\sigma_i}$ (i.e. a hyperplane or an open halfspace) we have (cf. [19], S. 3-2)

$$(H^{\sigma_i})^x = \begin{cases} H^{\sigma_i}, & \text{if } f_i(x) = o \\ \mathbf{R}^d, & \text{if } \operatorname{sig} f_i(x) = \sigma_i \neq o \\ \emptyset, & \text{otherwise} \end{cases} .$$

For C as the intersection of all $H_i^{\sigma_i}$ $(i=1,\ldots,n)$ we get because of $C^x = \bigcap \{(H_i^{\sigma_i})^x : i=1,\ldots,n\}$ (see [19], Satz 3;13)

$$C^x = \begin{cases} \emptyset, & \text{if there exists a } f_i \text{ with} \\ & f_i(x) \neq 0,\ \operatorname{sig} f_i(x) \neq \sigma_i \\ \bigcap \{H_i^{\sigma_i} : f_i(x) = 0\}, & \text{otherwise} \end{cases} .$$

Finally, for Q itself we get Q^x as the union of all C^x with $C \in \mathbf{C}(H,Q)$ (cf. (09) and [19], Satz 3;14). We arrive at the following algorithm:

Let $T_o(H,Q) := \{t \in T(H,Q): f_i(x) = o \text{ or } \operatorname{sig} f_i(x) = \sigma_i \text{ for all } i\}$.
If $T(H_o,Q) = \emptyset$ then $Q^x = \emptyset$.
Otherwise, let $H_x := \{H_i : f_i(x) = 0\}$. If $H_x = \emptyset$ then $Q^x = \mathbf{R}^d$.
Otherwise, put $T(H_x,Q^x) := \{t|H_x : t \in T_o(H,Q)\}$.

P7: Find all Q^x

Input: $T(H,Q)$. **Output:** $T(H_x,Q^x)$ for all $x \in \mathbf{R}^d$.

Let C be a cell $\in \mathbf{C}(H)$ represented by the tuple $t \in T(H)$. For every $x \in C$ we have $\operatorname{sig} f_i(x) = t(i)$ $(i=1,\ldots,n)$, hence it suffices to solve **P6** once for every cell $\in \mathbf{C}(H)$. In general, some of the resulting $T(H_x,Q^x)$ may represent identical pyramids. This has to be tested (see **E2**) and, if necessary, adjusted.

P8: Find clos Q

Input: $T(H,Q)$. **Output:** $T(H,\text{clos } Q)$.

According to (08), we must determine all cells $\in C(H)$ which lead to a $Q^x \neq \emptyset$. This is an immediate application of P6 and P7.

7. References

[1] Aho, A.V., Hopcroft, J.E., Ullman, J.D.: Data structures and algorithms. Reading: Addison-Wesley 1983.

[2] Bentley, J.L., Ottmann, T.A.: Algorithms for reporting and counting geometric intersections. IEEE Trans. Comput. C-28, 643-647 (1979).

[3] Bieri, H.: Eine Charakterisierung der Polyeder. Elemente Math. 35, 143-144 (1980).

[4] Bieri, H., Nef, W.: A recursive sweep-plane algorithm, determining all cells of a finite division of R^d. Computing 28, 189-198 (1982).

[5] Bieri, H., Nef, W.: A sweep-plane algorithm for computing the volume of polyhedra represented in Boolean form. Linear Algebra Appl. 52/53, 69-97 (1983).

[6] Bieri, H., Nef, W.: A sweep-plane algorithm for computing the Euler-characteristic of polyhedra represented in Boolean form. Computing 34, 287-302 (1985).

[7] Bieri, H.: Wechselwirkung zwischen der Computergrafik und der Theorie der Polyeder. Informatik-Fachberichte 126, 441-455. Berlin: Springer 1986.

[8] Brüderlin, B.D.: Rule-based geometric modelling. Dissertation, ETH Zürich. Zürich: Verlag der Fachvereine 1988.

[9] Bruggesser, H.: Ein Programmsystem für die graphische Darstellung von Polyedern. Dissertation, Universität Bern 1975.

[10] Chazelle, B., Dobkin, D.P.: Intersection of convex objects in two and three dimensions. J.ACM 34, 1-27 (1987).

[11] Edelsbrunner, H.: O'Rourke, J., Seidel, R.: Constructing arrangements of lines and hyperplanes with applications. SIAM J. Comput. 15, 341-363 (1986).

[12] Edelsbrunner, H.: Algorithms in combinatorial geometry. Berlin: Springer 1987.

[13] Hertel, S., Mäntylä, M., Mehlhorn, K., Nievergelt, J.: Space sweep solves intersection of convex polyhedra. Acta Informatica 21, 501-519 (1984).

[14] Laidlaw, D.H., Trumbore, W.B., Hughes, J.F.: Constructive solid geometry for polyhedral objects. ACM SIGGRAPH'86 Proc., 161-170.

[15] Maibach, B.: MATIP - Eine Benutzersprache und ein Interpreter für mathematische Anwendungen. Dissertation, Universität Bern 1982.

[16] Mehlhorn, K., Simon, K.: Intersecting two polyhedra one of which is convex. Lecture Notes in Computer Science 199, 534-542. Berlin: Springer 1985.

[17] Meier, A.: Methoden der grafischen und geometrischen Datenverarbeitung. Stuttgart: Teubner 1986.

[18] Muller, D.E., Preparata, F.P.: Finding the intersection of two convex polyhedra. Theor. Comput. Sci. 7, 217-236 (1978).

[19] Nef, W.: Beiträge zur Theorie der Polyeder, mit Anwendungen in der Computergraphik. Bern: Herbert Lang 1978.

[20] Preparata, F.P., Shamos, M.I.: Computational geometry - An introduction. Berlin: Springer 1985.

[21] Requicha, A.A.G.: Representations for rigid solids: Theory, methods, and systems. ACM Comput.Surv. 12, 437-464 (1980).

[22] Schmidt, P.M.: Algorithm for constructing a sweep-plane which is in general position to a given point set. Manuskript, Friedrich-Schiller-Universität Jena 1987.

[23] Shamos, M.I., Hoey, D.: Geometric intersection problems. 17th Annual IEEE Symp. Foundations of Comput Sci. 1976, 208-215.

[24] Six, H.W., Wood, D.: Counting and reporting intersections of d-ranges. IEEE Trans. Comput. C-31, 181-187 (1982).

[25] Vogel, V.: Mathematische Modelle für die Geometrieverarbeitung - mengentheoretisch-algebraische Grundlagen und ein (Fleisch, Haut)-Modell. Technische Universität Dresden, Sektion Mathematik, Nr. 07-07-84.

A Divide–and–Conquer Algorithm for Computing 4–Dimensional Convex Hulls

C. E. Buckley
Integrated Systems Laboratory
Federal Institute of Technology at Zürich

Abstract

This paper contains a description an algorithm for computing four dimensional convex hulls of point sets using the divide–and–conquer paradigm. The algorithm features minimal asymptotic time and memory complexity with respect to the size of its input point set. It is based upon a fully–dual four–dimensional boundary representation (BREP) data structure called Hexblock, also developed by the author, which was inspired by Guibas' and Stolfi's quadedge data structure.

The algorithm was developed in order to quickly compute three–dimensional Delaunay triangulations of large numbers of points. It has been implemented. Also implemented for comparison purposes was a more conventional algorithm for computing such triangulations due to Sever. Preliminary tests suggest that the implementations in fact perform commensurate with theoretical expectations.

1 Introduction

The calculation of convex hulls plays an integral role in many geometric applications. Through appropriate incidence–preserving transformations, techniques for computing convex hulls may be used to solve other problems.

In this particular instance, the problem motivating this work was an expressed need to compute *Delaunay triangulations.* A Delaunay triangulation of a set of points in some d–dimensional Euclidean space is a mapping of those points into a simplicial complex such that for each simplex of $d+1$ points, no other point in the complex finds itself inside the sphere defined by these points.

In the case of two dimensions, this triangulation is the one in which the minimum angle of each triangle (simplex) is maximized. Mock and Sever have made theoretical studies relating the generation of finite elements using Delaunay triangulation to good behavior of the resulting matrix of simulation equations [Sev86]. Other authors have also expressed interest in basing their simulation codes on elements generated through Delaunay triangulation.

No attempt was made in this work to determine the suitability of Delaunay triangulations for this purpose. Rather, it was simply sought to facilitate the computation of such triangulations for potentially large input point sets (on the order of a million).

2 Delaunay Triangulations and Convex Hulls

The transformation which "maps" the problem of computing Delaunay triangulation into one of computing convex hulls is a well-known [GS85],[Ede87]. It runs as follows.

Let a generic point in R^d be represented by p_n, where n is an index held by that point alone. The d coordinates of the point may be represented by $\{p_{ni},\ i = 1, \cdots, d\}$.

Now let S describe the set of $d+1$ test points $\{p_{n_i},\ i = 1, \cdots, d+1\}$ defining a simplicial element in R^d, and let p_t describe an additional point to be tested for inclusion in the d-dimensional sphere defined by the points in S. Throughout this paper, it is assumed that all points lie in general (nondegenerate) position — the implementation mechanics used to back up this assumption are described in section 6.

Correspondingly, let S' define a set of $d+1$ points $\{p'_{n_i},\ i = 1, \cdots, d+1\}$ in R^{d+1} obtained by subjecting each of the points in S to the following coordinate mapping PR:

$$\begin{aligned} p'_{ni} &= p_{ni}, \qquad i = 1, \cdots, d \\ p'_{ni} &= \sum_{j=1}^{d} p_{nj}^2, \quad i = d+1. \end{aligned} \tag{1}$$

Likewise, map p_t into p'_t using the same method. Intuitively, it is as if, prior to the mapping, the points lay in a "horizontal" d–dimensional hyperplane in R^{d+1}, with last coordinate 0. The mapping simply lifts each point vertically upward until it lies in the paraboloid of revolution about the vertical $d+1$st coordinate axis defined by the equation

$$\sum_{j=1}^{d} p_{\cdot j}^2 = p_{\cdot d+1} \tag{2}$$

as shown in Figure 1.

The equivalence between the two problems can now be succinctly stated:

> The condition that point p_t lie in the hypersphere defined by the points in S is exactly the same as the point p'_t lying below the hyperplane in R^{d+1} defined by the points in S'.

Therefore, to calculate the Delaunay triangulation of a point set I, it suffices to apply the mapping PR to each of the points in I, calculate the convex hull of the resulting point set I', and then remap the points back to their original state by dropping their last coordinate.

Actually, only the faces of the convex hull which face "downward" correspond to elements of the Delaunay triangulation[1], and so the "upwards" pointing faces must be removed or ignored.

[1]To be specific, the Delaunay triangulation is the dual of the *nearest-point Voronoi partition* of the points, and the "upwards" pointing faces are actually the dual of the *furthest-point Voronoi partition* [Ede87].

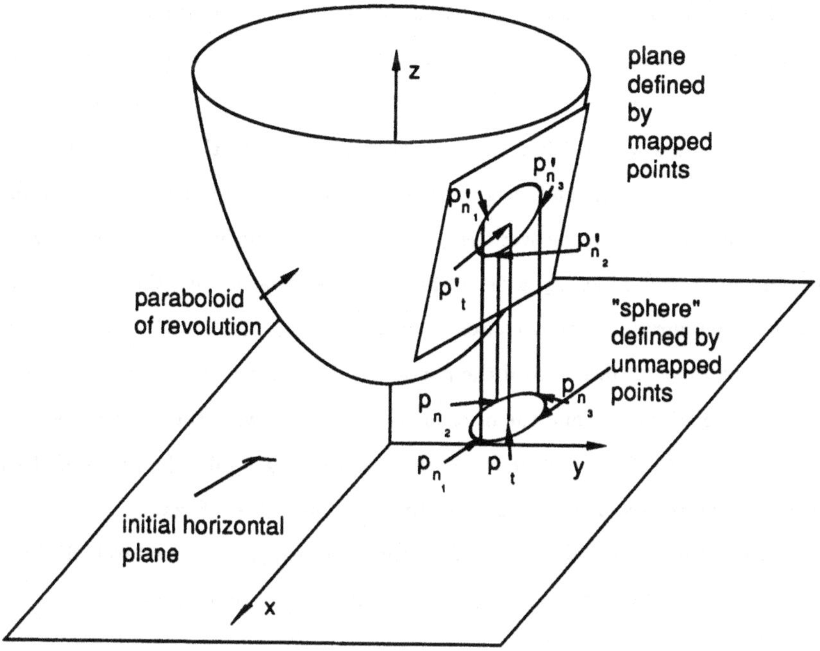

Figure 1: The 3–Dimensional Version of the Parabolic Mapping

This removal process can be simplified by including a single extra point $e \in R^{d+1}$ at infinity with coordinates $(0, \cdots, 1, 0)$ in the input set. After having computed the convex hull, this point, and all faces and edges separating the d–facets adjacent to it may be removed using purely topological operations. This will leave only the desired downward–pointing faces.

3 Computing Convex Hulls: State of the Art

Of course, the advantage of transforming the Delaunay problem into one of convex hull calculation is that more results are available for the latter. For the general case of computing convex hulls in Euclidean spaces of arbitrary dimension, essentially three algorithms exist, based upon three different conceptual ideas.

3.1 The Gift–Wrapping Method

The first and oldest of these, called the *gift–wrapping method*, is due to Chand and Kapur [CK70], although its asymptotic complexity was not analysed for general dimension until much later [Swa85]. The idea behind it is simple — the points for which a convex hull is to be computed are considered as hanging immovable against all forces in R^d. The idea is to take an infinitely stretchable, weightless piece of $d-1$-dimensional wrapping foil and wrap it around the point set.

Reported out of the algorithm are simply the $d-1$-facets of the convex hull, identified by groupings of the d points which define their space of affine completion (a. k. a. tangent hyperplane). Since all facets are simplicial, neighboring facets have all but one point in common. This is the key to the algorithm — to discover the group of points identifying a $d-1$-facet neighboring one already known. i. e. to identify a facet sharing a $d-2$-facet with the current one. To do so, each point of the current facet is chosen in turn to "leave the basis"[2]. Each $d-2$-facet is shared by exactly two $d-1$-facets, so if the particular $d-2$-facet under consideration has been seen before, this case is dropped and the next considered. Otherwise, the point for which a vector from any point in the $d-2$-facet to it forms the smallest angle with the plane of the previous face is taken with the points of the $d-2$-facet to form a new $d-1$-facet.

Having sketched the general idea of the algorithm, the reader is referred to [PS85] for further details, stating further only that the asymptotic (input) time complexity is

$$C_{\text{gift}} = O(n^{\lfloor \frac{d}{2} \rfloor + 1}) + O(n^{\lfloor \frac{d}{2} \rfloor} \log n) \approx O(n^{\lfloor \frac{d}{2} \rfloor + 1}), \tag{3}$$

where d is the dimension of the ambient space, and n is the number of input points.

3.2 The Beneath–Beyond Method

The idea behind this method is to incrementally construct the convex hull by adding a point at a time. After the addition of the jth point, a partial convex hull of the first j points can always be output. The form in which the convex hull is output is an *incidence graph*, in which nodes represent facets of all orders, from points up to nodes whose spaces of affine completion are d-dimensional hyperplanes. There are fields in each node point to adjacent faces of lower and higher order, as well as fields to hold non-topological properties.

When a new point is merged with a prior convex hull, either one of two cases apply:

1. The point lies inside prior hull, in which case the prior hull is output unchanged.
2. The point lies outside the prior hull, in which case the hull will be changed to reflect its addition.

Without loss of generality, we can assume:

- that the previous hull is always at least a d-dimensional simplex (meaning that the first d points are treated specially), and

[2]just like in linear programming

- that points to be added are sorted so that each additional point added lies outside of its immediately preceding partial hull.

In this case, with respect to the new point, each facet on the previous hull is either:

1. totally visible
2. totally invisible
3. partly visible (its relative interior is invisible, but certain subfacets on its boundary may be visible)

In the new convex hull, the totally visible facets of the previous hull will disappear, and the others will remain, except that new faces will be created from totally visible subfaces of partially visible ones and the new point. The complexity of the merge step consists of classification of the facets, creating the new facets, and updating the incidence graph to reflect the presence of the new point.

Again we stop after having given a qualitative description, referring the reader to [Ede87] for details. The asymptotic time input complexity is

$$C_{\mathrm{bb}} = O(n \log n + n^{\lfloor \frac{d+1}{2} \rfloor}) \approx O(n^{\lfloor \frac{d+1}{2} \rfloor}). \tag{4}$$

3.3 The Shelling Method

This is the newest of the convex hull methods for arbitrary dimension, due to Seidel [Sei87]. It also outputs a facial incidence graph, but the idea behind its construction is different.

Central to the algorithm is that d points in R^d describe a hyperplane, and that this hyperplane intersects a unique *shelling line* passing "through" the input point set in exactly one (projective) point. These intersection points induce an ordering of facets.

The algorithm proceeds by moving along continuously the shelling line until, passing through infinity and approaching from the other direction, the starting point is discovered again. Adjacent facets in the shelling line ordering will differ in their defining sets by single points, and this property induces an enumeration equivalent to the gift-wrapping method. Special data structures and linear programming techniques are used to facilitate the task (see [Sei87] for details). The input time complexity is reported as

$$C_{\mathrm{shell}} = O(n^{\lfloor \frac{d}{2} \rfloor} \log n). \tag{5}$$

3.4 Dimension–Specific Algorithms

The above algorithms shared in common being of roughly equivalent time and memory complexity, and *that the input point sets must be treated as a single, undifferentiated mass.* In the case of the beneath–beyond method, they must be treated sequentially, in the case of the other two methods, all points must be present when answering queries about neighboring facets.

This latter aspect essentially enforces processing sequentiality — having more processors available to work for you does no good, since you have to wait for the previous step to complete.

In lower dimensions, there exist algorithms which work according to the *divide–and–conquer paradigm* which avoid this sequentiality. Under this paradigm, the input point set is first sorted along a given dimension so that the convex hulls of arbitrary divisions are guaranteed to be separate. Then, the input set is separated recursively down into subsets of equal size until the last separation produces a point set of a size which can be treated canonically.

In the case of convex hull calculation, canonical treatment can be carried out in the following two cases:

1. If there are less than d points, these points are simply passed upward in the recursion until the following case is true.
2. If there are at least d but less than $2d$ points, the first d of these are formed into a simplex by copying a pre–stored output result graph and stuffing the points into it. Any remaining points are merged into this structure sequentially.

With divide–and–conquer, there need be no interaction between two halves of a subdivision until their resulting convex hulls are to be merged together. To support this, in essence one need only write an algorithm capable of merging:

1. two partial convex hulls
2. one partial convex hull with a single point[3]

Divide–and–conquer algorithms have been previously reported for the computation of convex hulls of points in R^2 and in R^3. This paper reports a new algorithm of this type for calculating convex hulls of points in R^4, which can consequently be used for computing Delaunay triangulations of points in R^3.

4 Describing Partial Hull Boundaries

Quite a number algorithms of minimal asymptotic complexity using various approaches have been published for computing convex hulls in R^2. Credit for the first of these goes to Graham [Gra72].

[3]The merge algorithm for the beneath–beyond method satisfies this case.

Preparata and Hong are credited with unifying results for R^2 and R^3 in the divide–and–conquer paradigm. Edelsbrunner reports evolved versions of their algorithms having time complexities of $O(n \log n + n) \approx O(n \log n)$ [Ede87]. These fit into a general complexity framework of

$$C_{\mathbf{gen}} = C_{\mathbf{sort}} + C_{\mathbf{merge}} = O(n \log n) + O(n^{\lfloor \frac{d}{2} \rfloor}) \tag{6}$$

in which the cost of presorting the points is treated separately from the cost of merging them afterwards. Having made such a separation, the best that can be expected for the merge phase is a complexity on the order of the number of nodes in the output graph.

The divide–and–conquer algorithms of Preparata and Hong and their derivatives all depend on the use of graph–like data structures which describe the boundary of the partial hulls. In the two dimensional case, such a data structure is easy to manage, since there are only 0–facets (points) and 1–facets (edges). These are transparently connected together in doubly linked lists.

In three dimensions, there are also 2–facets to contend with. To manage these, Edelsbrunner uses an obvious derivative of the *quadedge data structure* reported by Guibas and Stolfi [GS85]. The main pointer block of this structure represents half of an edge (or dually, half of a dual edge[4]), and is shown in Figure 2[5].

The utility of such a structure in computing convex hulls is that, given a pointer to an edge, the following can be obtained in constant time:

- the vertex data of the edge origin,
- the vertex data of the edge destination, and
- pointers to the faces on the left and right of the edge.

In addition, pointers to all of the edges emanating from a vertex can be obtained in time linear in the output set.

Guibas and Stolfi found it particularly valuable that the structure automatically delivered at the same time a dual structure describing the relationship between faces and dual edges. This was because they could accomplish their goal, the computation of Voronoi diagrams, simply by computing a Delaunay triangulation, and then returning the dual graph formed as a by–product fro a result. In Edelsbrunner's version, the boundary data structure has been redefined to provide the equivalent functionality without carrying the dual graph, which halves the amount of memory required.

[4]an edge having as its origin and destination *faces* instead of *vertices*

[5]This is the oriented manifold version of the structure, and the letter "e" has been appended to the front of each pointer name given by Guibas and Stolfi, to avoid confusion with pointers of similar functionality in the three–dimensional boundary structure to be introduced shortly.

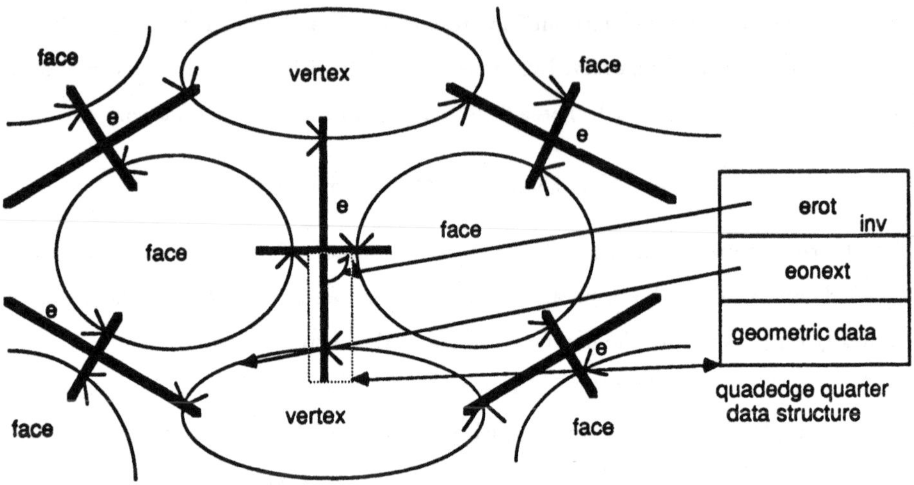

Figure 2: Basic Topological Data Block of Quadedge Data Structure

Prior to hearing of Edelsbrunner's results, the author had developed the equivalent of his three–dimensional convex hull algorithm using Guibas and Stolfi's data structure. In doing so, the duality inherent in the structure also played an important role, quite apart from the describing the Voronoi/Delaunay relationship. In fact, this dual role is central to the extension of the algorithm to four dimensions. It is treated in the next section.

5 Convex Hulls and the Gaussian Sphere

The idea behind divide–and–conquer convex hull algorithms is to "wrap" the *a priori* discrete partial hulls generated previously during upward recursion with pieces of $d-1$–dimensional hyperplanes which support both hulls. The pieces are bounded on either side by facets of even lower dimension which belonged to the prior hulls.

In two dimensions, these hyperplane pieces are simply line segments, and there are two of them — one above and one below (see Figure 3). Because the two are in general separate from one another, they must be treated individually.

In three dimensions, the hyperplanes are 2–dimensional, and taken together, they form a sort of "sleeve" around the previous hulls. In this case it is the number of neighbors which are enumerated: each 2–facet has exactly two neightbors. The merge step can be broken into three parts:

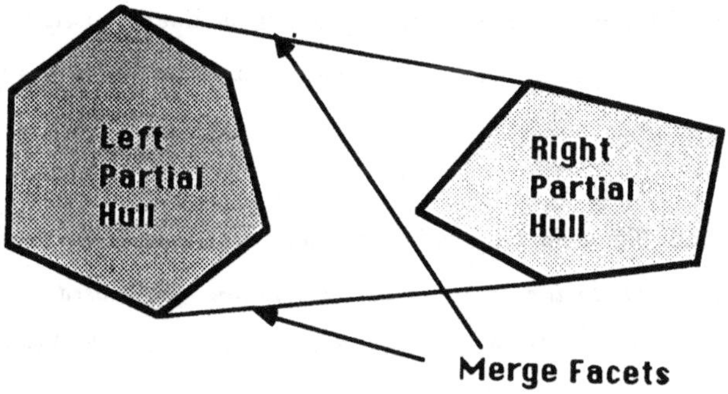

Figure 3: 2–Dimensional Divide–and–Conquer Algorithm Merge Step

1. identification of an initial facet,
2. pivoting along in one direction until a neighbor facet is found. This is done repeatedly until the initial facet is found again, at which point
3. the facets are merged into the original hulls on both sides.

While this process is already intuitively easy to understand *per se*, it takes on an even more compelling significance when the dual of the boundary graph is mapped onto the unit Gaussian ball such that each facet maps onto the set of support directions for which it (the primal facet) is the support set.

5.1 3–Dimensional Convex Hulls

In three dimensions, this means that:

- *Facets* map to *vertices* whose geometric coordinates represent the normalized outward–pointing face normal,
- *Vertices* map to (spherical) *facets*, and
- *Edges* map to (dual) *edges*.

This dual imbedding is invariant to coordinate scaling of the input polyhedron. Figure 4a shows a two–dimensional Delaunay triangulation calculated as a three–dimensional convex hull. To be quite precise, is shows the resulting convex hull, but viewed from below, so that the results are indistinguishable. Figure 4b shows a view of the convex hull from the side, in order to give an idea of the input points having been mapped onto the paraboloid of equation 2. The unconnected edges leading away from the hull lead towards the point e at infinity, and will be removed once the hull has been finished. Figure 4c shows a subdivision of the dual Gaussian ball corresponding the the view direction of Figure 4a, in which each dual point corresponds to a primal facet, etc., as described above.

In preparation for the merge step, the dual mappings of both partial hulls are coembedded in the same Gaussian ball. Then, *in this dual space, the calculation of the "sleeve" of the three-dimensional algorithm is exactly the same as the calculation of a closed chain of dual edges.* Where these dual edges slice a previously existing dual edge edge in a dual point corresponds to the generation of a primal plane face. The duals of these dual edges are the primal edges which span between the two partial hulls. The dual faces of each of the partial hulls through which the dual edge passes correspond to the vertices of the two prior hulls between which the edge passes.

The direction of the dual edge corresponds to a portion of the great circle whose polar is the direction cosine vector of the primal edge. Every time a previously existing dual edge *is* crossed, this corresponds to a change in the vertex of the primal edge on that side. As a graphic demonstration of this principle, Figure 5 shows the output of a demonstration program of this algorithm.

Figure 5a show a pair of partial hulls part way through the process of being merged together. Already-generated crossing edges are also shown. The original for this figure had color-coded lines, and was not reproducible here. Instead, edges have been marked as follows:

○ indicates edges of one partial hull,

△ indicates edges of the other partial hull, and

— (crosshatching) indicates a crossing edge.

Figure 5b shows the corresponding dual subdivision, viewed from the same direction, in which dual edges are marked with the same symbols as the primal edges to which they correspond. The ends of the continuous chain of crossing edges is mmarked with a heavy black spot (•), and is grown from one end towards the other. When the two ends meet, all crossing edges will have been generated.

As far is complexity of the algorithm is concerned, once an initial dual point on the intersection manifold is found, no dual edge (and consequently no primal edge) is visited more than twice. Likewise, if the same data structure is used to form the intersection manifold, pointers to the cut edges may be set into the non-topological fields of the vertices of the dual intersection edges, and so the unzipping and zipping can be accomplished in time linear in the number of dual intersection edges.

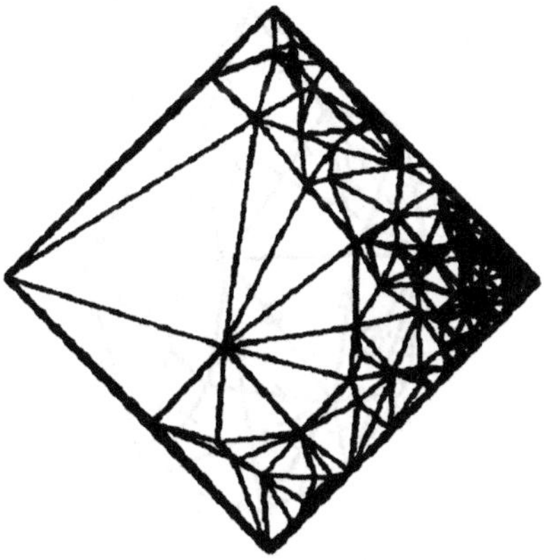

(a)

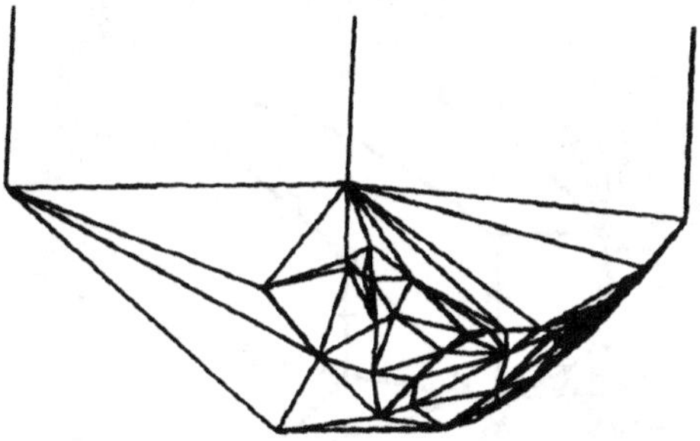

(b)

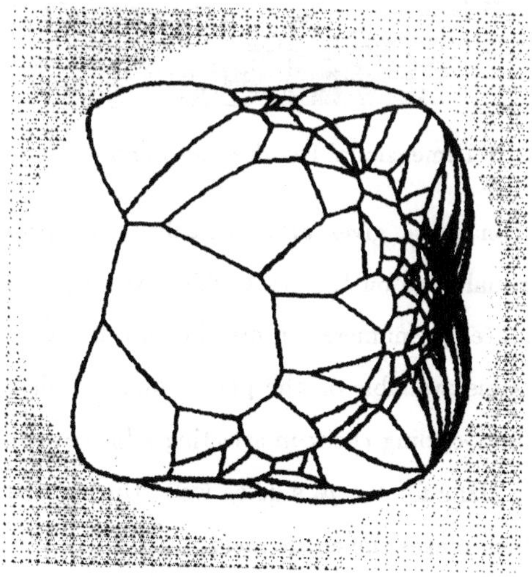

(c)

Figure 4: A 3–D Delaunay/Convex Hull with Dual Gaussian Subdivision

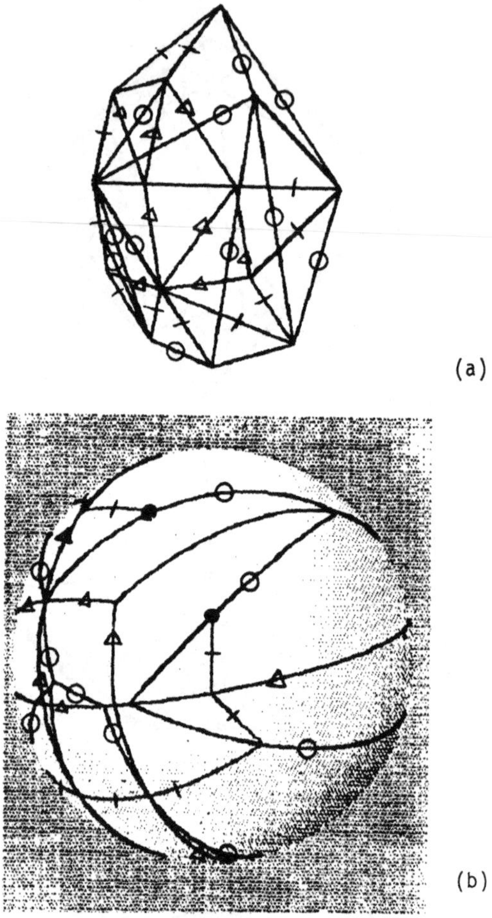

Figure 5: 3–Dimensional Divide–and–Conquer Merge Step

For the initial coordination phase, given an arbitrarily chosen primal vertex p_0 on one prior hull, it is easy to find a primal edge on the right which, when taken with p_0 forms a support plane for the other prior hull (edge enumeration may be used if need be). If, on the other hand, not only a point but an edge is available on the partial hull containing p_0, then its two primal vertices can be used in an anti–cycling check in selecting adjacent edges on the other side until the hyperplane is a support plane for the p_0 side as well. During this phase, no edge is visited more than twice as well, which makes the entire merge step of linear complexity in the edges.

Through the two–dimensional Euler characteristic, using the fixed degree of 3 for each dual vertex implied by the assumption of points in general position if you like, this is relatable to a merge complexity linear in the number of vertices belonging to the two hulls to be merged, and by standard divide–and–conquer complexity arguments to a bottom–to–top linear merge complexity.

5.1.1 On–Line Variant

The phrase "on–line" is applied to convex–hull algorithms capable of supporting the addition of a single point to a previously existing hull. This case hardly requires special consideration in the context of the algorithm described above: the dual subdivision for the partial hull which is only a point consists only of a single dual 2–facet covering the entire Gaussian ball. Dual edges are cut only for the dual subdivision corresponding to the non–point prior hull.

5.2 Extension to Higher Dimensions

Apart from bringing a new intuitive twist to the convex hull algorithm, this dual interpretation is also important in that ***it is completely extendable to arbitrary higher dimension.*** In each case, the k–facets of the two prior convex hull boundaries are mapped onto $d-k-1$–facets on the unit Gaussian ball in d–space, a closed $d-1$–dimensional manifold. The merge step consists of the elaboration of a closed $d-2$–dimensional dual intersection manifold which cuts each of the prior–hull $d-1$–manifolds into two. These manifolds are then subsequently "zipped apart" and the outside halves are "zipped onto" the dual intersection manifold, which becomes a part of the resulting dual subdivision.

The merge step remains subdividable into three parts:

1. identification of an initial facet (corresponding to finding an initial point on the dual intersection manifold),
2. the contiguous evolution of this point along all directions on the intersection manifold until a closed manifold is generated, and
3. the use of this manifold as a guide to slice the previously existing dual subdivisions and to weave the halves together again.

5.3 The 4–Dimensional Case

Of course, in order for this to work, a data structure is required which:

- can describe a boundary of the appropriate dimension,
- can reflect the necessary duality, and
- can preserve this duality relationship through modifications made to one or the other of the mutually dual subdivisions.

5.3.1 The Hexblock Data Structure

Guibas' quadedge structure fulfilled this need perfectly for the case two–dimensional boundary manifolds. What was needed was simply an extension one dimension higher. That is, the duality correspondances required were:

- primal *vertices* must correspond to 3–facets in the unit Gaussian ball (call them dual *voids*),
- primal *edges* must correspond to dual 2–facets (call them dual *faces*),
- primal *faces* must correspond to dual 1–facets (dual *edges*), and
- primal *voids* must correspond to 0–facets (dual *vertices*).

Having so phrased the problem, such a structure was not difficult to develop, based on the observation that the neighborhood description of a vertex must be isomorphic to the boundary description of a void. The structure was called Hexblock, to indicate its quadedge lineage[6]. The fragment "block" was used because it was no longer possible relate a basic data structure block to any classical geometric entity.

A paper describing the Hexblock structure is in preparation. Here we content ourselves with presenting Figure 6, showing the topological pointers of the basic data structure block. Salient features of the Hexblock structure are:

- it requires storage linear in the edges used (worst case quadratic in the input vertices),
- topological neighbors to any geometric entity may be enumerated in time linear in the size of the output set, and
- as with Guibas' quadedge structure, all structural modifications may be expressed in terms of a single, self–inverting modifier operation called `splice`, which leaves primal and dual subdivisions consistent. In this case, `splice` has two interpretations. Either
  1. it takes two edges on the boundaries of two voids and "zips" them together (or apart), or
  2. it takes two faces, addressed from two points, and melds them into one (or unmelds them into two).

After the fact, it was determined that similar data structures had been independently developed and reported by Laszlo [Las87], and also developed but unpublished by Rajan[7]. Rajan would not divulge details of his structure, but an initial comparison of Hexblock with the structure reported by Laszlo indicated:

- the orientation of the boundary manifold was treated by Laszlo, whereas Hexblock was written to work for oriented manifolds only, and

[6] and to reflect a then–current hypothesis that 2^d topological pointers would be necessary to describe boundaries of d–dimensional boundary. This has not turned out to reflect the way things have since developed ···

[7] private communication

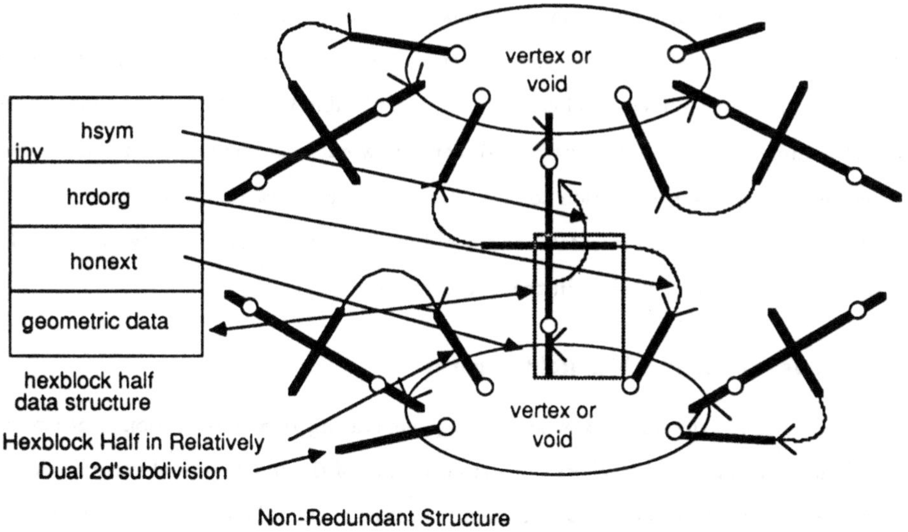

Figure 6: The Basic Topological Pointer Blocks of the Hexblock Data Structure

- as observed by Laszlo, his modification primitives can leave the dual structure inconsistent ("exquisite garbage" was the term used [DL87]), whereas Hexblock's splice avoids this problem.
- Otherwise, the structures seem to perform the same function.

At any rate, Hexblock has proven to be an adequate tool for the implementation of the four-dimensional version of a divide-and-conquer convex-hull algorithm, as described following.

5.3.2 The Merge Step

In four dimensions, the dual intersection manifold is two-dimensional. Each dual face of the manifold corresponds to a crossing edge separating two primal vertices on each of the prior hulls. The boundaries of the dual intersection manifold faces are determined by the dual faces of the prior hulls that are cut. When a left and right prior-hull exist, the dual intersection face is an intersection of the faces that would exist if one or the other of the prior hulls were only a point. Currently these are computed separately and then intersected, although a speedup could be obtained if these two steps were merged. Programmed either way, this operation is linear in the dual faces, and consequently linear in the primal edges and worst case quadratic in the vertices.

Its not hard to see that all of the other aspects of the merge process and all of the other rubrics applicable to the three–dimensional case[8] are also applicable to the four–dimensional case, each time with complexity linear in the edges (and therefore worst–case quadratic in the input vertices).

There is an important exception to this, which concerns how to splice together the various faces of the dual intersection manifold, and avoid generating the same face twice. This deserves special attention, which it receives in the next section.

5.3.3 "Shoestringing" a Manifold Together

In three dimensions, determining when to stop generating dual intersection edges was not a problem, since one only had to check against encounter of the initial primal vertex pair.

In four dimensions, each dual face of the intersection manifold corresponds uniquely to a pair of primal vertices (the endpoints of its corresponding primal crossing edge). However, in order to be able to splice such faces together, it is also necessary to select one of a variable number of edges on the face. If this were done in an array, the storage complexity would rise to $O(n^3)$, which is no longer commensurate with the potential output complexity of the structure.

But, an *occupancy array* in which each cell is only a single Boolean bit is commensurate with this bound, and so it is possible to determine whether or not a dual intersection face has been previously generated. To match up edges to be spliced together, during dual face generation, and before any of these faces are spliced together, a "shoestring" is wrapped around each face, leaving only a small opening at one end of the edge between the new face and the face which prompted its generation. The shoestring data structure is none other than the two–dimensional counterpart to quadedge in three dimensions and Hexblock in four. Such a process is shown in Figure 7.

Each edge of the shoestring contains a pointer to the edge of the dual intersection facet it surrounds, which in turn contains a pointer to the dual face of the prior hull being cut. Whenever an additional dual intersection face is brought into the shoestring perimiter, both sides of the small opening (call it a "vee") are pushed onto a vee–stack, which is used to drive the splicing together of the faces after all have been generated. This proceeds as follows:

1. An entry is removed from the vee stack.
2. If the cut dual face ultimately pointed to by the shoestring edges on both sides of the vee are the same, then:
   (a) the two shoestring edges and the shoestring vertex corresponding to the vee are spliced out of the shoestring,

---

[8]including that discussed in Section 5.1.1

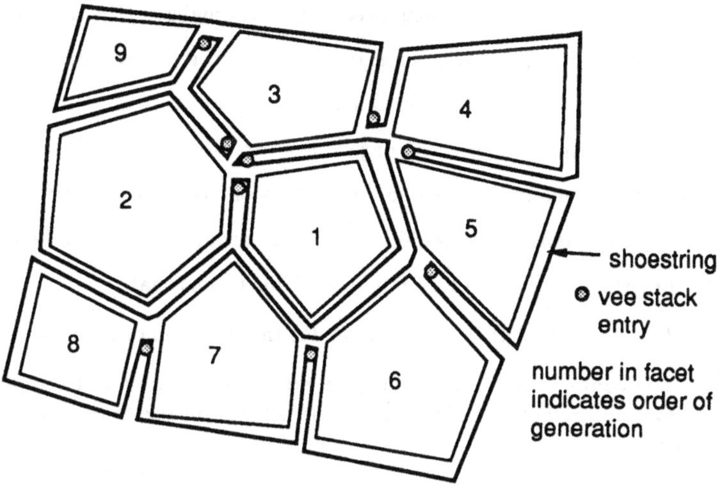

Figure 7: "Shoestringing" 2–Dimensional Facets Together

(b) the two Hexblock edges pointed to by the vee edges are spliced together, and

(c) the newly formed vee vertex (at the other end of the just–removed shoestring edges) is pushed on the vee stack.

Otherwise, the vee-stack entry is simply discarded.

3. the "un–shoestringing" stops when the vee stack is empty.

This entire un–shoestringing process can be carried out using time and with storage linear in the number of dual intersection facet edges, and is thus commensurate with the overall complexity bounds.

6 The Topological/Geometric Interface

Up until this point, several things have been taken for granted which should not be:

1. that points in the input point set were in general position, and
2. that dual edges and faces could be tested for intersection.

Both of these assumptions must be disposed of with care.

First of all, it is highly undesirable to calculate the support directions which make up the dual vertex coordinates at all, especially when numbers are represented in floating point form. However, it is perfectly reasonable to represent primal vertex coordinates in fixed–point form, and if one accepts that support directions need not be normalized (the dual unit Gaussian ball is actually a projection of a positively homogeneous space), it is even possible to calculate the normals using operations which are closed under the integers.

This alone, though, does not prevent degeneracies in the input set from causing problems. It becomes necessary to maintain some sort of "ancestor" relationship with the primal points from which these dual quantities were derived. To get around this, the scheme of "conceptual perturbations" described in [Ede87] was employed. This means that:

- Each point is "tagged" with a unique index for the duration of its presence in the dataset.
- All geometric computational steps must be expressed in terms of hyperplane tests, which return a Boolean predicate stating whether or not a point lise to one side or another (never on) a hyperplane.
- The hyperplane test must always be defined in terms of computing the exact determinant of a matrix formed from homogeneous coordinate representations of the d tagged primal points defining the hyperplane, and the test point.

The duality information inplicit in the data structures used plays an important role here: given a pointer to a dual vertex (whose coordinates are never actually calculated and stored), the coordinates of the primal points which define it are always available in constant time.

In the implementation, only two real potential difficulties were encountered:

1. The fixed-point multiplications required in computing the determinant produced very long fixed–point numbers. Fortunately, the computer on which the development was done[9] just happened to automatically support the representation of such numbers without additional programming effort, but this is not true of conventional architectures.
2. Even assuming that big numbers are no problem, and that multiplication is a constant–time operation, calculating determinants requires $O(d!)$ operations. Using the conceptual perturbations scheme, not only is the calculation of a single determinant required, but also the recalculation of certain specific minors.

   In initial tests, it was observed that a substantial fraction of the total execution time is given over to performing hyperplane tests. As an answer to this, initial feasibility studies indicate that even using the most mundane of the currently available hardware technologies, a reasonably priced embedded system could be built which would perform such tests in $O(db)$, where b is the number of bits in the fixed–point representation of a point coordinate.

7 Implementation of a Comparison Algorithm

Performing the implementation on a Symbolics machine turned out to quite worthwhile, since:

- several features available "for free" in Lisp were invaluable in this implementation:
  - painless dynamic memory allocation

[9] a Symbolics Lisp Machine workstation

  - automatic big number representation
  - typeless data

- These features, and a good interactive working environment really speed development. For example, it took longer to reimplement part of the Hexblock structure alone in C++, working from commented Lisp source code, than it did to:

  1. conceive of the Hexblock structure,
  2. implement it in Lisp,
  3. conceive and implement the the 3–D convex hull algorithm, and
  4. get well into developing the 4–D algorithm.

However, because the Symbolics remains a relatively unusual machine, no published figures for comparison algorithms may be fairly used. It was therefore necessary to especially implement another algorithm for comparison. The algorithm selected for implementation was that published by Sever [Sev86]. In published form, the algorithm is cast in terms of computing Delaunay triangulations. When recast in terms of computing convex hulls, it turns out to be very similar to the beneath–beyond method mentioned previously.

Sever's algorithm works as follows. Without loss of generality, it may be assumed that the first d points may be formed into a simplex in a canonical fashion, and subsequent points are merged in one at a time. During the merge–in of each additional point:

1. Each d–facet of the previous hull is checked for visibility or invisibility with respect to the new point[10]. Two types of dual edges were collected into two different lists during this enumeration:

   (a) dual edges between two visible dual vertices

   (b) dual edges between a visible and non–visible dual vertex

2. The dual edges (primal faces) in the first list were spliced out of the prior hull structure.

3. One entry in the second list was used as a pointer to the opened-up primal void. New faces or edges (depending on the dimension) exactly corresponding to the new facets added in the beneath–beyond method are then created and spliced into the structure.

Thus, the only real difference between the Sever algorithm and the beneath–beyond method is that with the former, visible faces of the prior hull are individually removed, while with the latter, it is possible to work along the boundary between visible and invisible faces without having to delve into the middle.

The time complexity of the Sever algorithm ***per point*** is $O(n)$ in the three–dimensional case, and $O(n^2)$ in the four–dimensional case, leading to overall time complexities of $O(n^2)$ and $O(n^3)$, respectively.

[10]The requirement only to do hyperplane tests in accordance with Section 6 precluded using structure walks, just as is true in the coordination phase in the divide–and–conquer algorithms.

8 Results of the Implementations

For the Symbolics machine in particular, and all Lisp implementations in general, allocation of dynamic memory (for data structure blocks, for example) is directly and naturally expressed as part of the programming language syntax (the **cons** function).

Dynamic memory *deallocation*, however, is only implicitly expressed. Although Symbolics offers certain "black–magic" functions which partially mitigate this, and modern-day Lisp compilers make a serious attempt to allocate as much temporary storage as possible on a push–down list reminiscent of the return stack in "normal" languages with reentrant subroutine capability (like C), it is simply not possible to keep from using up all dynamically available memory in the course of computing substantial triangulations without running a garbage collector in the background.

In practice, this takes place transparent to the user, *except* when benchmarking, where it is hard to keep the activity of the garbage collector from influencing results obtained. To minimize this influence, it becomes necessary to run tests repeatedly in order either:

- to allow detection of results unduly skewed from the others, and/or
- to average out the influence of the garbage collector on the results.

As this article had to go to press, an unexplained yet fatal interaction between this same garbage collector and the system software which supports the representation of big integers prevented the battery of tests designed to compensate for the effects of the garbage collector from running to completion. They will be included in a later version of the article.

Based on the few tests which *did* complete, the following *qualitative* observations may be guardedly made:

1. For small input point sets, the Sever algorithms clearly run faster than the divide–and–conquer algorithms, in both three and four dimensions.
2. As input point sizes grow, the time required by the Sever algorithms grows faster than that required by the divide–and-conquer algorithms. This is in accordance with theoretical predictions.

The following questions remain to be answered:

1. At what approximate input set size do the Sever and the divide–and–conquer algorithms deliver equivalent performance?
2. Are the asymptotic power curve behaviors predicted by the theory actually borne out in practice?
3. What sort of difference in behavior may be expected:
   (a) for worst–case input data (neighborly polytopes)?
   (b) for uniformly distributed input data?
   (c) for "real" input data, in which point density varies spatially?

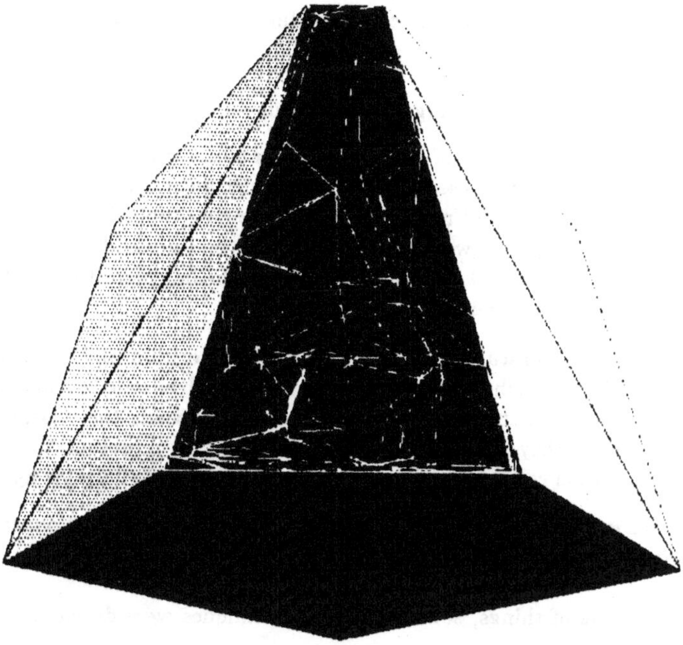

Figure 8: Cutaway View of a Completed Delaunay Triangulation

9 Open Problems and Future Work

Practically speaking, doubt remains that global Delaunay triangulation of a set of free–standing input points alone has much of a future as a basis for mesh–generation for simulation problems — other methods seem equally promising, faster and certainly easier to implement. Figure 8[11] shows a cut–through view of a Delaunay triangulation generated by the algorithm of a set of points consisting of:

- the eight corners of the unit cube and
- 100 randomly distributed points in the center of the cube, which were drawn according to a distribution which biases their location towards the corner which has been most cut away.

We observe that, setting aside the problem of the generation of extremely flat elements along the boundary, curable through the introduction of nodes *on* the boundary, that extremely flat elements also arise in the center of the cube. However, increasing the input point density reduces

[11]The software for producing these cutaway figures is described in [DP88].

the uneveness, but at a increased element cost. As stated previously, evaluation of mesh quality was not a goal of this work.

Concerning open questions pertaining to convex hull computation *per se*:

- The evaluation of the efficiency of this algorithm in terms of output set complexity remains to be examined. In a way, these divide–and–conquer algorithms can be considered as instances of the beneath–beyond method in which the new "point" to be added is just more complicated topologically. However, the beneath–beyond method has been shown to compare poorly in terms of output set complexity to the shelling algorithm. Is this also true for the divide–and–conquer variant of the algorithm?

- As published in the literature, the beneath–beyond algorithm and the shelling algorithm make use of incidence graphs to report their output, which encode less information than the boundary representation structures described here. How would the relative merits of these algorithms change if they were recast in terms of boundary representation structures? The study of this question could be pursued immediately for all dimensions up to 4.

- The efficacious generation of duality–describing boundary representations for $d > 4$ remains for the moment elusive.

On the practical side of things, several valuable techniques were developed in the course of this work which are being applied elsewhere. Companion to these come unsolved issues:

- The Hexblock structure itself has been found to be useful independent of the merits of the work reported here. There are a number of issues pertaining to the use of the Hexblock and its relatives. Perhaps the most important of these pertain to their use outside the rather forgiving Lisp programming environment. Explicitly,

  1. can the primitives used to manipulate such structures be expanded to return freed–up memory to the dynamic store in an explicit way without increasing complexity? If not,
  2. is it still worth it to accept the increased costs of making the return of no longer needed memory explicit, as opposed to simply burying the problem in the depths of a difficult–to–analyze garbage collector scheme?

- The "conceptual perturbation" technique was indispensible to the success of the algorithms presented here. However, as previously observed, it will require explicit treatment in hardware to make it manageable to use. Such work has yet to be undertaken.

- Finally, although the divide–and–conquer was in particular sought after because of its capacity for parallelization, the implementation described here was done on a sequential machine. Issues pertaining to its implementation on a parallel machine remain to be treated. Only when this has been done will it be possible to experimentally evaluate the degree to which the parallelism sought after in undertaking this work can be achieved.

10 Conclusion

A convex–hull algorithm for points in R^4 based on the divide–and–conquer paradigm has been developed and implemented. The algorithm meets asymptotic complexity expectations raised by experience with algorithms of a similar type for points in lower dimensions. Key to the algorithm was the development of a duality–bearing data structure for describing 4–dimensional boundary representations.

Preliminary evaluations of the performance of a Lisp implementation of the algorithm suggest qualitatively better asymptotic performance than an algorithm of higher theoretical complexity, albeit with higher per–unit manipulation costs.

Acknowledgement

The author would like to thank Prof. Dr. Wolfgang Fichner, head of the Integrated Systems Laboratory of the Federal Institute of Technology at Zürich, for financial support and for use of computer facilities.

References

[CK70] D. R. Chand and S. S. Kapur. An algorithm for convex polytopes. *Jounal of the ACM*, 17:78–86, 1970.

[DL87] D. P. Dobkin and M. J. Laszlo. Primitives for the manipulation of three–dimensional subdivisions. In *ACM Siggraph Symposium on Computational Geometry*, pages 86–99, Waterloo, Ontario, June 1987.

[DP88] P. Dornier and S. Paschedag. The animation of simulation results in three dimensions. July 1988. ETH Abt. IIIB Semesterarbeit.

[Ede87] H. Edelsbrunner. *Algorithms in Combinatorial Geometry.* Springer–Verlag, Heidelberg, 1987.

[Gra72] R. L. Graham. An efficient algorithm for determining the convex hull of a finite planar set. *Information Processing Letters*, 1:132–133, 1972.

[GS85] L. J. Guibas and J. Stolfi. Primitives for manipulation of general subdivisions and the computation of voronoi diagrams. *ACM Transactions on Graphics*, 4(2):74–123, 1985.

[Las87] M. J. Laszlo. *A Data Structure for Manipulating Three–Dimensional Subdivisions.* Technical Report CS–TR–125–87, Princeton University ComputerScience Department, 1987. thesis.

[PS85] F. P. Preparata and M. I. Shamos. *Computational Geometry: an Introduction.* Springer–Verlag, New York, 1985.

[Sei87] R. Seidel. *Output-size Sensitive Algorithms for Constructional Problems in Computational Geometry.* PhD thesis, Cornell University, Ithaca, New York, January 1987.

[Sev86] M. Sever. Delaunay partitioning in three dimensions and semiconductor models. *COMPEL*, 5(2):75–93, 1986.

[Swa85] G. Swart. Finding the convex hull facet by facet. *Journal of Algorithms*, 6:17–48, 1985.

Triangulating a Monotone Polygon in Parallel

Hubert Wagener

Technische Universität Berlin, Germany (FRG)
Franklinstr. 28/29, FR 6-2, D-1000 Berlin 10
huwa@tub.bitnet ...!pyramid!tub!huwa

Abstract: Given a simple n-sided polygon, the triangulation problem is to partition the interior of the polygon into $n-2$ triangles by adding $n-3$ nonintersecting diagonals. We propose an $O(\log n)$-time algorithm for triangulating monotone n-sided polygons using only $n/\log n$ processors in the CREW-PRAM model. This improves on the previously best bound of $O(\log n)$ time using n processors.

1. Introduction

Let P be a simple n-sided polygon, given by a list $(v_0, v_1, \cdots, v_{n-1})$ of its vertices as they are encountered in a clockwise traversal of the boundary. Without loss of generality we assume throughout this paper that v_0 is the vertex of P with minimal x-coordinate, and that the vertices have distinct x-coordinates as well as distinct y-coordinates. We choose the x-axis as preferred direction, and call a simple polygon $P = (v_0, v_1, \cdots, v_{n-1})$ monotone (with respect to the x-axis, if there exist a vertex v_k of P such that the vertices $v_0, \cdots, v_k$ are in increasing order by x-coordinates and the vertices $v_k, \cdots, v_{n-1}, v_0$ are in decreasing order by x-coordinates.

The edges of P are the open line segments whose endpoints are v_i and v_{i+1}, for $0 \leq i < n$ (indices are taken mod n). The diagonals of P are the open line segments joining two nonadjacent vertices of P and lying entirely in the interior of P. A triangulation of a simple polygon P is a partition of the interior of P into $n-2$ triangles, whose sides are either edges or diagonals of P not intersecting each other. Note that $n-3$ diagonals are required to triangulate a n-sided simple polygon.

Recently, Tarjan and van Wyk showed that triangulation of simple polygons is not as hard as sorting by presenting an $O(n \log\log n)$-time algorithm for this problem [Tarjan

& van Wyk '87]. One of the foremost open problems in computational geometry is to determine whether triangulation can be done in linear time. On the other hand linear-time algorithms for triangulating special classes of polygons are known, as for e.g. monotone polygons [Garey et al. '78], starshaped polygons [Fournier & Montuno '84], polygons with few concave vertices [Hertel & Mehlhorn '83], etc. Here we are interested in parallel algorithms for the triangulation problem. As model of parallel computation we use the concurrent-read exclusive-write parallel random access machine, CREW-PRAM for short. In [Atallah et al. '86] an algorithm for computing the trapezoidal decomposition of a set of n nonintersecting line segments is given. This algorithm runs in time $O(\log n)$ using n processors in the CREW-PRAM model. From this result it follows that a simple n-sided polygon can be triangulated within the same time- and processor-bounds. Note that the number of operations performed by such an algorithm is $\Omega(n \log n)$. A natural question in this context is the following:

> For which of the special classes of polygons allowing linear-time triangulation there exist parallel triangulation algorithms that run in logarithmic time, and perform only a linear number of operations?

In the following we propose an $O(\log n)$-time algorithm for triangulating monotone n-sided polygons using $O(n/\log n)$ processors of a CREW-PRAM, only. Our algorithm decomposes a monotone polygon into structurally simpler polygons; first into unimonotone ones, and then into so called fully decomposable ones. For these latter polygons a trapezoidal decomposition is computed, inducing immediately a triangulation.

Section 2 of this paper gives some preliminary results concerning parallel computation. In section 3 the nearest dominator problem is defined, and an optimal solution for this problem is given. This solution will form the core of the trapezoidal decomposition procedure for fully decomposable polygons, and is of interest in its own right. Section 4 presents the algorithm for triangulating monotone polygons.

2. Preliminary Results

In this section we collect some results concerning parallel computation which are used in the triangulation algorithm.

Proposition 2.1 [Merging]
Given two sorted sequences A and B, each stored in an array, and having a total of n elements, the sorted sequence C containing the elements of $A \cup B$ can be computed and stored into an array in time $O(n/p + \log n)$ using p processors in the CREW-PRAM model.

For a proof see [Kruskal '83]

Proposition 2.2 [Prefix Computation]
Let $y_1, \cdots, y_n$ be n elements from a semigroup with associative operation $+$. If $y_1, \cdots, y_n$ are stored in an array in the given order, the n prefix sums $\sum_{j=1}^{i} y_j$, $1 \leq i \leq n$, can be computed in time $O(n/p + \log n)$ using p processors in the CREW-PRAM model.

For a proof see [Ladner & Fischer '80].

An easy corollary of Proposition 2.2 is the following:

Corollary 2.1

i) Given a sequence $A = (a_1, \cdots, a_n)$ stored in an array, any subsequence $a_{s(1)}, \cdots, a_{s(k)}$ of marked elements can be stored in consecutive order in an array of size k in time $O(n/p + \log n)$ using p processors in the CREW-PRAM model.

ii) Given the sequence $A = (a_1, \cdots, a_n)$ and a subsequence $a_{s(1)}, \cdots, a_{s(k)}$ of marked elements from A, the $k + 1$ consecutive subsequences $A_1 = (a_1, \cdots, a_{s(1)})$, $A_2 = (a_{s(1)+1}, \cdots, a_{s(2)}), \cdots, A_{k+1} = (a_{s(k)+1}, \cdots, a_n)$ can be stored in $k + 1$ arrays in time $O(n/p + \log n)$ using p processors such that only $O(n)$ space is required.

We will refer to the operation on sequences described in Corollary 2.1 ii) as splitting the sequence at marked elements.

Proposition 2.3 [List Ranking]
Let L be a linked list with n elements. The rank of a list element is one more than the number of elements properly preceding it in the list. For each element l in the linked list L, the rank $r(l)$ within L can be computed in time $O(n/p + \log n)$ using p processors.

For a proof see [Cole & Vishkin '86].

3. The Nearest Dominator Problem

Let $A = (a_1, \cdots, a_n)$ be a sequence of elements taken from a total order. The left nearest dominator of a_i is the element a_k of A with the following properties:

i) $k < i$,

ii) $a_i \leq a_k$,

iii) $a_i > a_j$, for $k < j < i$.

Analogously, the right nearest dominator of a_i is defined.

The nearest dominator problem now asks for the right and left nearest dominator of each element a_i of a given sequence A, if these dominators exist. Figure 1 demonstrates the intimate relation between the nearest dominator problem and the horizontal visibility problem for vertical halflines. Each $a_i \in A$ is interpreted as a vertical halfline extending downward from the point $p_i = (i, a_i)$. Then the right or left nearest dominators of a_i correspond to halflines visible in horizontal direction from p_i.

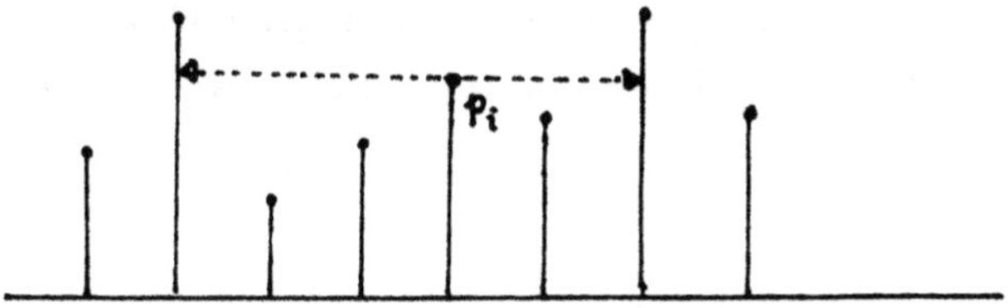

Figure 1

Lemma 3.1
The nearest dominator problem can be solved sequentially in linear time.

Proof: Let $A = (a_1, \cdots, a_n)$. We show how to compute the right nearest dominator for each $a_i \in A$; the problem of computing the left nearest dominator is symmetric. Our algorithm finds the right nearest dominators by considering the elements $a_1, a_2, \cdots, a_n$ in turn. It uses a stack $q_0, q_1, \cdots, q_t$ to store the elements for which the right nearest dominator are yet unknown; q_t is the top of the stack. We initialize the stack with $q_0 = a_1$.

(1) $s \leftarrow 2$
(2) while $s \leq n$
(3) do
(4) while $q_t \leq a_s$
(5) do pop q_t from the stack, right-dom(q_t) $\leftarrow a_s$ od
(6) push a_s
(7) $s \leftarrow s + 1$
(8) od
(9) $q_0, q_1, \cdots, q_t$ have no right dominator

Before the loop body (lines (3) to (8)) is executed the following invariant holds:

i) $q_0, \cdots, q_t$ are those elements of A having no right dominator in $(a_1, \cdots, a_{s-1})$
ii) $q_i > q_{i+1}$, for $0 \leq i \leq t$
iii) for all elements of $(a_1, \cdots, a_{s-1})$ not stored in the stack the right nearest dominator has been found.

This invariant can easily be shown by induction, thus establishing the correctness of the algorithm. Since each element of A is pushed exactly once into the stack, and poped from the stack at most once, the linear time-bound follows.

◇

The method for solving the nearest dominator problem given above is inherently sequential. In the following we first give a sequentially slower algorithm which is easy to parallelize. Then the optimal $O(\log n)$-time algorithm using $O(n/\log n)$ processors will be presented. This final algorithm will use the sequential as well as the following procedure as subroutines.

Lemma 3.2
The nearest dominator problem for $A = (a_1, \cdots, a_n)$ can be solved in time $O(\log n)$ using $O(n)$ processors in the CREW-PRAM model.

Proof: Again, we show how to compute the right nearest dominators. First, a binary tree with n leaves is initialized. With the i-th leaf in left-to-right order the element a_i of A is associated. With each inner node N we associate the maximal value stored in the leaves of N's left and right subtree, respectively. These values are denoted by leftmax(N) and rightmax(N). For each element a_i of A the right nearest dominator is determined by the following search-procedure:

(1) $N \leftarrow$father(l), where l is the leaf associated with a_i
(2) <u>while</u> rightmax(N) $< a_i$ or l is in the right subtree of N
(3) <u>do</u> $N \leftarrow$father(N) <u>od</u>
(4) <u>while</u> N is no leaf
(5) <u>do</u>
(6) <u>if</u> leftmax(N) $\geq a_i$ and l is not in the left subtree of N
(7) <u>then</u> $N \leftarrow$leftson(N)
(8) <u>else</u> $N \leftarrow$rightson(N)
(9) <u>od</u>
(10) right-dom(a_i) $\leftarrow a(N)$, where $a(N)$ is the element of A associated with leaf N.

The correctness of the algorithm is obvious. To analyze the complexity first note that the binary tree together with the required information (i.e. rightmax- and leftmax-values) can be computed in time $O(\log n)$ using n processors. Then the n processors are used to determine for all a_i in parallel (one processor per element of A) the corresponding right nearest neighbors. Such an search requires $O(\log n)$ time, because each of the two loops (lines (2), (3) and lines (4)-(9)) are performed at most h times, where h is the height of the binary tree.

◇

Now we are able to give the final O($\log n$)-time algorithm solving the nearest dominator problem with only O($n/\log n$) processors in the CREW-PRAM model:

Let $A = (a_1, \cdots, a_n)$ be given with a_i taken from a total order. For convinience we assume that A is stored in an array, and that $a_i \neq a_j$ for $i \neq j$.

(1) Decompose A into $\lceil n/\log n \rceil$ consecutive subsequences A_i, $1 \leq i \leq \lceil n/\log n \rceil$, of size $\leq \lceil \log n \rceil$ each.

(2) Solve the nearest dominator problem for each A_i by assigning one processor per subsequence A_i, and solving the problem for all A_i in parallel with the algorithm given in Lemma 3.1. Report all nearest dominators found so far. Obviously, the nearest dominators with respect to A_i are nearest dominators with respect to A, too.

(3) Now for each sequence A_i two subsequences L_i and R_i are computed. L_i contains the elements of A_i having no left dominator in A_i, and R_i contains the elements of A_i having no right dominator in A_i. To compute for $1 \leq i \leq \lceil n/\log n \rceil$ the subsequences L_i, allocate one processor per sequence A_i, and scan A_i from left to right. During this scan collect all elements having no left dominator in A_i. Analogously, the subsequences R_i are computed (here scan from right to left). Note that the subsequences L_i and R_i are sorted in increasing order.

(4) For $1 \leq i \leq \lceil n/\log n \rceil$ compute $m_i := \max\{a | a \in A_i\}$, and create the sequence $M = (m_1, m_2, \cdots, m_{\lceil n/\log n \rceil})$.

(5) Solve the nearest dominator problem for M using the algorithm given in Lemma 3.2 . Note that the dominators found in this step are in general not dominators with respect to A; but this information will help to find the nearest dominators with respect to A.

(6) In this step we rearrange the information gained in step (5) into a suitable form. For $1 \leq i \leq \lceil n/\log n \rceil$ compute the subsequences $\overline{L_i}$ of M containing all elements of M having m_i as left nearest dominator with respect to M. Similarily, compute $\overline{R_i}$, the subsequences of M having m_i as right nearest dominator. In order to compute $\overline{L_i}$ for $1 \leq i \leq \lceil n/\log n \rceil$, one processor is allocated to every element m_j of M. If m_j and its right nearest dominator in M have both the same left nearest dominator in M, then a link from m_j to its right nearest dominator is created. Now we have the following situation: If $m_{i+1} > m_i$, then $\overline{L_i} = \emptyset$. Else, $\overline{L_i}$ is now given as a linked list, whose first element is m_{i+1}. By the construction of $\overline{L_i}$ it is clear, that the elements of $\overline{L_i}$ are sorted in increasing order. Applying the list ranking algorithm (cf. Proposition 2.3) to the obtained linked lists yields for each element its rank in the corresponding list. Now these lists can easily be stored into arrays. Analogously, the subsequences $\overline{R_i}$ of M are computed such that they are sorted in increasing order, too.

Before continuing the description of the algorithm we give a lemma that supports the understanding of the remaining part of the algorithm.

Lemma 3.3

i) Let $a \in R_i$, let m_k and m_j be two successive elements of $\overline{L_i}$ with $m_k > a > m_j$, then the right nearest dominator of a in A is an element of A_k.

ii) If $a \in R_i$ with $a > m$ for all $m \in \overline{L_i}$, and if the right nearest dominator of m_i in M is m_k, then the right nearest dominator of a in A is an element of A_k.

iii) Let $a \in L_i$, let m_k and m_j be two successive elements of $\overline{R_i}$ with $m_k > a > m_j$, then the left nearest dominator of a in A is an element of A_k.

vi) If $a \in L_i$ with $a > m$ for all $m \in \overline{R_i}$, and if the left nearest dominator of m_i in M is m_k, then the left nearest dominator of a in A is an element of A_k.

Proof: We only give a proof for part i), the proofs for the other parts are similar. First note that $i < j < k$ holds in the situation described in i). Since $a < m_k$, the nearest right dominator of a in A is in some subsequence A_s with $i < s \leq k$. Thus it remains to show that the subsequences A_s with $i < s < k$ contain no elements dominating a. Now recall that m_j has m_i as left nearest dominator, this is because $m_j \in \overline{L_i}$ holds. This implies that $m_s < m_j$ holds for $i < s < j$. Since m_j and m_k are successive elements of $\overline{L_i}$, $m_s < m_j$ or $m_s > m_k$ holds for $j < s < k$. Clearly $m_s < m_k$ follows from $m_k \in \overline{L_i}$. Alltogether we have shown that $a > m_j \geq m_s$ holds for $i < s < k$. Thus in A_s, $1 < s < k$, there are no elements dominating a.

◇

Lemma 3.3 suggests a method to identify for each $a \in L_i$ (R_i, respectively) the subsequence A_k containing the left (right, respectively) nearest dominator of a in A. After this identification it remains to find the nearest dominator within this subsequence of size $\leq \log n$. How this can be achieved efficiently is described in the remaining part of the algorithm.

(7) For all i, $1 \leq i \leq \lceil n/\log n \rceil$, in parallel merge L_i with $\overline{R_i}$, and R_i with $\overline{L_i}$. In order to merge L_i with $\overline{R_i}$ allocate $|\overline{R_i}|$ processors to this task, and use the algorithm of Proposition 2.1..

(8) Split the sequences $MERGE(L_i, \overline{R_i})$ obtained in step (7) at the elements from $\overline{R_i}$ into subsequences L_{ij}, $1 \leq j \leq |\overline{R_i}|$ (see Proposition 2.2 and Corollary 2.1). Similarily, split the sequences $MERGE(R_i, \overline{L_i})$ into subsequences R_{ij}. Note that L_{ij} contains the elements of L_i having their left nearest dominator with respect to A in the same subsequence A_k (by Lemma 3.3).

(9) Now we can find all missing, but existing left nearest dominators in A as follows:

Allocate to each sequence L_{ij} one processor and for all these sequences do the following in parallel: Let m_k be the maximal element in L_{ij}. If $k \neq i$, then let $group(i,j) = k$; else, let $group(i,j) = s$, where m_s is the left nearest dominator of m_i in M. Note that by Lemma 3.3 $A_{group(i,j)}$ contains the left nearest dominator in A for all elements of L_{ij}. Now concatenate $A_{group(i,j)}$ and L_{ij} into $A_{group(i,j)} \circ L_{ij}$, and solve the nearest dominator problem for $A_{group(i,j)} \circ L_{ij}$ sequentially using the algorithm of Lemma 3.1. Report the left nearest dominators found. In order to compute the missing right nearest dominators proceed analogously.

The above discussion guaranties the correctness of this algorithm. It remains to analyze its run time.

By Corollary 2.1 step (1) can be performed in time $O(\log n)$ using $O(n/\log n)$ processors. In steps (2), (3), and (4) each processor is allocated to a job of size $O(\log n)$. Each of these $O(n/\log n)$ jobs can be performed in linear time. Thus $O(\log n)$ time suffice, if $O(n/\log n)$ processors are available. By Lemma 3.2 step (5) requires $O(\log n)$ time with $O(n/\log n)$ processors, and by Proposition 2.3 step (6) can be performed within the same time- and processor-bound. In step (7) we have to merge several pairs of sequences. For each pair the size of one sequence is bounded by $\lceil \log n \rceil$; let k be the size of the other sequence. Note that we allocate in this case k processors to the merge task. Thus, by Proposition 2.1 the time required is $O(\log n)$. Then number of processors allocated in this step is bounded by $\sum_{i=1}^{\lceil n/\log n \rceil} |\overline{L_i}|$ and $\sum_{i=1}^{\lceil n/\log n \rceil} |\overline{R_i}|$, respectively. Since the sets $\overline{L_i}$ are disjoint subsets of M, the two above sums can be bounded by $O(n/\log n)$. Finally, Corollary 2.1 and Lemma 3.1 show that step (8) and (9) can also be performed in $O(\log n)$ time using $O(n/\log n)$ processors. Note that in step (9) $O(n/\log n)$ different nearest neighbor problems of logarithmic size have to be solved.This is because $\lceil n/\log n \rceil$ sequences L_i (R_i, respectively) were split at $\leq n/\log n$ positions, yielding at most $O(n/\log n)$ sequences.

The result of this section can be summarized as:

Theorem 3.1
Let $A = (a_1, \cdots, a_n)$ be a sequence of elements taken from a total order, the nearest dominator problem can be solved in time $O(\log n)$ using $O(n/\log n)$ processors in the CREW-PRAM model.

4. Triangulating a Monotone Polygon in Parallel

A polygon $P = (v_0, v_1, \cdots, v_{n-1})$ is called monotone (with respect to the x-axis), if there exist a vertex v_k such that $v_0, \cdots, v_k$ are sorted in increasing order with respect to x-coordinates, and that $v_k, \cdots, v_{n-1}, v_0$ is sorted in decreasing order with respect to x-coordinates. We call $v_0, \cdots, v_k$ the upper chain of P, and $v_k, \cdots, v_{n-1}, v_0$ the lower chain of P. P is called unimonotone, if the lower or the upper chain consists of a single edge. The chain of a unimonotone polygon P consisting of a single edge is called baseedge of P. A unimonotone polygon is said to be fully decomposable, if the endvertices of the baseedge either have both larger y-coordinates than all other vertices of P, or both have smaller y-coordinates than all other vertices.

Our triangulation algorithm first decomposes a monotone polygon into unimonotone ones; these unimonotone polygons are decomposed again. The resulting subpolygons are either fully decomposable, or a specified rotation of these subpolygons yield fully decomposable ones. Finally, the obtained fully decomposable polygons are triangulated.

4.1 Decomposing a Monotone Polygon into Unimonotone Ones

The sequential standard procedure to decompose a monotone polygon into unimonotone ones can easily be parallelized.

Let $P = (v_0, \cdots, v_{n-1})$ be a monotone polygon with upper chain $v_0, \cdots, v_k$.

(1) Merge $v_0, \cdots, v_k$ and $v_k, \cdots, v_{n-1}$, into a sequence Π of vertices $(v_{\pi(0)}, \cdots, v_{\pi(n-1)})$ sorted by x-coordinates.

(2) Decompose $\Pi = (v_{\pi(0)}, \cdots, v_{\pi(n-1)})$ between $v_{\pi(i)}$ and $v_{\pi(i+1)}$, if the line segment $\overline{v_{\pi(i)}v_{\pi(i+1)}}$ is no side of P. For each subsequence $\Pi_i = (v_{\pi(i+1)}, \cdots, v_{\pi(k)})$ obtained in this way, add $v_{\pi(i)}$ as first and $v_{\pi(k+1)}$ as last element, provided these elements exist.

We claim that the line segment $\overline{v_{\pi(i)}v_{\pi(i+1)}}$, which are no sides of P, are diagonals of P, and that these diagonals do not intersect each other. Consider the vertical strip bounded by the vertical lines through $v_{\pi(i)}$ and $v_{\pi(i+1)}$, respectively. The intersection of P and this vertical strip is a trapezoid, and two of its vertices are $v_{\pi(i)}$ and $v_{\pi(i+1)}$. Thus $\overline{v_{\pi(i)}v_{\pi(i+1)}}$ is either a side or a diagonal of this trapezoid, and therefore either a side or a diagonal of P. Furthermore, since Π forms a monotone polygonal chain, no two of these line segments intersect each other. Moreover, one easily verifies that step (2) corresponds to decomposing P by adding the diagonals $\overline{v_{\pi(i)}v_{\pi(i+1)}}$. The construction implies that each subpolygon of this decomposition is unimonotone.

By Proposition 2.1 and Corollary 2.1 both steps can be performed in time $O(\log n)$ using $O(n/\log n)$ processors in the CREW-PRAM model.Thus we have

Lemma 4.1

A monotone polygon $P = (v_0, \cdots, v_{n-1})$ can be decomposed into unimonotone ones in time $O(\log n)$ using $O(n/\log n)$ processors in the CREW-PRAM model.

4.2 Decomposing a Unimonotone Polygon into Fully Decomposable Ones

Let $P = (v_0, \cdots, v_{n-1})$ be a unimonotone polygon. Without loss of generality we assume that (v_0, v_{n-1}) is the baseedge of P, and that $y(v_0) \geq y(v_{n-1})$ holds, where $y(v_i)$ denotes the y-coordinate of vertex v_i.

To decompose $P = (v_0, \cdots, v_{n-1})$, P is processed as follows:

(1) For $0 \leq i < n$, compute prefixmin$(i) := \min\{y(v_j) | 0 \leq j \leq i\}$

(2) For $0 \leq i < n$, mark v_i, if $y(v_i) <$prefixmin$(i-1)$; furthermore mark v_0.

(3) Form the subsequence Δ of P, consisting of all marked vertices. Let $\Delta = (v_{\delta(0)}, v_{\delta(1)}, \cdots, v_{\delta(k)})$. Note that $v_{\delta(0)} = v_0$ and $v_{\delta(k)} = v_{n-1}$.

(4) Decompose P into subpolygons by adding the line segments $\overline{v_{\delta(i)}v_{\delta(i+1)}}$ for $0 \leq i < k$, if $\overline{v_{\delta(i)}v_{\delta(i+1)}}$ is no side of P. To obtain the standard representation of these subpolygons, just split $P = (v_0, \cdots, v_{n-1})$ between $v_{\delta(i)}$ and its immediate successor in P, provided $\overline{v_{\delta(i)}v_{\delta(i+1)}}$ is no side of P. To each obtained subsequence $(v_{\delta(i)+1}, \cdots, v_{\delta(i+j)})$ add $v_{\delta(i)}$ as first element. Let us denote such an sequence $(v_{\delta(i)}, \cdots, v_{delta(i+j)})$ by $P_{\delta(i)}$. Now the sequences $P_{\delta(i)}$ and Δ having more than two vertices give the standard representation of the subpolygons.

Again, one easily verifies that the added line segments are diagonals, and that these line segments do not intersect each other. Moreover, by constuction the obtained subpolygons $P_{\delta(i)}$ are fully decomposable; indeed, $y(v_{\delta(i+j)}) < y(v_{\delta(i)}) < y(v_j)$ for $\delta(i) < j < \delta(i+j)$. Clearly, the subpolygon Δ is not fully decomposable. But note, that Δ is unimonotone with respect to the x-axis as well as with respect to the y-axis. In both cases (v_0, v_{n-1}) is the baseedge of Δ. This implies that Δ is unimonotone with respect to any line parallel to its baseedge. Thus rotating Δ such that the baseedge becomes parallel to the x-axis yields a polygon Δ^{rot} that is unimonotone with respect to the x-axis. Since obviously any unimonotone polygon with a baseedge parallel to the x-axis is fully decomposable, Δ^{rot} is fully decomposable.

To analyze the run time of this procedure, note that step (1) requires time $O(\log n)$ with $O(n/\log n)$ processors by Proposition 2.2. To perform step (2) within this complexity bound is trivial. For steps (3) and (4) we can achieve this bound by Corollary 2.1.

Thus we can summarize as

Lemma 4.2
Let $P = (v_0, \cdots, v_{n-1})$ be a unimonotone polygon. P can be decomposed into fully decomposable ones in time $O(\log n)$ with $O(n/\log n)$ processors in the CREW-PRAM model.

4.3 Triangulating Fully Decomposable Polygons

To triangulate a fully decomposable polygon $P = (v_0, \cdots, v_{n-1})$ with baseedge (v_0, v_{n-1}) and $y(v_0) > y(v_{n-1})$, we proceed as follows:

(1) For all i, $0 \leq i < n$, compute $a_i := y_{max} - y(v_i)$, where y_{max} is the maximal y-coordinate of the vertices, and form the sequence $A = (a_0, \cdots, a_{n-1})$.

(2) Solve the nearest dominator problem for A. Let $v(a_i), v_{left}(a_i)$, and $v_{right}(a_i)$ denote the vertices corresponding to a_i, to the left nearest dominator of a_i, and to the right nearest dominator of a_i, respectively.

(3) For all i, $0 \leq i < n$, add $\overline{v(a_i)v_{left}(a_i)}$ as diagonal to P, provided this line segment is no edge of P. Analogously, proceed with the segments $\overline{v(a_i)v_{right}(a_i)}$.

Now one easily verfies that all added line segments are either sides or diagonals of P, and that these line segments do not intersect each other. So it remains to show that all obtained subpolygons are indeed triangles. Recall that each simple polygons are triangulated by adding $n-3$ nonintersecting diagonals. Now a simple counting argument suffice to complete the proof of correctness. Note that for $0 < i < n-1$ $a_0 > a_i$ and $a_{n-1} > a_i$ holds by the assumptions we made for P. Therefore each element of A, except a_0 and a_{n-1} have a left and a right dominator in A. Thus, step(2) yields $2n-3$ line segments, n of which are (can be) sides of P. Consequently, the remaining $n-3$ segments are diagonals.

To perform steps (1) and (3) in time $O(\log n)$ with $O(n/\log n)$ processors is easy, and step (2) can be performed within the stated time- and processor-bound by Theorem 3.1.

Lemma 4.3
Let $P = (v_0, \cdots, v_{n-1})$ be a fully decomposable polygon. P can be triangulated in time $O(\log n)$ using $O(n/\log n)$ processors of a CREW-PRAM.

A triangulation of a n-sided monotone polygon P can now be obtained by decomposing P into unimonotone ones, allocating a suitable fraction of the processors to each unimonotone polygon, decomposing the unimonotone ones into fully decomposable ones, reallocate the processors, and then triangulate the fully decomposable polygons. Each of these tasks can be performed in time $O(\log n)$ with a total of $O(n/\log n)$processors. Thus we have shown:

Theorem 4.1
A n-sided monotone polygon P can be triangulated in time $O(\log n)$ using $O(n/\log n)$ processors in the CREW-PRAM model.

5. References

[Tarjan & v. Wyk '87] R. E. Tarjan and C. van Wyk. An $O(n \log \log n)$-time algorithm for triangulating simple polygons. 1986. Manuscript, submitted to *SIAM Journal on Computing.*

[Garey et al. '78] M. R. Garey, D. S. Johnson, F. P. Preparata, and R. E. Tarjan. Triangulating a simple polygon. *Information Processing Letters*, 7(4):175-179, 1987 .

[Fournier & Montuno '84] A. Fournier and D. Y. Montuno. Triangulating simple polygons and equivalent problems. *ACM Transactions on Graphics*, 3(2):153-174, 1984.

[Hertel & Mehlhorn '83] S. Hertel and K. Mehlhorn. Fast triangulation of simple polygons. In *Proceeding of the Conference on Foundations of Computing Theory*, pages 207-218, Springer-Verlag, Berlin, 1983.

[Atallah et al. '86] M. J. Atallah and M. T. Goodrich. Efficient plane sweeping in parallel. In Proceedings of the 2nd Symposion on Computational Geometry, pages 216-225, 1986.

[Kruskal ,83] C. P. Kruskal. Searching, merging, and sorting in parallel computation. *IEEE Transactions on Computers*, C32(10):942-946, 1983.

[Cole & Vishkin '86] R. Cole and U. Vishkin. Approximate and exact parallel scheduling with applications to list, tree and graph problems. *Proceeding of the 27th FOCS*,pages 478-491, 1986.

Abstract Voronoi Diagrams and their Applications (Extended Abstract)

Rolf Klein*

Abstract

Given a set S of n points in the plane, and for every two of them a separating Jordan curve, the abstract Voronoi diagram $V(S)$ can be defined, provided that the regions obtained as the intersections of all the "halfplanes" containing a fixed point of S are path–connected sets and together form an exhaustive partition of the plane. This definition does not involve any notion of distance. The underlying planar graph, $\widehat{V}(S)$, turns out to have $O(n)$ edges and vertices. If $S = L \bigcup R$ is such that the set of edges separating L–faces from R–faces in $\widehat{V}(S)$ does not contain loops then $\widehat{V}(L)$ and $\widehat{V}(R)$ can be merged within $O(n)$ steps giving $\widehat{V}(S)$. This result implies that for a large class of metrics d in the plane the d–Voronoi diagram of n points can be computed within optimal $O(n \log n)$ time. Among these metrics are, for example, the symmetric convex distance functions as well as the metric defined by the city layout of Moscow or Karlsruhe.

Keywords: Voronoi diagram, metric, computational geometry.

1 Introduction

Generally, the Voronoi diagram of a set of sites p_i in a space M partitions M into regions, one to each site, such that the region of p_i contains all those points of M that are closer to p_i than to any other site p_j. Here we assume that M is endowed with a distance measure that allows to compute the distance between any point in M and any of the sites p_i. If only two sites, p and q, are considered then M is decomposed into the region of all points closer to p than to q, the region of all points closer to q than to p, and the set of those points that are of equal distance to both, the *bisector* of p and q. Intuitively, the Voronoi diagram of the whole site set consists of pieces of such bisectors that separate the regions of the sites from one another.

The Voronoi diagram belongs to the most useful data structures in computational geometry. Many proximity problems (closest pair, all nearest neighbors, nearest neighbor) can be solved efficiently once the Voronoi diagram for the underlying set of sites is constructed, see [PrSh] for an overview in the Euclidean case. Moreover, Voronoi diagrams and their generalizations have become important tools in motion planing, see, for example, [DuYa].

The first Voronoi diagrams studied were those of n point sites in the real plane, based on the standard Euclidean metric. Optimal $O(n \log n)$ algorithms have been presented in [ShHo] and, later, in [Br] and [F]. Later on, Voronoi diagrams based on various non–Euclidean metrics in the plane have been studied ([WiWuWo], [ChDr], [L], [Ar], [LeWo], [H], to mention but a few). Most of the algorithms presented for the computation of such Voronoi diagrams are of divide–and–conquer type. They have the remarkable feature in common that they are only working on (a subset of) *the set of*

*Institut für Informatik, Universität Freiburg, Rheinstr. 10–12, 7800 Freiburg, Fed. Rep. of Germany

bisectors of the given sites and are not making use of any other properties of the underlying metric than such expressible in terms of bisectors.

Motivated by this observation we show in this paper how to define and to compute Voronoi diagrams *in terms of bisecting Jordan curves* without using the notion of distance. Instead, we assume that the regions of such abstract Voronoi diagrams are path–connected sets, forming a partition of the plane. Surprisingly, connectedness of the Voronoi regions is an assumption strong enough to make the (counter–)clockwise scan principle work, the very core of Shamos/Hoey's divide–and–conquer algorithm, even in our general situation. This enables us to merge two abstract Voronoi diagrams within a linear number of steps, provided that their common bisector contains no loops.

On the one hand, our result clarifies what is really essential in standard divide–and–conquer algorithms for the computation of Voronoi diagrams (convexity of the regions, for example, is not!). On the other hand, it provides us with a "universal" tool: Whenever a metric in the plane, d, is such that its bisectors do have the properties needed in our abstract model, we infer that the d–Voronoi diagram can be computed efficiently. Different sufficient criteria for metrics to have these properties are given in [KlWo] and [Kl]. As an application, in section 4 the metric defined by the city of Moscow (or Karlsruhe) is studied, where the streets are either circles around the center or rays emanating from the center, see Figure 1.

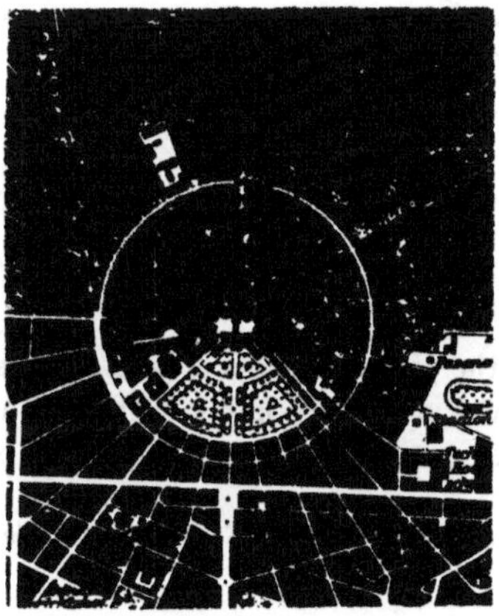

Figure 1: The city of Karlsruhe

2 The definition of abstract VDs

Let $S = \{p_1, \ldots, p_n\}$ be a set of n points in the plane. For each pair $p, q \in S$ let $B_0(p,q) = B_0(q,p)$ be a simple, piecewise smooth Jordan curve tending to infinity at both ends [1] that separates p from q. Assume that the intersection of two such curves B_0, B_0' consists of at most finitely many connected components.

After removing $B_0(p,q)$ the plane is dissected into two domanis, each of them containing one of p, q. In order to avoid borderline conflicts the curve $B_0(p,q)$ itself should belong to the region of one of the points. A consistent choice can be obtained as follows. We fix an arbitrary order "$\prec$" of all points in the plane, for example the lexicographic order, and define

[1] i. e., after applying stereographic projection the curve becomes a simple closed Jordan curve on the surface of the sphere that passes through the pole.

Definition 2.1

1. *If $p \prec q$ then (see Figure 2)*
   $R(p,q) :=$ $B_0(p,q) \cup$ *domain containing* p
   $R(q,p) :=$ *domain containing* q
2. $R(p,s) := \bigcap_{q \in S, q \neq p} R(p,q)$ *(Voronoi region of* p *)*
3. $V(S) := \bigcup_{p \in S} \partial R(p,S)$ *(Voronoi diagram)*

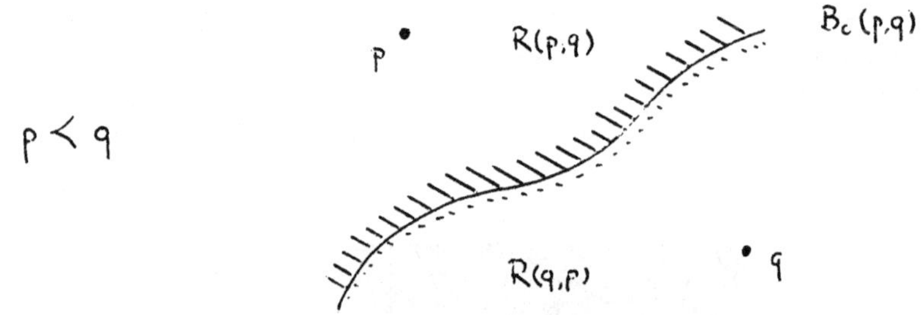

Figure 2

Note that we could get rid of the physical points p_i (and consider p_i, p_j as indices) if we assumed instead that each set $R(p_i, S)$ has a nonempty interior (then points p_i could be picked from these sets, if necessary). So, our abstract model is not restricted to the point site case.

Definition 2.2 *The system* $(S, \{B_0(p,q); p \neq q \in S\})$ *is admissible if both of the following conditions hold for each subset* $S' \subseteq S$.

1. $R(p,S')$ *is path–connected for each* $p \in S'$
2. $\Re^2 = \bigcup_{p \in S'} R(p,S')$.

The system shown in Figure 3a is not admissible since condition 2 is violated. The system of Figure 3b is admissible if and only if $p \prec q$, r, because otherwise the two components belonging to p-land are not connected.

In order to compute the Voronoi diagram, we are less interested in the set $V(S)$ than in the graph structure of the region boundaries. This graph appears once the thin parts of the regions are pumped up, see Figure 4. Note that no conflict arises here since each point of the plane belongs to only one region. Furthermore, the connectedness of the regions is preserved. But a curve segment of $V(S)$ can be represented by two edges of $\hat{V}(S)$, and a point of $V(S)$ can give rise to many vertices of $\hat{V}(S)$, as Figure 4 indicates. However, from Euler's formula follows

Theorem 2.3 *The underlying planar graph* $\hat{V}(S)$ *has* n *faces and* $O(n)$ *edges and vertices.*

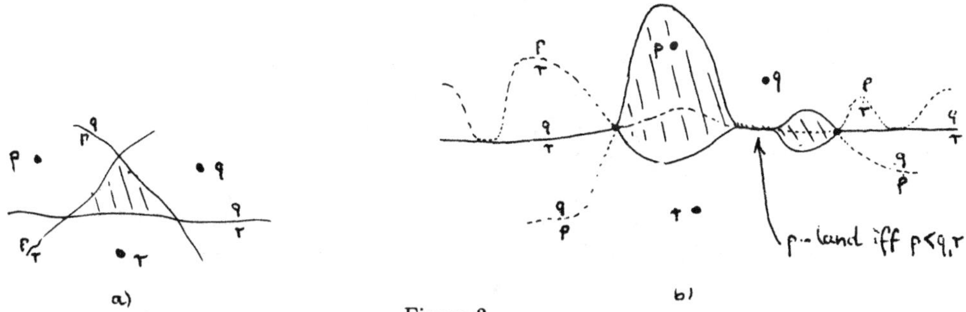

Figure 3

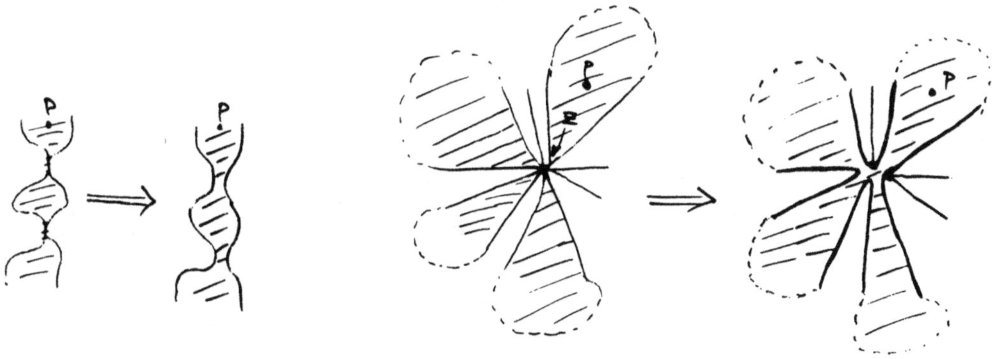

Figure 4

3 The computation of abstract VDs

In order to apply a divide–and–conquer approach we have to concentrate on the merge phase. Assume that $S = L \bigcup R$, where $|S| = n$. If it is possible to merge $\hat{V}(L)$ and $\hat{V}(R)$ within $O(n)$ steps giving $\hat{V}(S)$ then we obtain an $O(n \log n)$ overall performance, provided that in the recursion step S is always split into subsets of about equal cardinality .

Merging $\hat{V}(L)$ and $\hat{V}(R)$, the crucial part is in computing the bisector of L and R, $\widehat{B_0}(L,R)$. It consists of all " bisecting " edges of $\hat{V}(S)$ that are common border of both an L– and an R–face. $\widehat{B_0}(L,R)$ can consist of several connected components, among them cycles, as shown in Figure 5. The problem is how to detect all the components during the computation of $\widehat{B_0}(L,R)$. Those who are known to be adjacent to a fixed vertex can easily be found. Therefore, we assume

A: *For all subsets $L' \subseteq L$ and $R' \subseteq R$ the bisector $\widehat{B_0}(L',R')$ contains no cycle.*

As was first noticed by [ChDr] the non–cyclic components of $\widehat{B_0}(L,R)$ are unbounded and can be detected at infinity–which is just another vertex of the compactification of $\hat{V}(S)$ on the sphere. From now on we assume that the bisecting edges in $\widehat{B_0}(L,R)$ are oriented in such a way that the adjacent L–faces are on their left–hand side. Now $\widehat{B_0}(L,R)$ can be decomposed into chains in the following way. We begin with an arbitrary bisecting edge starting from infinity. Whenever we reach an endpoint v of such an edge different from ∞, we must determine if the point of the plane

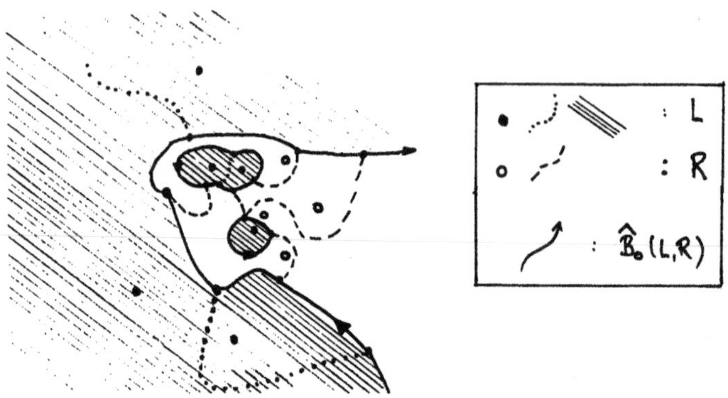

Figure 5

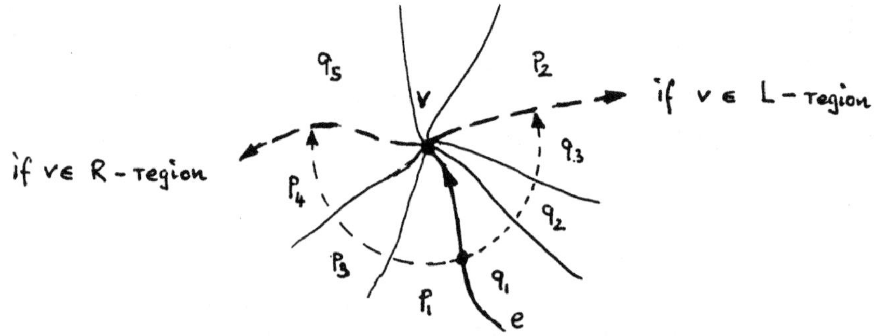

Figure 6: Possible continuations of bisecting edge e

represented by the vertex v of $\widehat{V}(S)$ belongs to an L–region or to an R–region of $V(S)$. In the first case we continue with the first bisecting edge in counterclockwise order around v that starts from vertex v; in the second case we choose the first bisecting edge in clockwise order with initial point v, see Figure 6. Since, by assumption A, there are no loops in $\widehat{B_0}(L,R)$, this procedure brings us back to ∞. We continue until all bisecting edges incident to ∞ are encountered.

Let $K_1,\ldots,K_m$ denote the chains $\widehat{B_0}(L,R)$ consists of. For each K_l let L_l be the union of all regions $R(t,S)$ such that the t–face of $\widehat{V}(S)$ lies of the left–hand side of K_l. Due to the above chain definition we can prove

Lemma 3.1 *Let $p \in L$.*

1. *$R(p,L) \bigcap L_l$ is path–connected for each $l \leq m$*
2. *$R(p,S) = R(p,L) \bigcap \bigcap_{p \in L_l} L_l$.*

An analogous statement holds for the right-hand sides of the chains. During the merge phase we compute the continuation of bisecting edges to bisecting chains much the same way. However, we first have to determine the *endpoint* of each bisecting edge in $\widehat{V}(S)$ from the subdiagrams $\widehat{V}(L)$ and $\widehat{V}(R)$. Suppose a piece of $B_0(p,q)$, e, running in $R(p,L) \bigcap R(q,R)$ is known to belong to a dividing chain. In order to determine it's endpoint in $\widehat{V}(S)$ we have to look for the first nontrivial intersection of $B_0(p,q)$ with the boundary of $R(p,L)$(or $R(q,R)$). In the standard Euclidean case this is achieved by scanning the boundary of $R(p,L)$ counterclockwise (and the boundary of $R(q,R)$ clockwise) for the first intersection with $B_0(p,q)$, knowing that there is exactly one such intersection since the regions are convex and $B_0(p,q)$ is straight, see Figure 7. These properties are, of course,

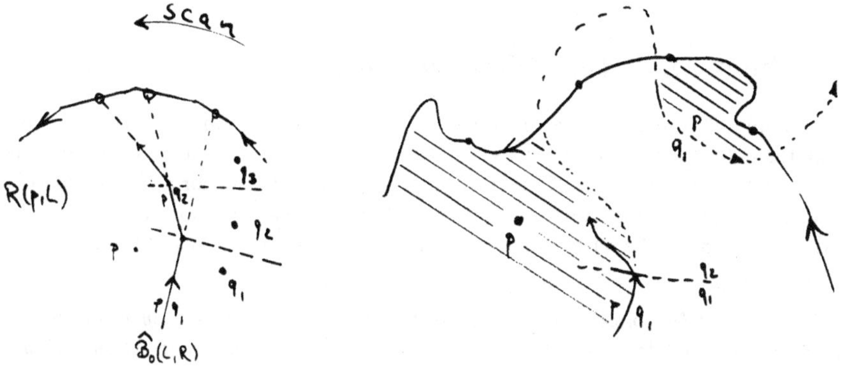

Figure 7

not guaranteed in our general case. However, a situation as shown in Figure 7b cannot occur because the two shaded areas would both belong to $R(p,L) \bigcap R(p,q_1) = R(p, L \bigcup \{q_1\})$, in contradication to the connectedness of the latter ! The main tool in proving that the scan principle is still applicable is the following Lemma.

Lemma 3.2 *Let C_1, C_2 be two simple, closed, oriented curves on the surface of the sphere. Assume that all intersections of C_1 and C_2 are proper crossings. Let D_1, D_2 denote the domains "to the left" of C_1 and C_2, correspondingly. If $D_1 \bigcap D_2$ is connected then the points of $C_1 \bigcap C_2$ appear on C_1 in the same (cyclic) order as on C_2.*

See Figure 8 for illustrations. Another problem is how to efficiently determine the continuation of a

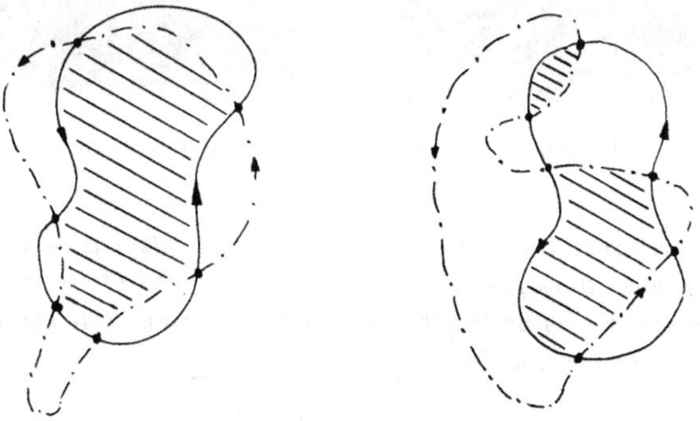

Figure 8: (a) connected intersection (b) disconnected intersection

bisecting edge whose endpoint is a vertex in $\widehat{V}(L)$ (and $\widehat{V}(R)$) that has buddy vertices all representing the same point of the plane. However, one can show

Theorem 3.3 *Let $(S, \{B_0(p,q); p \neq q \in S\})$ an admissible system that fulfils the assumption A. Let $S = L \bigcup R$. Then $\widehat{V}(L)$ and $\widehat{V}(R)$ can be merged within $O(|S|)$ steps giving $\widehat{V}(S)$.*

Here one step may involve one elementary curve operation like, for example, determining on which side of B_0 a given point z lies, or determining the first intersection with B_0' on B_0 in a given direction.

4 Applications. The Moscow metric

In order to apply Theorem 3.3 to a single metric d (or to classes of metrics) we first have to make shure that the d–bisectors $B_0(p,q)$ for a given set of points form an admissible system in the sense of Definition 2.2. It was shown in [Kl] and [KlWo] that this holds true for all "nice" metrics that are subject to four simple axioms. Next, one has to verify that the assumption A of Section 3 is fulfilled for suitable partitions of the set of points, S. This can be done by means of the following Theorem also shown in [Kl].

Theorem 4.1 *Let d be a nice metric. Assume that all d-circles are simply–connected, and that a set of points is divided into subsets L,R by a separating curve l whose intersection with an arbitrary d-circle is empty or connected. Then $\widehat{B_0}(L,R)$ contains no loop.*

To give an application, we consider the Moscow metric that reflects a city layout where the streets are either straight lines radially emanating from a fixed center, or circles around this center. This pattern can be found in the map of numerous cities, among them Moscow and Karlsruhe, see Figure 9.

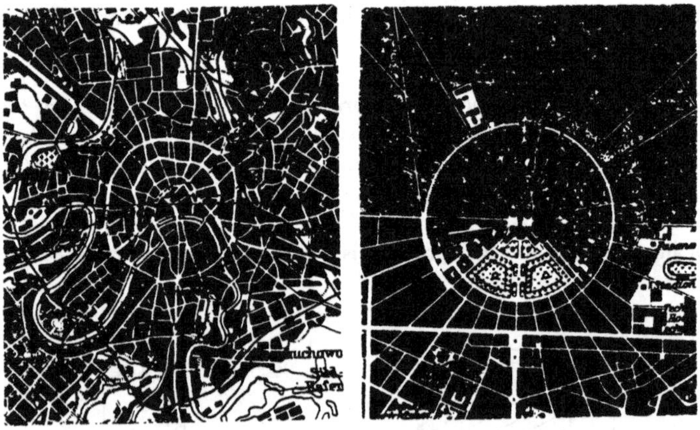

Figure 9: The maps of Moscow and Karlsruhe

Definition 4.2 *A curve connecting two points, a and b, in the plane is called admissible if it consists only of segments of straight lines passing through the origin, and of segments of circles centered at the origin. The minimum Euclidean length of all admissible connecting curves is called the distance of a and b in the Moscow metric, $d_M(a,b)$.*

Figure 10a displays an admissible curve. Clearly, the d_M–distance of two points is not invariant under translations.

Straight (=shortest) curves in d_M can be characterized as follows.

Lemma 4.3 *A curve connecting a and b is d_M-shortest iff it is part of the unbounded curves depicted in Figure 10b and 10c.*

Thus, if the angle α between a and b is greater than 2 ($\simeq$ 114.59... degrees) then the shortest connecting path follows the radii through the origin. If $\alpha < 2$ then the shortest path goes a roundabout way along the circle of the point closer to 0. If $\alpha = 2$ then infinitely many shortest paths from a to b exist. Next we address the *d_M-circles.*

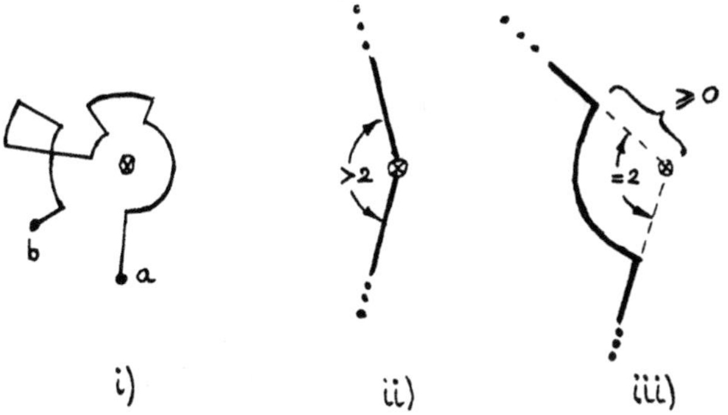

Figure 10: An admissible curve and shortest path extensions

Lemma 4.4 *Let l be a straight line through the origin. Then for each circle $C_r^{d_M}(v)$ the intersection $C_r^{d_M}(v) \cap l$ is connected (or empty)*

This assertion is illustrated by Figure 11.

Note, however, that d_M-circle are in general not *d_M-convex*: the unique shortest path connecting the points a and b of the innermost circle in Figure 11 is not itself fully contained in this circle. Similarly, Figure 12b shows that the Voronoi regions need not be d_M-convex. Figure 11 also illustrates that the *d_M-bisectors*

$$B(v,w) = \{z; d_M(v,z) = d_M(w,z)\}$$

are curves, not regions (as happens in L_1 for points at diagonal vertices of a square), due to the shape of the circles.

Theorem 4.5 *The Voronoi diagram of n points in the Moscow metric can be computed in optimal $O(n \log n)$ time, using linear space.*

Proof: It is easy to check that the Moscow metric is nice. The d_M-circles have no holes because each d_M-straight curve can be prolonged towards infinity, by Lemma 4.3. Given n points, we compute their polar coordinates and sort them in lexicographic order. Then we run a divide-and-conquer algorithm on the polar angles, using straight lines through the origin as separators. If L and R are point sets separated this way then $\widehat{V}(L)$ and $\widehat{V}(R)$ can be merged in time $O(|L \cup R|)$, by Lemma 4.4, Theorem 4.1, and Theorem 3.3. The Voronoi diagram of k sorted points of the same polar angle can be computed in time $O(k)$, see Figure 12a. Figure 12b depicts the Voronoi diagram of 7 points in the Moscow metric. □

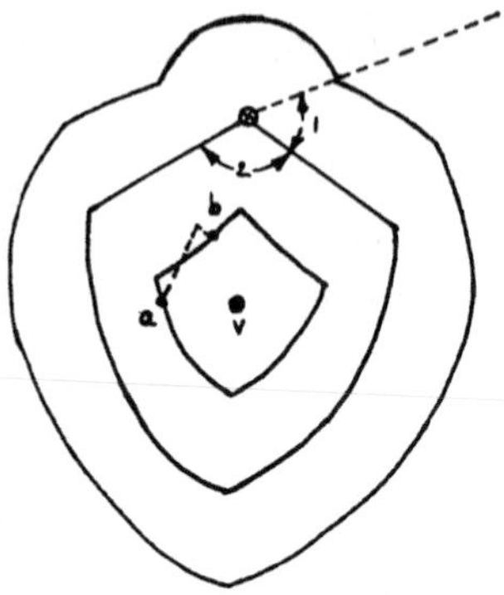

Figure 11: d_M–circles expanding from v

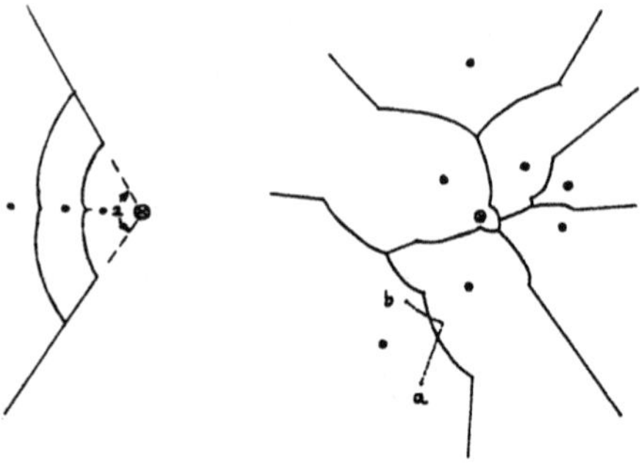

Figure 12: Voronoi diagrams in the Moscow metric

5 Concluding Remarks

We have introduced the notion of abstract Voronoi diagrams that are not based on a distance measure but on systems of Jordan curves dissecting the plane, together with certain connectedness assumptions. Then we have shown how to compute such diagrams by using a divide–and–conquer technique, using elementary operations on the given set of curves. This settles one of the three subproblems in computing Voronoi diagrams based on distance functions in the plane, namely the construction of the diagram from the bisectors. The second problem is how to ensure that the bisectors of a given distance function are (or can be chosen) such that our abstract model applies. For metrics some sufficient conditions were given in [KlWo] and [Kl]. The third–and still open–problem is how to carry out efficiently the elementary bisector operations for classes of interesting metrics.

References

[Ar] B. Aronov, "On the geodesic Voronoi diagram of point sites in a simple polygon", *Proc. 3rd ACM Symposium on Computational Geometry*, Waterloo, 1987, pages 39–49.

[Br] K. Q. Brown, "Voronoi diagrams from convex hulls", *Inf. Proc. Lett.* 9, pages 223–228, 1979.

[ChDr] L. P. Drysdale, III, "Voronoi diagrams based on convex distance functions", *Proc.1st ACM Symposium on Computational Geometry*, Baltimore, 1985, pages 235–244.

[DuYa] C. ó'Dúnlaing and C. K. Yap, " A retraction method for planning the motion of a disc" , in J. Schwartz, M. Sharir, and J. Hopcroft (eds.), *Planning, Geometry, and Complexity of Robot Motion*, Ablex Publishing Corp., Norwood, NJ, 1986.

[F] S. Fortune, "A sweepline algorithm for Voronoi diagrams", *Algorithmica 2 (2)*, 1987, pages 153–174.

[H] F. K. Hwang, "An $O(n \log n)$ algorithm for rectilinear minimal spanning trees", *JACM 26*, 1979, pages 177–182.

[Kl] R. Klein, "Voronoi diagrams in the Moscow metric", Technical Report No 7, Institut für Informatik, Universität Freiburg, to be presented at WG '88, Amsterdam.

[KlWo] R. Klein and D. Wood, "Voronoi diagrams based on general metrics in the plane", in R. Cori and M. Wirsing (eds.), *Proc. 5th Annual Symposium on Theoretical Aspects of Computer Science (STACS)*, Bordeaux, France, 1988, LNCS, pages 281–191.

[L] D. T. Lee, "Two–dimensional Voronoi diagrams in the L_p metric", *JACM 27*, 1980, pages 604–618.

[LeWo] D. T. Lee and C. K. Wong, "Voronoi diagrams in L_1 (L_∞ metrics with 2–dimensional storage applications", *SIAM J. COMPUT.9*, 1980, pages 200–211.

[PrSh] F. Preparata and I. Shamos , "*Computational Geometry* " *An Introduction*, Springer, 1985.

[ShHo] M. I. Shamos and D. Hoey, "Closest- point problems", *Proc.6th IEEE Symposium on Foundations of Computer Science*, 1975, pages 151–162.

[WiWuWo] P. Widmayer, Y. F. Wu, and C. K. Wong, "Distance problems in computational geometry for fixed orientations ", *Proc. 1st ACM Symposium on Computational Geometry*, Baltimore, 1985, pages 186–195.

Geometric Modeling of Smooth Surfaces

by

Hans Hagen*

Universität Kaiserslautern

FB Informatik

Abstract

The methods of Computer Aided Geometric Design have arisen from the need of efficient computer representation of practical curves and surfaces used in engineering design. The generation of smooth surfaces from a set of three-dimensional data points is a key problem in this field. The purpose of this paper is to present algorithms for designing and testing smooth free-form surfaces.

Keywords

geometric modeling, computer aided geometric design, design of curves and surfaces, automatic smoothing

Introduction

Designing curves and surfaces plays an important role in the construction of quite different products such as car bodies, ship hulls, airplane fuselages and wings, propeller blades, etc., but also in the description of chemical, physical, geological and medical phenomena.

The choice of the surface form depends upon the application. The two main techniques are the so called Bezier- and B-Spline methods and the Gordon-Coons-type surfaces. Chapter 1 and 2 are devoted to the details of these fundamental techniques. The topic of chapter 3 is the design of "smooth" surfaces, that means how to specify the inner control points of Bezier- and B-Spline-surfaces or the twist vectors of Gordon-Coons patches in order to achieve smooth surfaces.

In constructing surfaces supported by a graphics system it is necessary to control the "quality" of the surface. Such "quality testing" algorithms or "surface interrogation" algorithms are covered in chapter 4.

(1) Bezier- and B-Spline Surfaces

The curves and surfaces now known as Bezier curves and surfaces were independently developed by **P. de Casteljau** and by **P. Bezier**. The underlying mathematical theory, based on the concept of Bernstein polynomials, was first introduced by **R. Forrest** (see [**Forrest, 72**]). The fundamental idea of this approach is to evaluate and manipulate the curves and surfaces by a (small) number of "control points". Since we build up Bezier-surfaces from curve-net works, we first consider Bezier-curves.

A Bezier-curve is a segmented curve. The segments $X_l(u)$; $l = 0, \dots, k$ of a Bezier-curve of degree m over the parameter intervall $u_l \leq u \leq u_{l+1}$ are:

$$X_l(u) := \sum_{i=0}^{m} b_{l \cdot m+i} \cdot B_i^m\left(\frac{u-u_l}{u_{l+1}-u_l}\right) \tag{1.1}$$

The Bernstein polynomials $B_i^m(t) := \binom{m}{i}(1-t)^{m-i}t^i$; $0 \leq t \leq 1$ are used as blending functions.

Bernstein polynomials are special degenerated B-Splines (see [**Boehm, Farin, Kahmann, 84**]). If we use B-Splines as blending functions instead of Bernstein polynomials, we can generalize the whole concept to so called B-Spline curves and surfaces (see [**Gordon-Riesenfeld, 74**]). B-Spline curves are similar to Bezier curves in that a set of blending functions combine the effect of $n+1$ control points

$$Y(u) := \sum_{j=0}^{n} d_j \cdot N_{j,M}(u) \tag{1.2}$$

The most important difference is the local support property of the B-Spline blending functions $N_{j,k}(u)$. Both curve-typs have the convex hull and the variation diminishing property (for more details see [**Boehm et al., 84**]).

* this research was supported by a NATO Collaborative Research Grant under contract number 0097/88

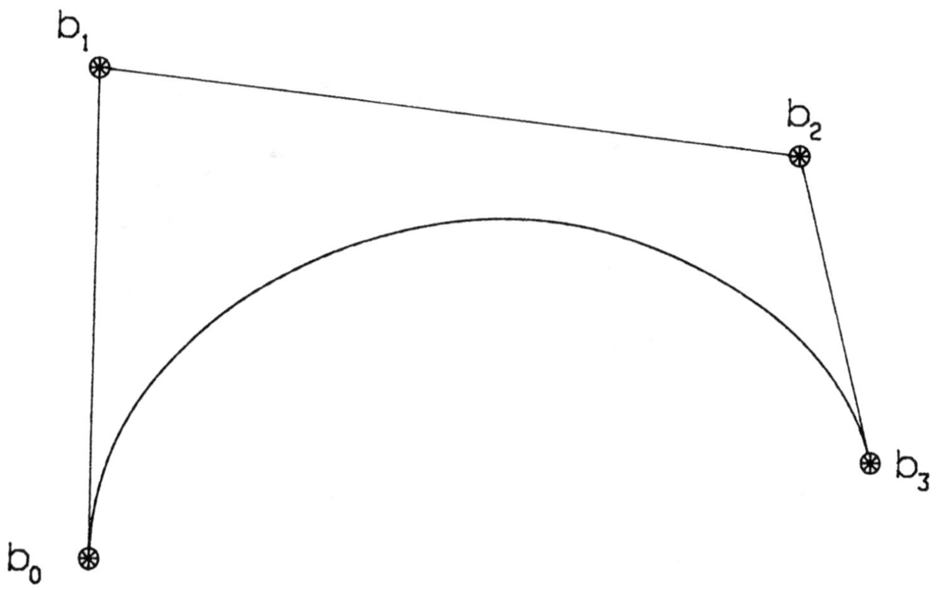

Figure 1: Bezier curve

Bezier- and B-Spline curves can be contructed by repeated linear interpolation.

(1.3) Bezier-segment-algorithm (de Casteljau)

Input: M = order of the curve; interval $[a, b]$;
control polygon $\{b_0^{(0)}, \ldots, b_m^{(0)}\}$, $m := M - 1$

$FOR\ p := 0\ TO\ N\ DO$
 $BEGIN$
  $t := \frac{p}{N} \cdot (b - a) + a$
  $FOR\ k := 1\ TO\ m\ DO$
   $BEGIN$
    $FOR\ j := k\ TO\ m\ DO$
     $BEGIN$
      $b_j^{(k)} := \frac{t-a}{b-a} \cdot b_j^{(k-1)} + \left(1 - \frac{t-a}{b-a}\right) \cdot b_{j-1}^{(k-1)}$
     END
   END
  $X(t) := b_m^{(m)}$
 END

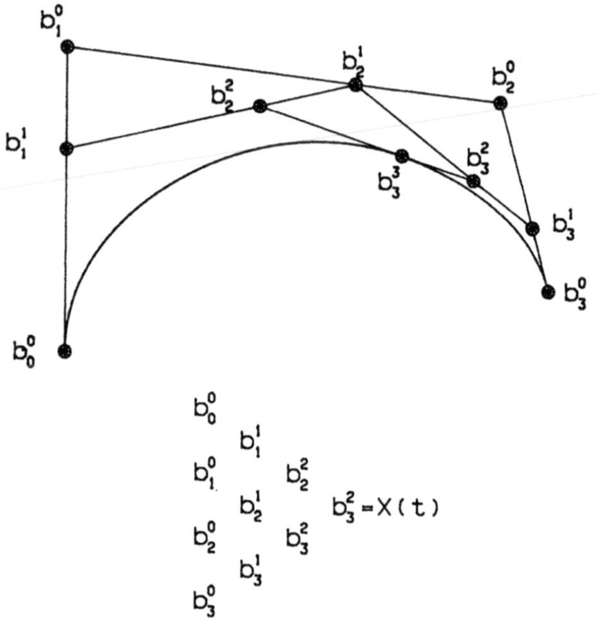

Figure 2: Bezier-segment-algorithm (de Casteljau algorithm)

(1.4) B-Spline-algorithm (de Boor)

Input: M = order of the curve;
control polygon $\{d_0^{(0)}, \ldots, d_n^{(0)}\}$, $n \geq M-1$
KV (knot vector) open or closed

IF $KV =$ closed
  THEN
    $\{x_0, \ldots, x_{n+1}\} := (i - M \text{ div } 2) \bmod (n+1)$
  ELSE
    $\{x_0 \ldots, x_{n+M}\} := \{\underbrace{0, \ldots, 0}_{M}, 1, 2, \ldots, \underbrace{n-M+2, \ldots, n-M+2}_{M}\}$
FOR $p := 0$ *TO* N *DO*
  BEGIN
    $s := \frac{p}{N} \cdot A\,;\quad A := n+1 \text{ or } A := n-M+2$
    FOR $k := 1$ *TO* $M-1$ *DO*
      BEGIN
        $i \leq s$
        FOR $j := i-M+k+1$ *TO* i *DO*
          BEGIN
            $d_j^{(k)} := \frac{s-x_j}{x_{j+M-k}-x_j} \cdot d_j^{(k-1)} + \left(1 - \frac{s-x_j}{x_{j+M-k}-x_j}\right) \cdot d_{j-1}^{(k-1)}$
          END
      END
    $Y(s) := d_i^{(M-1)}$
  END

A Bezier-surface is a segmented surface. The segments $X_{pq}(u,w)$; $p = 0,\ldots,k$; $q = 0,\ldots,r$ of a Bezier surface of degree m, n over the rectangular paramienter domain $u_p \le u \le u_{p+1}$; $w_q \le w \le w_{q+1}$ are:

$$X_{pq}(u,w) := \sum_{i=0}^{m}\sum_{j=0}^{n} b_{p\cdot m+i/q\cdot n+j} \cdot B_i^m\left(\frac{u-u_p}{u_{p+1}-u_p}\right) \cdot B_j^n\left(\frac{w-w_q}{w_{q+1}-w_q}\right) \tag{1.5}$$

Instead of a control polygon a Bezier-surface (-segment) has a control polyhedron.

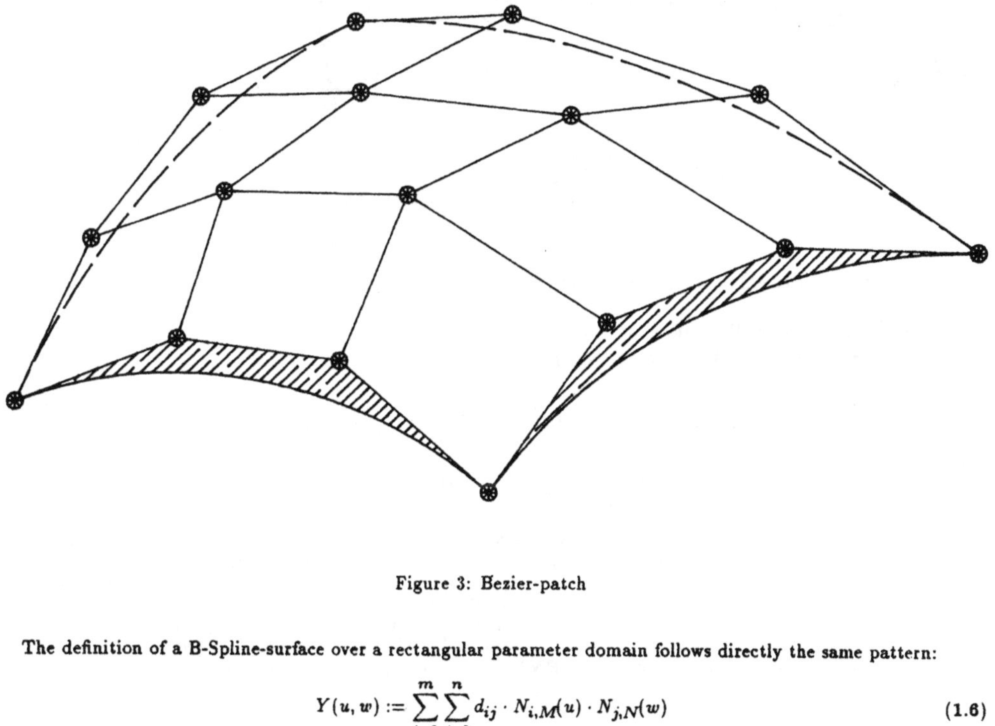

Figure 3: Bezier-patch

The definition of a B-Spline-surface over a rectangular parameter domain follows directly the same pattern:

$$Y(u,w) := \sum_{i=0}^{m}\sum_{j=0}^{n} d_{ij} \cdot N_{i,M}(u) \cdot N_{j,N}(w) \tag{1.6}$$

These so called tensor-product-surfaces can be easily generated by applying the de Casteljau algorithm or the de Boor-algorithm twice.

A triangular Bezier patch is defined by

$$X(u,w) := \sum_{I} b_{i,j,k} \cdot B_{i,j,k}^n(r(u,w), s(u,w), t(u,w)) \tag{1.7}$$

where r, s, t are local barycentric coordinates of the triangular parameter domain and "I" denotes summation over all $i, j, k \ge 0$, $i + j + k = n$. The $B_{i,j,k}^n$ are generalized Bernstein polynomials of degree n given by:

$$B_{i,j,k}^n(r,s,t) := \frac{n\,!}{i\,!\,j\,!\,k\,!} \cdot r^i \cdot s^j \cdot t^k$$

These Bernstein polynomials have properties very much like the univariate ones. Therefore they are appropriate blending functions (see [**Farin, 79**]).

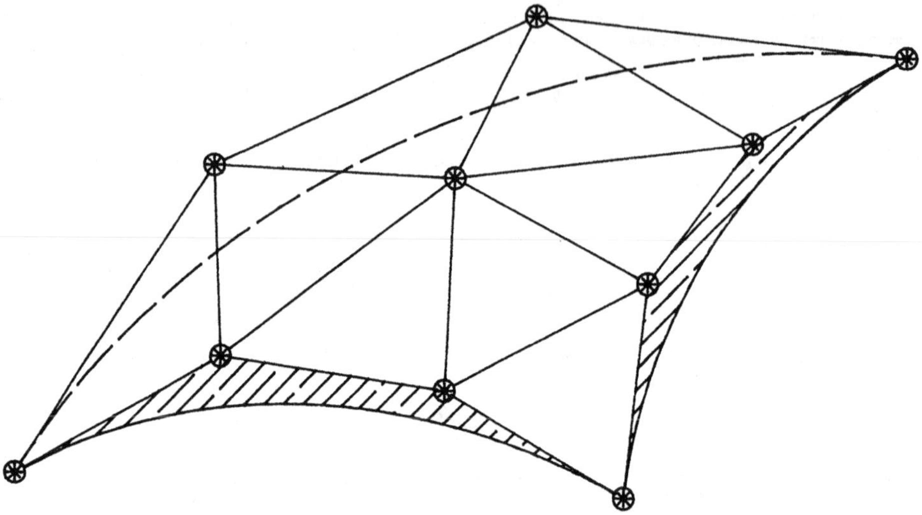

Figure 4: triangular Bezier-patch

(1.8) Triangular-Bezier-patch-algorithm (Farin)

Input: M = order of the surface; $m := M - 1$
$\frac{(m+1)\cdot(m+2)}{2}$ control polygon $\{b_{i,j,k}\}$

$FOR\ l := 1\ TO\ m\ DO$
$\quad BEGIN$
$\quad\quad FOR\ j := 0\ TO\ m - l\ DO$
$\quad\quad\quad BEGIN$
$\quad\quad\quad\quad FOR\ i := 0\ TO\ m - l - j\ DO$
$\quad\quad\quad\quad\quad BEGIN$
$\quad\quad\quad\quad\quad\quad k := m - l - i - j$
$\quad\quad\quad\quad\quad\quad b^{(l)}_{i,j,k} := u \cdot b^{(l-1)}_{i+1,j,k} + v \cdot b^{(l-1)}_{i,j+1,k} + w \cdot b^{(l-1)}_{i,j,k-1}$
$\quad\quad\quad\quad\quad END$
$\quad\quad\quad END$
$\quad END$

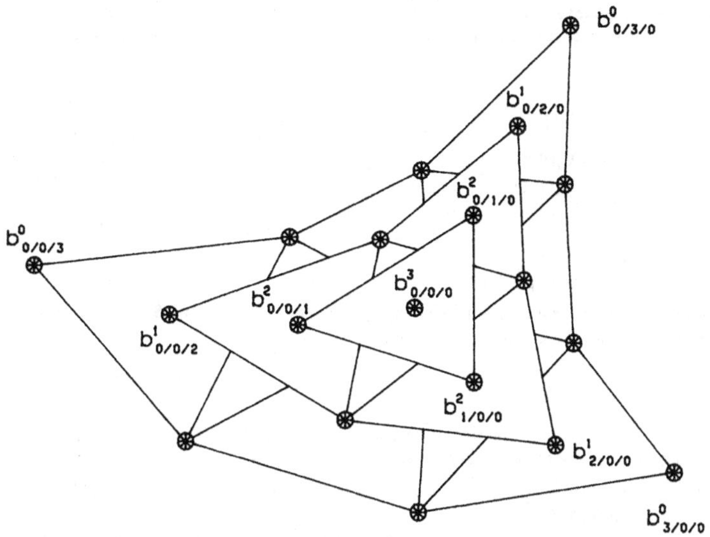

Figure 5: triangular Bezier-patch algorithm

The boundary control points of these patch-types can be supplied by a smooth network of curves. In chapter 3 we present a method how to specify the "inner control points" of each patch in an efficient way to generate smooth surfaces.

(2) Generalized Gordon-Coons Patches

This surface-patch-technique was introduced by **S. Coons** (see [**Coons, 64**]) and **W. Gordon** pointed out (see [**Gordon, 69**]), that such a patch can be written as a boolean sum of projectors, which are themselves interpolants to lower dimensional information. A bicubic Coons' patch can be written as a Boolean sum of two projectors:

$$P_1 := \sum_{i=0}^{1} X(i,w)\cdot H_i(u) + X_u(i,w)\cdot \tilde{H}_i(u)$$

$$P_2 := \sum_{j=0}^{1} X(u,j)\cdot H_j(w) + X_w(u,j)\cdot \tilde{H}_j(w)$$

$(X_u := \frac{\partial X}{\partial u}, \ldots)$ H_i, $\tilde{H}_i$ are the cubic Hermite basis functions. It is easy to generalize this concept to the biquintic case:

$$P_1 := \sum_{i=0}^{1} X(i,w)\cdot H_i(u) + X_u(i,w)\cdot \overline{H}_i(u) + X_{uu}(i,w)\cdot \overline{\overline{H}}_i(u)$$

$$P_2 := \sum_{j=0}^{1} X(u,j)\cdot H_j(w) + X_w(u,j)\cdot \overline{H}_j(w) + X_{ww}(u,j)\cdot \overline{\overline{H}}_j(w)$$

H_i, $\overline{H}_i$, $\overline{\overline{H}}_i$ are the quintic Hermite basis functions.

The position $(X(i,w)$ and $X(u,j))$; the tangent $(X_u(i,w)$ and $X_w(u,j))$ and the second derivative-information $(X_{uu}(i,w)$ and $X_{ww}(u,j))$ can be supplied by a smooth network of curves.

Using this technique, there were two major problems: Appropriate input for the twist vectors (cross partial derivatives) and twist incompatibilities. To remove the incompatibility of the cross partial derivatives we can use either **Gregory's** square (see [**Barnhill, 74**] and [**Barnhill-Gregory, 75a**]) or **Nielson's** convex combination extension of the boolean sum scheme (see [**Nielson, 79**]):

$$X(u,w) := P_1 + P_2 - \alpha P_1 P_2 - \beta P_2 P_1$$

The functions α and β have to fulfill:

$$\alpha + \beta = 1\ ;\quad \alpha,\beta \geq 0\ ;\quad \alpha(i,w) = 1\ ;\quad \beta(i,w) = 0\ ;\quad \alpha(u,j) = 0\ ;\quad \beta(u,j) = 1$$

Since we are interested in curvature input, but not in partial derivatives of curvature functions, we set $X_{uuw} = 0$; $X_{uww} = 0$; $X_{uuww} = 0$. If we replace $X(i,w)$, $X_u(i,w)$, etc. by their quintic Hermite interpolants, we get the following matrix representation of this patch:

$$\begin{bmatrix} H_0(u) \\ H_1(u) \\ \overline{H}_0(u) \\ \overline{H}_1(u) \\ \overline{\overline{H}}_0(u) \\ \overline{\overline{H}}_1(u) \end{bmatrix}^T \cdot \begin{bmatrix} X(0,0) & X(0,1) & X_w(0,0) & X_w(0,1) & X_{ww}(0,0) & X_{ww}(0,1) \\ X(1,0) & X(1,1) & X_w(1,0) & X_w(1,1) & X_{ww}(1,0) & X_{ww}(1,1) \\ X_u(0,0) & X_u(0,1) & \tilde{X}_{uw}(0,0) & \tilde{X}_{uw}(0,1) & 0 & 0 \\ X_u(1,0) & X_u(1,1) & \tilde{X}_{uw}(1,0) & \tilde{X}_{uw}(1,1) & 0 & 0 \\ X_{uu}(0,0) & X_{uu}(0,1) & 0 & 0 & 0 & 0 \\ X_{uu}(1,0) & X_{uu}(1,1) & 0 & 0 & 0 & 0 \end{bmatrix} \cdot \begin{bmatrix} H_0(w) \\ H_1(w) \\ \overline{H}_0(w) \\ \overline{H}_1(w) \\ \overline{\overline{H}}_0(w) \\ \overline{\overline{H}}_1(w) \end{bmatrix} \quad (2.1)$$

In this scheme there are no twist incompatibilities.

The author presented recently a solution for twist input problem (see [**Hagen, 88a**]). The twist vectors $\tilde{X}_{uw} = \alpha X_{uw} + \beta X_{wu}$ are replaced using the Gauß-frame-technique:

$$\tilde{X}_{uw} = <\tilde{X}_{uw}, N> \cdot N + <\tilde{X}_{uw}, X_u> \cdot X_u + <\tilde{X}_{uw}, X_w> \cdot X_w$$

$<,>$ is the scalar product and the scalar functions $<\tilde{X}_{uw}, N>$, $<\tilde{X}_{uw}, X_u>$ and $<\tilde{X}_{uw}, X_w>$ can be supplied by an automatic smoothing process described in chapter 3.

In the last couple of years interpolating triangular patches became more and more important. We give here a short overview, for details see [**Hagen, 88b**].

The first results of blending functions interpolation on triangles were presented by **Barnhill**, **Birkhoff** and **Gordon** (see [**Barnhill et al., 73**]), where rational interpolation functions are derived from Boolean sum combinations. To remove the incompatibility of the cross partial derivatives we can use **Gregory**'s square (see [**Barnhill, 74**] and [**Barnhill-Gregory, 75a**]). Rational blending functions can be replaced by polynomial blending functions (see [**Barnhill-Gregory, 75a**]).

Convex combination schemes were introduced in geometric modeling by **Gregory** (see [**Gregory, 74**] and [**Gregory, 78**]) and generalized by **Gregory** and **Charrot** (see [**Gregory-Charrot, 80**] and [**Gregory, 83**]). Using convex combination schemes with rational Hermite projectors there are no twist problems. The methods mentioned above are based upon the combination of interpolation operators consisting of univariate interpolation along lines parallel to the sides of a triangle. **Nielson** presented a side vertex method for interpolation in triangles, where the fundamental operators consist of univariate interpolation along lines joining a vertex and his opposing side (see [**Nielson, 79**]). The author presented 1986 a convex combination scheme for interpolating function values, derivatives and curvature values on the boundary of arbitrary triangles (see [**Hagen, 86**]): This method is based upon a geometric Hermite-operator and is a generalization of Nielson's side vertex method.

(3) Automatic Smoothing

The functional $\int_S (k_1^2 + k_2^2)dS$ (k_1 and k_2 are the principal curvatures of surface S) is a standard fairness criterion for surfaces in engineering (see [**Nowacki-Reese, 83**]), because it is equivalent to the strain energy of flexure and torsion in a thin rectangular elastic plate of small deflection.

Hagen and **Schulze** used this functional for a variation formulation and solution of the surface fairing problem (see [**Hagen-Schulze, 87**]) under the assumption of orthogonal parameter lines, just recently **Farin** and **Hagen** proved without any regularity constraints (see [**Farin-Hagen, 88**]):

A surface is smooth in the sense of

$$\int_S (k_1^2 + k_2^2)dS \longrightarrow \min$$

if

$$h_{12} = \frac{g_{12}(g_{11}h_{22} + g_{22}h_{11})}{g + 2g_{12}^2} \quad (3.1)$$

This gives smooth normal components of twist vectors as an output of an automatic smoothing process. This information can be used to specify the inner control points of Bezier and B-Spline patches and can be used as a curvature input for generalized Gordon-Coons-patches (2.1) and for the triangular patch given in [**Hagen, 86**].

The following hairdryer was constructed using this method.

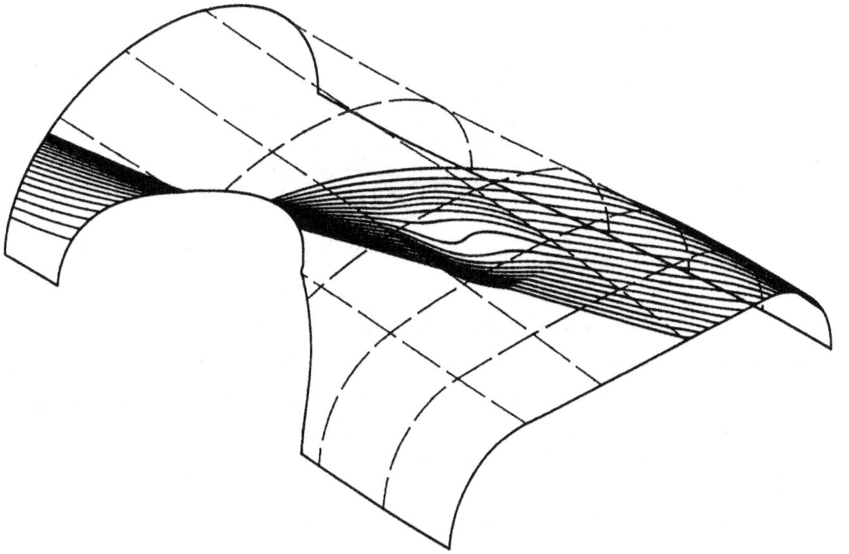

Figure 7: reflection-line-simulation

Another method is to use isophotes, lines of equal light intensity on a surface illuminated by parallel beams. Shadow outlines are special isophotes. A C^r-cont. surface has C^{r-1} cont. isophotes. This is the key for applying isophotes for quality testing of surfaces. This fact can be visualized in even simple cases. The following solid is only C^0-continous and so the isophotes have gaps.

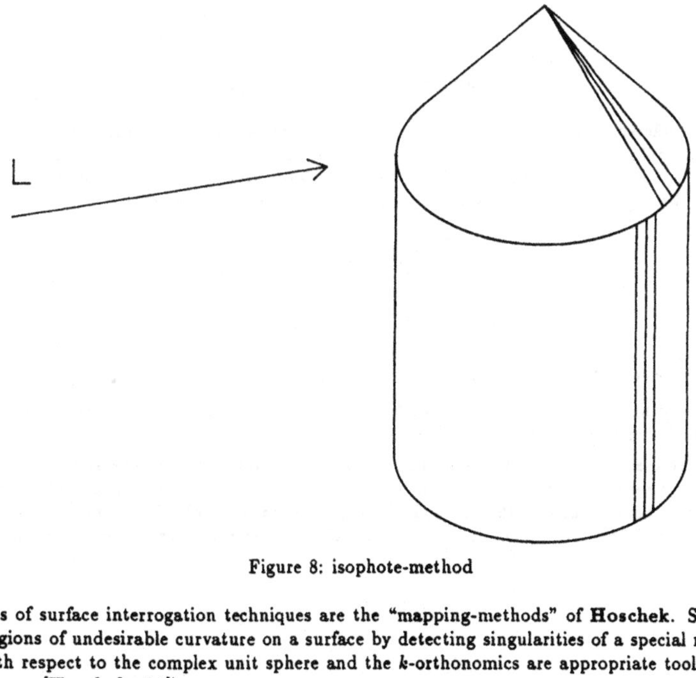

Figure 8: isophote-method

Another class of surface interrogation techniques are the "mapping-methods" of **Hoschek**. Such a method "recognizes" the regions of undesirable curvature on a surface by detecting singularities of a special map of the surface. The polarity with respect to the complex unit sphere and the k-orthonomics are appropriate tools for this approach (for more details see [**Hoschek, 84**]).

It is also possible to use isolines (lines of constant curvature, intersection curves,...) to detect surface irregularities. Such techniques are sucessfully used by **Nowacki** (see [**Hartwig-Nowacki, 82**]).

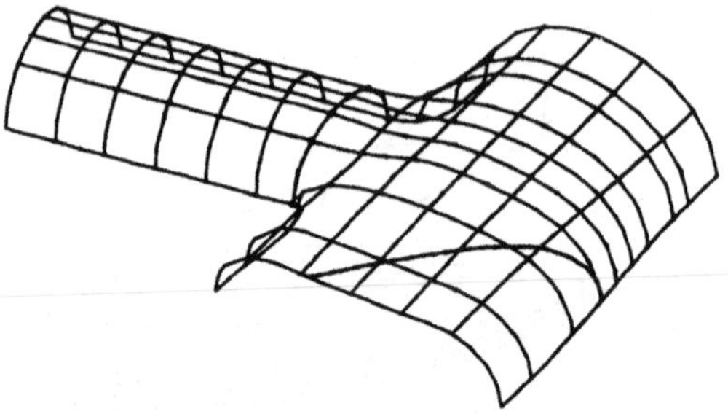

Figure 6. Curvature-plot-analysis

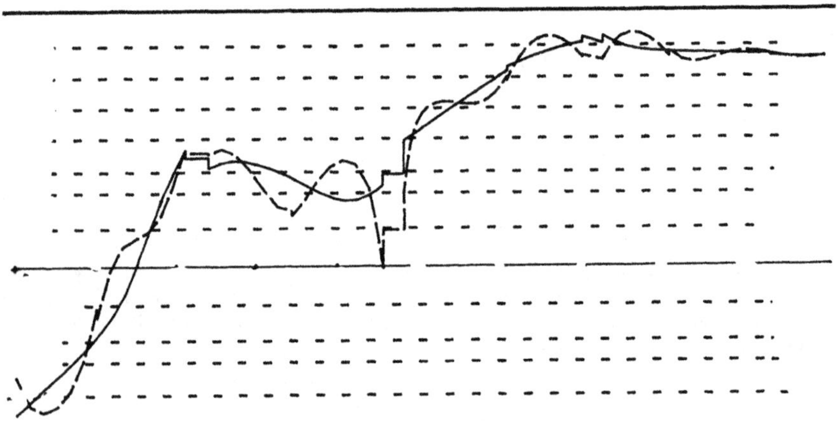

Figure 6: curvature-plot-analysis

To visualize the difference between zero twist input and optimal twist input we intersect with an arbitrary plane and plot the curvature-lines in both cases: you can not see the difference between the two curves by just looking on the surface plot. The dashed curvature-line belongs to the surface curve corresponding to zero twist input and the solid curvature-line belongs to the optimal smoothing parameter h_{12} in the sense of

$$\int_S (k_1^2 + k_2^2)\,dS \longrightarrow \min .$$

(4) Surface Interrogation Methods

The curvature-plot-technique is not the only surface-testing-method. We give here a short overview over surface interrogation methods, for more details see [**Hagen-Hoschek, 88**].

A technique often used in the car industry is the reflection line simulation (see [**Klass, 80**] and [**Farin, 85**]). The reflection pattern of a series of parallel fluorescent light bulbs is simulated on the surface. If this reflection pattern is "nice", then the esthetic quality of the car body is considered satisfactory. It is very helpfull to simulate these reflection patterns during the design phase in order to detect surface irregularities before the surface is manufactured.

References

Barnhill (1974): *Smooth interpolation over triangles* in Barnhill-Riesenfeld (eds.): Computer Aided Geometric Design, Academic Press, New York, p. 45–70

Barnhill-Birkhoff-Gordon (1973): *Smooth interpolation in triangles,* J. Approx. Theory 8, p. 114–128

Barnhill-Gregory (1975a): *Compatible smooth interpolation in triangles* J. Approx. Theory 15, p. 214–225

Barnhill-Gregory (1975b): *Polynomial interpolation to boundary data on triangles,* Math. Comp. 29, p. 726–735

Boehm-Farin-Kahmann (1984): *A survey of curve and surface methods in CAGD,* Computer Aided Geometric Design 1, p. 1–60

Coons (1964): *Surfaces for computer aided design,* Report MAC–TR–4; Project MAC, M.I.T.

Farin (1979): *Subsplines über Dreiecken,* Diss. TU Braunschweig

Farin (1985): *A modified Clough-Tocher interpolant* Computer Aided Geometric Design 2, p. 19–27

Farin-Hagen (1988): *Twist Estimation for Smooth Surface Design,* to be published

Forrest (1972): *Interactive interpolation and approximation by Bezier polynomials* Computer J. 15, p. 71–79

Gordon (1969): *Free-form surface interpolation through curve networks,* GMR–921, GM Research Labs.

Gordon-Riesenfeld (1974): *B-Spline curves and surfaces* in Barnhill-Riesenfeld (eds.): Computer Aided Geometric Design, Academic Press, New York, p. 95–126

Gregory (1974): *Smooth interpolation without constraints* in Barnhill-Riesenfeld (eds.): Computer Aided Geometric Design, Academic Press, New York, p. 71–88

Gregory (1978): *A blending function interpolant for triangles* in Handscomb (eds.): Multivariate Approximation, Academic Press, New York, p. 279–287

Gregory (1983): *C^1 rectangular and non rectangular surface patches* in Barnhill-Boehm (eds.): Surfaces in CAGD, North Holland, Amsterdam, p. 25–33

Gregory-Charrot (1980): *A C^1-Triangular Interpolation Patch for Computer Aided Geometric Design,* Computer Graphics and Image Processing 13. p. 80–87

Hagen (1986): *Geometric Surface Patches without Twist Constraints* Computer Aided Geometric Design 3, p. 179–184

Hagen (1988a): *Generalized Gordon-Coons' Patches,* to be published

Hagen (1988b): *Computer Aided Geometic Design — Methods and Applications* to be published in the proceedings of the conference on Engineering Graphics and Descriptive Geometry, Wien '88

Hagen-Schulze (1987): *Automatic smoothing with geometric surface patches* Computer Aided Geometric Design 4, p. 231–235

Hartwig-Nowacki (1982): *Isolinien und Schnitte in Coonsschen Flächen* in "Geometrisches Modellieren" Inf. Fachb. GS, Springer-Berlin, p. 329–344

Hoschek (1984): *Detecting regions with undesirable curvature* Computer Aided Geometric Design 1, p. 183–192

Klass (1980): *Correction of local surface irregularities using reflection lines* CAD 12/2, p. 73–77

Nielson (1979): *The side vertex method for interpolation in triangles,* J. Approx. Theory 25, p. 318–336

Nowacki-Reese (1983): *Design and Fairing of Ship Surfaces* in Barnhill-Boehm (eds.): Surfaces in CAGD, North Holland, Amsterdam, p. 121–134

Collision Avoidance for Nonrigid Objects

– Extended Abstract –

Stephan Abramowski
Department of Computer Science
University of Karlsruhe

Summary

The path existence problem and the collision detection problem for time-varying objects in a geometric scene are discussed. For a large class of spherical nonrigid objects, exact solutions of the path existence problem are developed based on decomposition techniques and graph traversal.

Keywords

Data Structures and Algorithms, Computational Geometry, Motion Planning, Robotics, Animation.

1 Introduction into Collision Avoidance Problems

Given a set of geometric objects, some of them altering their shape or position over time, a major question is for collisions of any objects in the scene. This question can be asked in several forms, leading to quite different problems. For example in robotics, given two positions of a manipulator's arm, the ***path existence problem*** is to search for the possibility of any motion between the two configurations without collision with the objects in a surrounding working space. Assuming the existence of a path, the ***optimal path detection problem*** is to calculate an optimal feasible path according to certain boundary conditions. A typical criterion is shortest time. The class of paths which is to be searched for an optimal path depends on the kinematic properties of the moving object. In the weakest case, an optimal path may be the geometrically shortest feasible path in the scene.

While for the path existence problem as well as for the path planning problem the question is to ask for a path, the *collision detection problem* assumes a path is already given. The question of the collision detection problem is whether the given path is feasible, i.e. avoids collisions. The collision detection problem has applications in robotics and animation. In both areas, programming languages exist for specifying motions (Magnenat-Thalmann, Thalmann (1984) and Blume, Jakob (1983)). An intelligent compiler or run-time system must be able to detect the feasibility of a specified motion. Another application is in automatic navigation where usually on-line and in real-time the next action of an object must be planned dependent on the current position in the environment. Also, certain optimality conditions may be obeyed in the latter case, for example moving always close to the surfaces of the objects of the scene, or moving in maximum distance. Shortest path problems were discussed by e.g. Sharir, Schorr (1984), Wu, Widmayer, Schlag, Wong (1985). Optimality considerations are excluded in the following, focusing mainly on the path extistence problem.

A very general version of a path existence problem is that of Sharir, Schwartz (1983). Sharir and Schwartz ask for two given sets of linked objects whether they can be deformed into each other. They construct a graph with $O((2k)^{3^{d+1}} \cdot n^{2^d})$ nodes which is searched for a path corresponding to the desired transformation. The feasible object configurations and the obstacles are described by n polynomials in d variables of degree at most k. The problem becomes more simple for only one moving *rigid* object in a given scene of static objects. Well investigated examples of this type are a moving line segment resp. a disk in a polygonal scene of size n in the plane in time $O(n^2 \log n)$ resp. $O(n \log n)$ (O'Dunlaing (1983)), and the translational motion of a convex polygon in a scene of arbitrary polygons in the plane in time $O(n \log^2 n)$ (Kedem and Sharir (1985)). For more complex motions without restriction on translation, several heuristics were developed. The free space approach is representing the free space in the scene by certain simple geometric primitives, for example cones, and moving the object within these primitives (Brooks (1983)). Another approach is discretization. For example, the rotational angle may rastered in equal sized steps. Then, translations are carried out for fixed angles with the possibility to switch from one angle to the next if enough free space for rotation is available (Lozano-Perez and Wesley (1979), Brooks and Lozano-Perez (1983)).

Among the *nonrigid* objects, the arm of a manipulator is the most important one. Planning motions of robot arms is usually done by simplifying them to moving chains of arm segments. By Reif (1979) resp. Hopcroft, Joseph, Whitesides (1985), motion planning of chains of line segments with an arbitrary, but fixed number of joints was shown to be *NP*-hard and *PSPACE*-complete in 3-space resp. 2-space. A consequence of this result is that the degree of the polynomial bounding the complexity of any algorithm is unlimited if the number of line segments grows, assuming $P \neq NP$. In the special case of feasible configurations forced to lie within a circle, the motion planning problem can be solved polynomially with a degree independent of the number of joints.

A path existence problem related to robot arms and its generalizations will be discussed in the following. Our geometric model is a sphere model. In this model, a set of spherical obstacles,

$$S := \{H_j = sp(m_j, r_j) \;:\; m_j \in R^3,\; r_j \in R,\; j = 1, ..., n\}$$

is given. $sp(x, r)$ denotes a sphere of radius r with center x. The moving object is a "worm" consisting of a sequence of spheres. The shapes of a worm have to satisfy the condition that the distance of two consecutive spheres is fixed, i.e.

$$d(P_i, P_{i+1}) = l_i,\; i = 0, ..., m-1, \tag{1}$$

$$P_i \text{ the center of sphere } i,\; l_i \in R \text{ fixed},\; i = 1, ..., m.$$

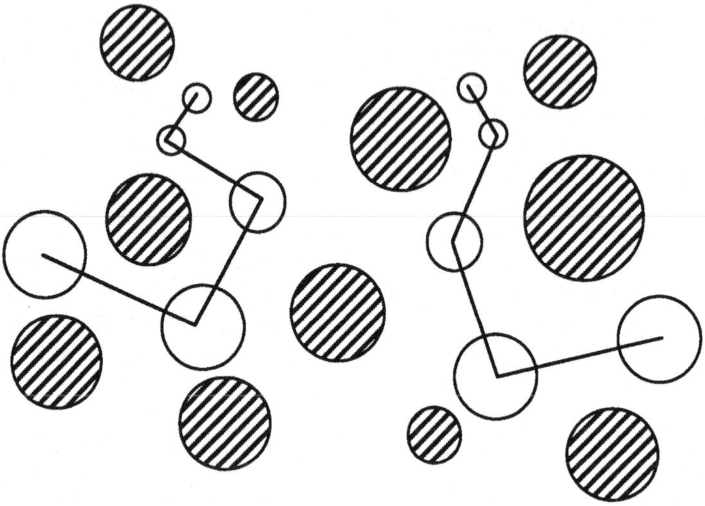

Figure 1. Two worm configurations in a scene of circles.

d is the Euclidian distance. A configuration of a worm in space is *feasible* if it does not intersect any objects in S, i.e.

$$d(P_i, m_j) \geq d_i + r_j, \ i = 0, ..., m, \ j = 1, ..., n. \tag{2}$$

d_i is the radius of the i-th sphere of the worm. Self-penetration of a worm is not forbidden, in order to simplify the discussion in the following.

Figure 1 shows an initial and a final 2-space configuration of a worm in a scene of disks.

Sphere models like ours are quite useful in applications. By Badler and Smoliar (1979), a sphere model of the human body was designed for computer animation. Sphere models are sometimes used as approximations by the robotics community leading to relatively efficient and easily implementable algorithms. In chemistry, atoms are traditionally represented by spheres. Motion planning problems as they may arise with molecular modeling have an immediate interpretation in the sphere model.

In the next section an exact solution of the path extistence problem is sketched for worms in a world of spheres, like defined above. "Exact" means that a path is found if and only if there is indeed a path. This differs from the heuristics usually applied in practice which do not guarantee this fact.

2 Path Existence or Worms

The path existence problem for worms is solved by constructing a graph composed by cell products. The cells are derived from a complex decomposition of the obstacle space. There is a (graph-theoretic) path between two nodes representing an initial and a final worm configuration in this so-called cell-configuration-graph iff there is a feasible (geometric) path between these two worm configurations. The solution is totally carried out in the space of obstacles. This avoids the explosion of the dimension inherent to algorithms proposed for related versions of the problems. The explosion of the dimension is a consequence of the usage of the so-called configuration space. For example, this space is constructed over the angles describing the relative or absolute orientations of the links of a nonrigid object.

Let P_i be the center of the i–th sphere of a worm in some position, $i = 1, ..., m$. Let be F_i the set of all feasible positions of P_i. For example,

$$F_1 := R^d - \bigcup_{j=1}^{n} sp(m_j, r_j + d_1), \quad d \in N. \tag{3}$$

In general, F_i consists of all points x for which there is a point in F_{i-1} with distance l_{i-1} to x, and x is inside none of the spheres $sp(m_j, r_j + d_i)$. Formally,

$$F_i := Disp_{l_{i-1}}(F_{i-1}) - \bigcup_{j=1}^{n} sp(m_j, r_j + d_i), \; i = 2, ..., m, \tag{4}$$

$$Disp_r(M) := \{x \; : \; \exists y \in M \text{ with } d(x,y) = r\}, \; M \subset R^d, \; r \in R, \; r \geq 0.$$

We are now interested in the subsets of $F = F_1 \times ... \times F_m$ corresponding to feasible configurations of the worm. For simplicity, the following discussion is restricted to the plane. Higher dimensions can be treated analogously.

In the following we always assume that our workspace carries the Euklidean metric and the corresponding topology.

We want to identify sets of feasible worm configurations such that two configurations in the same set can easily be transformed into each other via some feasible path. These sets of worm configurations are called configuration classes. They are essentially given as m-tuples $(c_1, \ldots, c_m)$ of cells c_i such that P_i is allowed to vary over c_i. It turns out that in the plane it is sufficient to consider cells which are bounded by circular arcs. The configuration classes are built upon so called segment classes, i.e. sets of feasible segments (P_i, P_{i+1}) which can be easily transformed into each other via some feasible path.

2.1 Segments

In the following we define segment classes and examine the neighbored segment classes. Then we introduce configuration classes and examine neighbored configuration classes.

Let be $1 \leq i < m$; a **segment class** is a subset of R^4 given as an ordered pair $(\check{c}_i, c_{i+1})$ such that

1. For $\check{c}_i$ there is a cell c_i of a partition of the plane induced by some set $M_i = M_i(c_i)$ of circles, where M_i contains the set of all obstacle circles blown up by d_i, i.e. $M_i \supseteq \{\partial sp(m_j, r_j + d_i) : 1 \leq j \leq n\}$. c_i must not be contained in any of the (blown up) obstacle circles.

2. c_{i+1} is a cell of a partition of the plane induced by some set M_{i+1} of circles, where M_{i+1} is a superset of the obstacle circles (blown up by d_{i+1}) and the offset $off_{l_i}(c_i)$ of cell c_i

$$M_{i+1} \supseteq off_{l_i}(c_i) \cup \{\partial sp(m_j, r_j + d_{i+1}) : 1 \leq j \leq n\}$$

The offset is given by

$$\begin{array}{rcll} off_l(c_i) & = & \{\partial sp(m, |r+l|) : & \partial sp(m,r) \in M(c_i) \ \} \ \cup \\ & & \{\partial sp(m, |r-l|) : & \partial sp(m,r) \in M(c_i) \ \} \ \cup \\ & & \{\partial sp(x, r) : & x \in V(c_i)\} \end{array}$$

$M(c_i)$ is the set of all circles "supporting" cell c_i, i.e. all circles of M_i which intersect the border of c_i. $V(c_i)$ is the set of nodes of c_i (i.e. the set of endpoints of the circular arcs bordering c_i). c_{i+1} must not be contained in any of the (blown up) obstacle circles.

3. By equation 1, P_i is constrained to lie on circle $\partial sp(P_{i+1}, l_i)$. By the above constraints, the decomposition of $\partial sp(P_{i+1}, l_i)$ does not change combinatorially as P_{i+1} varies over c_{i+1}; this decomposition consists of circular arcs and intersection points (1– and 0–dimensional cells). $\check{c}_i$ is a cell of this decomposition with $\check{c}_i \subseteq c_i$.

4. the segment class $(\check{c}_i, c_{i+1})$ is the subset of R^4 containing exactly the points (P_i, P_{i+1}) with $P_{i+1} \in c_{i+1}$ and $P_i \in \check{c}_i$.

See figure 2 for an example of a segment class.

Note that it is easy to feasibly transform segments of the same segment class into each other: simply find a path $p_{i+1}(t)$ for P_{i+1} from its start to its goal position, such that $p_{i+1}(t)$ does not leave the cell c_{i+1}, and construct a corresponding path $p_i(t)$ for P_i such that $p_i(t)$ does not leave its circular cell $\check{c}_i$.

Our next step is to consider feasible motions of segments which induce a change of the segment class. Segments are considered as points in R^4. It is sufficient to consider the case of a segment moving from the (relative) interior of its segment class to the border of the segment class. We consider segment classes $\check{c} = (\check{c}_i, c_{i+1})$ and $\check{c}' = (\check{c}'_i, c'_{i+1})$ whose cells are induced by the same sets M_i and M_{i+1} of circles. In this case, the segment classes either have nonempty intersection or one segment class is contained in the closure of the other. We examine the case that $\check{c}'$ is contained in the closure $cl(\check{c})$ of $\check{c}$. If $c_{i+1} = c'_{i+1}$, then we have $\check{c}' \subseteq cl(\check{c})$ iff $\check{c}'_i$ is one of the two arc cells neighboring $\check{c}_i$ in the decomposition of $\partial sp(c_{i+1}, l_i)$ by c_i. The other case is $c'_{i+1} \subseteq \partial c_{i+1}$. In order to treat this case, consider the cell decomposition of $\partial sp(x, l_i)$ for $x \in c_{i+1}$. The decomposition can be computed by intersecting $\partial sp(x, l_i)$ with all circles in M_i and consists of 0-dimensional cells (intersection points q_j) and 1-dimensional cells (circular arcs (q_j, q_{j+1}) excluding the endpoints). Obviously the cell decomposition can be represented by the tuple $(q_j)_{j=1...N}$ of intersection points ordered (e.g.) clockwise. Moving center x from c_{i+1} to the bordering cell c'_{i+1}, successive points q_j, q_{j+1}

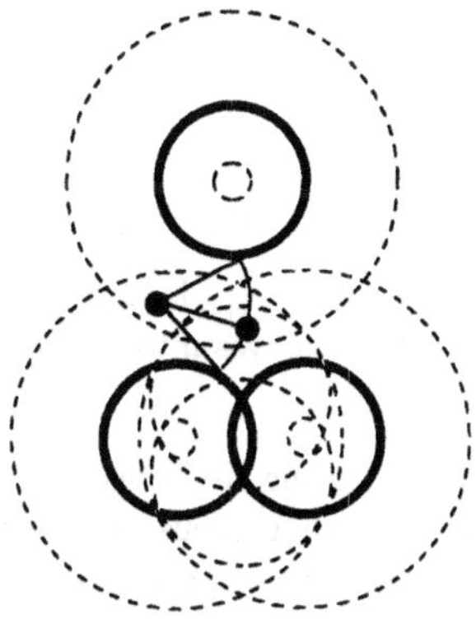

Figure 2.

Figure 2. A segment class. When P_{i+1} varies over one of the cells c_{i+1} drawn in broken line, the decomposition of $\partial sp(P_i, l_i)$ by the circles drawn in fat solid line into circular arc cells $\check{c}_i$ does not change combinatorially.

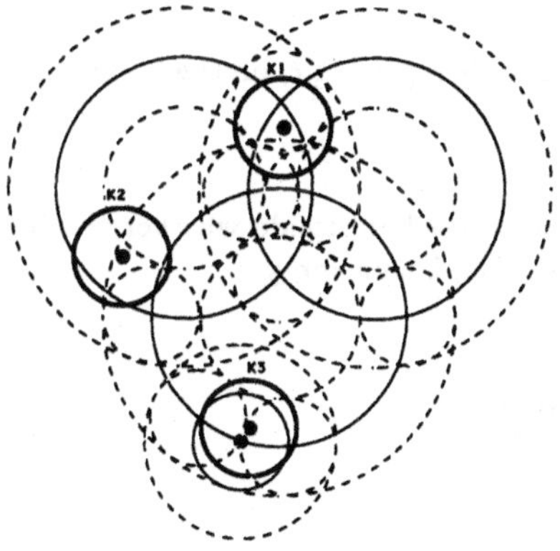

Figure 3.

Figure 3. The four cases of cell switching: The cells c_i are drawn in normal solid line; the cells c_{i+1} are drawn in broken line; $\partial sp(x, l)$, $x \in c_{i+1}$ are drawn in fat solid line (K_1, K_2, K_3 for three sample positions of x). Case 1: K_1 has four intersection points with c_i-cells; when K_1 moves south, the upper two intersection points do not change. Case 2: when K_2 moves south–east, its lower two intersection points merge into one. Case 3: when K_1 moves south, its lower two intersection points get a third point in between. Case 4: when K_3 moves south, two of its four intersection points vanish.

1. do not change, or
2. merge into one intersection point, or
3. are replaced by a sequence $(q'_{j,0}, q'_{j,1}, \ldots, q'_{j,M})$ with $q_j = q'_{j,0}$ and $q_{j+1} = q'_{j,M}$.
4. simply vanish; this can only happen when c'_{i+1} is just a point and $\partial sp(c'_{i+1}, l_i)$ is in M_i; the two points q_j, q_{j+1} vanish if center x moves into point cell c'_{i+1}.

Figure 3 shows circles K_1, K_2, K_3 demonstrating these cases. In all cases we can clearly define which cells of $\partial sp(c'_{i+1}, l_i)$ are **replaced** by which cells of $\partial sp(c_{i+1}, l_i)$. For example, consider case 3: the point cells q_j and q_{j+1} are replaced by $q'_{j,0}$ resp. $q'_{j,M}$ (i.e. by themselves) and the arc cell (q_j, q_{j+1}) is replaced by all point cells $q'_{j,k}$ for $0 < k < M$ and all arc cells $(q'_{j,k}, q'_{j,k+1})$ for $0 \leq k < M$.

When moving from c'_{i+1} to c_{i+1}, i.e. when moving to a neighboring cell of higher dimension, we have the inverse replacements.

Based on these observations it should now be obvious that $\check{c}' \subseteq cl(\check{c})$ iff

$c'_{i+1} = c_{i+1}$ and $\check{c}'_i$ is one of the two arc cells neighboring $\check{c}_i$ in the decomposition of $\partial sp(c_{i+1}, l_i)$ by c_i.

or

$c'_{i+1} \subseteq \partial c_{i+1}$ and $\check{c}'_i$ is one of the cells replacing $\check{c}_i$ on the boundary of the moving circle during transition from c_{i+1} to c'_{i+1}

The conditions given above can be used to decide whether the intersection of the closure of arbitrary segment classes $cl(\check{c}) \cap cl(\check{c}')$ is nonempty.

2.2 Worm configurations

2.2.1 Configuration classes

Based upon segment classes we now define configuration classes of worm configurations which can be easily transformed into each other.

Let H_{d_i} be the partition of the plane induced by the blown-up obstacle circles $\{\partial sp(m_j, r_j + d_i) : 1 \leq j \leq n\}$. In the following we have to deal with partitions D of the plane induced by sets M of circles. M is called the supporting set of circles of D. A superposition of two partitions D and D' is the partition induced by the union of the supporting sets of D and D'. D is a refinement of D' if the supporting set of D contains the supporting set of D'.

A **configuration class** is a subset of R^{2m} identified by an m-tuple $\check{c}$ such that there is an m-tuple D with

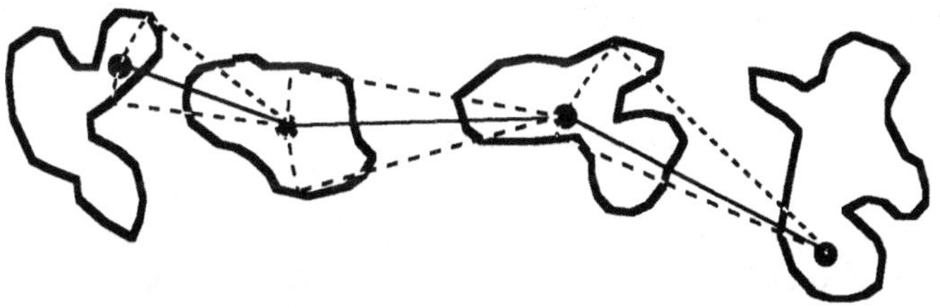

Figure 4. A configuration class $\check{c}$. The cells c_i are drawn in fat solid line, the circular arc cells $\check{c}_i$ are drawn in broken line. The figure shows a configuration $P = (P_1, \ldots, P_m) \in \check{c}$. We have $P_i \in c_i$ and $(P_i, P_{i+1}) \in (\check{c}_i, c_{i+1})$.

$\check{c}$	$=$	$(\check{c}_1, \ldots, \check{c}_m)$
D	$=$	$(D_1, \ldots, D_m)$
D_i	$=$	is a partition of the plane induced by a set of circles
D_1	$=$	H_{d_1}
D_{i+1}	$=$	a refinement of the superposition of the partition induced by $off_{l_i}(c_i)$ and $H_{d_{i+1}}$
c_i		is a cell of D_i which is not contained in any of the obstacle circles blown up by d_i $(1 \leq i < m)$
$\check{c}_m$	$=$	c_m
$\check{c}_i$		is a cell of the partition of $\partial sp(c_{i+1}, l_i)$ by the supporting circles of D_i $(1 \leq i < m)$

The configuration class $\check{c}$ consists of all worm configurations $w = (P_1, \ldots, P_m) \in R^{2m}$ for which the segment (P_i, P_{i+1}) is element of the segment class $(\check{c}_i, c_{i+1})$ $(1 \leq i \leq m)$.

See figure 4 for an example of a configuration class.

In (Abramowski, Müller 1988) a similiar approach is presented. The essential difference is, that here we allow different partitions D_i for different configuration classes, which yields a drastic improvement of the complexity of the algorithm.

We are especially interested in configuration classes for which partition D_{i+1} is only the superposition of the offset $off_{l_i}(c_i)$ and $H_{d_{i+1}}$ and no proper refinement of this superposition. Such configuration classes are called **maximal configuration classes.**

Every feasible worm configuration w is contained in exactly one maximal configuration class: c_1 is the cell of the partition induced by obstacle circles with $P_1 \in c_1$ and c_{i+1} is the cell induced by the obstacles and the offset $Disp_{l_i}(c_i)$ of c_i. Remember that the offset is defined such that if P_{i+1} varies over c_{i+1} the decomposition of $\partial sp(P_{i+1}, l_i)$ by c_i is combinatorially invariant, so the circular arc cell $\check{c}_i$ containing P_i is well defined.

Worm classes which are not maximal are only needed as intermediate stages when moving from one maximal configuration class to another.

Two worm configurations which belong to the same configuration class can be feasibly transformed into each other: find a path $p_m(t)$ for P_m from its start– to its goal–position such that $p_m(t)$ remains inside its cell c_m; given path $p_{i+1}(t)$ compute path $p_i(t)$ for P_i such that $p_i(t)$

remains inside its circular arc cell $\check{c}_i$.

2.2.2 Changing the configuration class

Now we examine how to move to a neighbored configuration class. Consider a feasible path $p(t)$, which transforms a start configuration s into a goal configuration g; $p(t)$ is a continous function $p(t) : [0,1] \to R^{2m}$ such that $p(0) = s$ and $p(1) = g$. Remember that for every t the worm configuration is contained in exactly one maximal configuration class. At some values t_i, $p(t)$ moves from its current maximal configuration class $\check{c}$ to a neighbored class $\check{c}'$. Because of the continuity of $p(t)$ we have $cl(\check{c}) \cap cl(\check{c}') \cap F \neq \emptyset$ (remember F is the set of feasible worm configurations).

Theorem 1 *If $p(t)$ is a feasible path of a worm configuration such that $\check{c}$ and $\check{c}'$ are consecutive elements of the sequence of maximal worm classes traversed by $p(t)$, then $cl(\check{c}) \cap cl(\check{c}') \cap F \neq \emptyset$.*

On the other hand, if $cl(\check{c}) \cap cl(\check{c}') \cap F \neq \emptyset$, then we can find a path $p(t)$ such that in the sequence of maximal configuration classes traversed by $p(t)$, $\check{c}$ and $\check{c}'$ are consecutive elements or separated by at most one further maximal configuration class.

Theorem 2 *Let $\check{c}, \check{c}'$ be maximal configuration classes. Then $cl(\check{c}) \cap cl(\check{c}') \cap F \neq \emptyset$ if and only if there is a nonempty configuration class $\check{c}''$ such that $(\check{c}''_i, c''_{i+1}) \subseteq cl((\check{c}_i, c_{i+1})) \cap cl((\check{c}'_i, c'_{i+1}))$ $(1 \leq i < m)$.*

We omit the proofs here. Theorem 2 allows to reduce the examination of the neighborhood of maximal configuration classes to the examination of the neighborhood of certain segment classes. Note that the configuration class $\check{c}''$ in the intersection of $cl(\check{c})$ and $cl(\check{c}')$ is not maximal in general.

2.2.3 The configuration graph

In this section the motion planning problem for worm configurations is reduced to the problem of searching a graph, the configuration graph. The nodes of this graph are the maximal configurations classes; If $\check{c}, \check{c}'$ are two maximal configuration classes, $\{\check{c}, \check{c}'\}$ is an edge if and only if $cl(\check{c}) \cap cl(\check{c}') \cap F \neq \emptyset$ (cf. theorem 1).

Theorem 3 *Given a start configuration s and a goal configuration g, there is a feasible path $p(t)$ transforming s into g if and only if the maximal configuration classes of s and g are in the same connected component of the configuration graph.*

Proof: (Sketch) If the maximal configuration classes of s and g are in the same connected component of the graph, then by theorem 1 there is a feasible path transforming s into g. On the other hand, if there is such a path $p(t)$, then for any pair $\{\check{c}, \check{c}'\}$ of maximal configuration classes successively traversed by $p(t)$, $\{\check{c}, \check{c}'\}$ is by construction an edge of the graph. $\square$

2.2.4 Complexity analysis

Theorem 4 *Let n be the number of obstacle circles, m the number of joints of the worm configuration.*

1. *The configuration graph (and its connected components) can be computed in* $O(2^{O(m^2)}n^{9m} \cdot \log n)$ *time using* $O(2^{O(m^2)} \cdot n^{6m})$ *space.*

2. *Having computed the configuration graph, the nodes (maximal configuration classes) corresponding to the start– and the goal–configuration have to be computed. This can be done using* $O(2^{O(m)}n \log n)$ *time and* $O(2^{O(m)}n)$ *space.*

3. *Having computed the configuration graph and the nodes corresponding to start– and goal configuration, the problem of the existence of a feasible path can be decided in constant time.*

4. *The configuration graph has* $O(2^{O(m^2)} \cdot n^{3m})$ *nodes and* $O(2^{O(m^2)} \cdot n^{6m})$ *edges. This also gives the maximal length of a path from start node to goal node and the maximal search time.*

5. *It requires* $O(2^{O(m^2)} \cdot n^{4m})$ *elementary motions to transform a start configuration into a goal configuration, if for an elementary motion* P_m *is allowed to move on a circular arc and* P_i *moves on a path compatible with the path of* P_{i+1} $(1 \leq i < m)$.

Proof: (Sketch) Consider the cells c_i of a configuration class $\check{c}$; let M_i be the supporting set of circles of the partition D_i (c_i a cell of D_i). M_1 is induced by the obstacle circles. So $|M_1| = O(n)$; M_{i+1} is obtained by "offsetting" the cell c_i; as c_i is induced by M_i, the size $|c_i|$ of c_i (the number of circular arcs on its boundary) is $O(|M_i|)$. Offsetting c_i enlarges the number of circles involved by some constant factor k' in the worst case, hence we get $|c_i| = O(2^{O(i)}n)$. A set of k circles induces a partition of the plane into $O(k^2)$ cells, so $|D_i| = O(2^{O(i)}n^2)$. Therefore there are $O(2^{O(1)}n^2) \cdot \ldots \cdot O(2^{O(m)}n^2) = O(2^{O(m^2)}n^{2m})$ cell tuples $(c_1, \ldots, c_m)$ of maximal configuration classes. Since c_i decomposes $\partial sp(P_{i+1}, l_i)$ $(P_{i+1} \in c_{i+1})$ into $O(|c_i|)$ circular arc cells, there are $O(|M_1|) \cdot \ldots \cdot O(|M_m|) = O(2^{O(m^2)}n^m)$ maximal configuration classes for a given cell tuple $(c_1, \ldots, c_m)$; so the total number of maximal configuration classes is $O(2^{O(m^2)}n^{3m})$, which is also the bound for the number of nodes of the configuration graph.

It seems to be hard to give a bound on the number of neighbored classes that is considerably better than quadratic (the quadratic bound is given in 4.). The reason is that the neighborhood conditions given in theorem 2 are quite complicated. For the same reason the construction of the edge set is done by checking all pairs $\{\check{c}, \check{c}'\}$ of maximal configuration classes $\check{c}, \check{c}'$ for configuration classes $\check{c}''$ contained in $cl(\check{c}) \cap cl(\check{c}')$. The number of intermediate classes $\check{c}''$ the algorithm may have to check before finding a nonempty $\check{c}'' \subseteq cl(\check{c}) \cap cl(\check{c}') \cap F$ is about as large as the number of maximal cell configurations, so $O(2^{O(m^2)}n^{9m})$ triples $(\check{c}, \check{c}', \check{c}'')$ have to be checked. This can be done withhin the bounds given in 1.

Given a point P and a set of k circles, the cell of the partition induced by the circles that contains P can be found in time $O(k \log k)$ using $O(k)$ space by a slight modification of an algorithm given in (Mehlhorn 1984). By an m–fold application of this algorithm we can construct the configuration class of a given worm configuration, which yields 2. 3. is trivial. 5. is nearly immediate from 4. □

3 Conclusion

We sketched an algorithm for the path planning problem for worm configurations among circular obstacles in the plane. The problem is solved exactly: if there is a path, the algorithm finds a path, otherwise the nonexistence of a path is detected.

We did not present any generalisations such as tree configurations instead of linear worm configurations (which provides a means to reduce the constants involved in the exponents in theorem 4), 3d– instead of 2d–workspace, restrictions posed on link angles, prohibition of self penetration etc. In most cases, these generalisations destroy the simple structure of the algorithm. The partitions of the workspace remain algebraic, but they are no longer induced by circles resp. spheres.

The algorithm avoids the doubly exponential growth (with respect to the degree of freedom of a configuration) of the fully general and much more complex algorithm given in (Schwartz, Sharir 1983). Its complexity is comparable to approximative grid techniques often found in practice and has the advantage of solving the problem exactly. It gives rise to the hope that there might be other specialized versions of the motion planning problem wich can be solved exactly and efficiently.

References

Abramowski S, Müller H (1988) Collision Avoidance for Nonrigid Objects to appear in Zeitschrift für Operations Research (ZOR)

Badler N I, Smoliar S W (1979) Digital Representation of Human Movement. ACM Computing Surveys 11: 19-38

Bentley J L, Ottmann T (1979) Algorithms for Reporting and Counting Geometric Intersections. IEEE Transactions on Computers 28: 643-647

Blume C, Jakob W (1983) Programming Languages for Industrial Manipulators (in German). Vogel-Verlag Würzburg

Brooks R A (1983) Solving the find-Path Problem by Good Representation of Free Space. IEEE Transactions on Systems, Man And Cybernetics: 190-197

Brooks R A, Lozano-Perez T (1983) A Subdivision Algorithm in Configuration Space for Find-path with Rotation. IJCAI: 799-806

Chazelle B (1985) Fast Searching in a Real Algebraic Manifold with Applications to geographic complexity. CAAP'85: 145-156

Hopcroft J, Joseph D, Whitesides S (1984) Movement Problems for 2-dimensional Linkages. SIAM Journal on Computing 13: 610-629

Hopcroft J, Joseph D, Whitesides S (1985) On the Movement of Robot Arms in 2-dimensional bounded Regions. SIAM Journal on Computing 14: 315-333

Kedem K, Sharir M (1985) An Efficient Algorithm for Planing Collisionfree Translational Motion of a Convex Polygonal Object in 2-dimensional Scene Amidst Polygonal Obstacles. 1. ACM Symposium on Computational Geometry: 75-80

Lozano-Perez T, Wesley M A (1979) An Algorithm for Planning Collision-Free Paths Among Polyhedral Obstacles. CACM 22: 560-570

Magnenat-Thalmann N, Thalmann D (1985) Computer Animation: Theory and Practice. Springer-Verlag Berlin

Mehlhorn K (1984) Data Structures and Algorithms III. Springer-Verlag Berlin

O'Dunlaing C, Sharir M, Yap C K (1983) Retraction: A New Approach to Motion-Planning. ACM Symposium on the Theory of Computing: 207-220

Reif J H (1979) Complexity of the Mover's Problem and Generalizations. IEEE FOCS: 421-427

Schwartz J T, Sharir M (1983) On the Piano Movers Problem. II. General Techniques for Computing Topological Properties of Real Algebraic Manifolds. Advances in applied Mathematics 4: 298-351

Sharir M, Schorr A (1984) On Shortest Paths in Polyhedral Spaces. ACM STOC: 144-153

Wu Y F, Widmayer P, Schlag M D F, Wong C K (1985) Rectilinear Shortest Paths and Minimum Spanning Trees in the Presence of Rectilinear Obstacles. IBM Research Report RC 11039(#49019)1/4/85

Yao A C, Yao F F (1985) A General Approach to d-dimensional geometric queries. ACM STOC: 163-168

On the detection of a common intersection of k convex polyhedra

Matthias Reichling *
Rechenzentrum, Universität Würzburg
Am Hubland, D–8700 Würzburg

Abstract

An algorithm is presented for the following problem: Given k convex polyhedra of n vertices each, determine whether they have a point in common. The running time of the algorithm is $O(k \log k \log^3 n)$.

1 Introduction

We consider the following problem:

Given k convex polyhedra of n vertices each, determine whether they have a common intersection.

For $k = 2$ Chazelle and Dobkin [2], [3] developed an algorithm with $O(\log^3 n)$ running time. Dobkin and Kirkpatrick [6] improved this result obtaining $O(\log^2 n)$ running time. A common intersection of k convex n-gons in the plane can be detected in $O(k \log^2 n)$ time [12].

We use the drum representation for polyhedra described in [6]. The *belt* of a drum is the ordered set of faces joining top and bottom face. In the algorithm we assume a preprocessing which allows us to obtain all necessary information about a polyhedron in constant time. All k polyhedra must be preprocessed in the same xyz coordinate system, divided into cross-sections parallel to the xy-plane. Therefore, we can decide in $O(\log n)$ time whether a point lies inside a polyhedron, and the vertex with minimal x-coordinate of a polyhedron can be obtained in $O(\log^2 n)$ time.

2 Preliminaries

Let $L_i = \{ s_1^i \leq s_2^i \leq \ldots \leq s_{n_i}^i \}$, $i = 1, \ldots, k$, be k sorted lists each stored in an array and let $z := \sum_{i=1}^{k} n_i$ be the number of all elements. We search for an "approximative median", i. e. for an element m which satisfies

*Research supported by the Stiftung Volkswagenwerk.

$$| \{ s_j^i \mid s_j^i \in L_i,\ s_j^i \le m \} | \ \ge\ cz$$

and

$$| \{ s_j^i \mid s_j^i \in L_i,\ s_j^i \ge m \} | \ \ge\ cz \qquad (*)$$

for some $c \in (0, \frac{1}{2}]$. In other words, there is a constant part of elements being not less than m and a constant part of elements being not greater than m.

For $z > 2k$ we developed the following algorithm: Only the lists which exceed a minimal size d are processed. Let $d := \lceil z/(2k) \rceil > 1$ and $I := \{ i \mid n_i \ge d \}$. We divide the lists L_i with $i \in I$ into $\lfloor n_i/d \rfloor$ parts of successive elements. Each part contains d elements, the last part between d and $2d-1$ elements. Letting $x := |I|$ and $s := \sum_{i \in I} n_i$ we have

$$s \ge z - (k-x)(d-1) \ge z - (k-x) \cdot \frac{z}{2k} > \frac{z}{2}.$$

For the number of parts we obtain

$$\sum_{i \in I} \left\lfloor \frac{n_i}{d} \right\rfloor \le \sum_{i \in I} \frac{n_i}{d} \le \sum_{i \in I} \frac{2kn_i}{z} \le \frac{2k}{z} \cdot \sum_{i=1}^{k} n_i = 2k$$

and because of $d < z/(2k) + 1 \Rightarrow z > 2k(d-1)$

$$\sum_{i \in I} \left\lfloor \frac{n_i}{d} \right\rfloor \ge \sum_{i \in I} \left(\frac{n_i}{d} - \frac{d-1}{d} \right) = \frac{s}{d} - x\frac{d-1}{d} \ge$$

$$\ge \frac{z-(k-x)(d-1)}{d} - x\frac{d-1}{d} = \frac{z-k(d-1)}{d} > k\frac{d-1}{d} \ge \frac{k}{2}.$$

The median of each of the parts can be determined in constant time (the elements are sorted and stored in an array, therefore we know the median's index). In linear time we can evaluate the median m of the at most $2k$ medians. With this technique m is less (greater) than half of the elements in half of the parts, hence less (greater) than at least $\frac{1}{2} \cdot z/(2k) \cdot \frac{1}{2} \cdot k/2 = z/16$ elements. m satisfies $(*)$ with $c = \frac{1}{16}$. The complexity of the algorithm is linear in k and independent of the numbers n_i.

*

Megiddo [9] introduced a technique for using a parallel algorithm to construct an efficient serial algorithm. In chapter 3 we need a special application hereof which is described below.

We consider the following problem:

Given real numbers a_i, b_i, $i = 1, \ldots, k$, and the lines g_i with $g_i(x) = a_i x + b_i$. Let $f : \mathbf{R} \to \mathbf{R}$ be a continuous function with the following properties: f is monotonous increasing, there exists an unknown $x \in \mathbf{R}$ with $f(x) = 0$, it requires $O(m)$ time to compute f at some point. We want to know which of the lines g_i is the median at a zero of f.

The order of the lines changes only at the $O(k^2)$ intersection points of two of them. Therefore a solution of the problem could be: Determine the x-coordinates s_j of all intersection points of two lines, compute $f(s_j)$ for all j and determine $s_g := \min_j \{ s_j \mid f(s_j) > 0 \}$, $s_l := \max_j \{ s_j \mid f(s_j) \le 0 \}$. The median of the values $g_i(\frac{1}{2}(s_g + s_l))$ gives the right answer. The complexity of the steps is $O(k^2)$, $O(k^2 m)$, $O(k^2)$ and $O(k)$ respectively, hence the algorithm needs $O(k^2 m)$ time.

With Megiddo's technique we can proceed as follows:
The parallel median algorithm of Cole and Yap [5] needs $O((\log\log k)^2)$ time on $O(k)$ processors. We apply this algorithm in a serialized form, hence it runs in $O(k(\log\log k)^2)$ time. During each of the $O((\log\log k)^2)$ phases we have to update the interval containing a zero of f. This requires $O(k + m \cdot \log k)$ time (see [9]). Hence the total complexity is only $O((k + m \cdot \log k)(\log\log k)^2)$.

Recently Cole [4] has improved Megiddo's technique. Applying his method to our problem we obtain $O((k+m)\log k)$ time. (It is considered that $m \ge O(k)$.) However, his algorithm is based on the sorting network of Ajtai-Komlos-Szemeredi [1] which causes a large constant factor.

*

The algorithm of Dobkin and Kirkpatrick [6] detecting an intersection of two convex polygons can be modified to obtain not only a point common to both polygons, but the leftmost point $pq_{\min}$ of the common intersection. Assuming for simplicity that the polygons P and Q do not have edges parallel to the y-axis, we can partition them into upper and lower chains P_u, P_l and Q_u, Q_l by determining the vertices with minimal and maximal x-coordinate (denoted by $p_{\min}, p_{\max}, q_{\min}, q_{\max}$). If $P \cap Q \neq \emptyset$, one of the following holds for $pq_{\min}$:

i) $pq_{\min} = p_{\min}$ iii) $pq_{\min} \in P_u \cap Q_l$
ii) $pq_{\min} = q_{\min}$ iv) $pq_{\min} \in Q_u \cap P_l$.

Especially $pq_{\min} \notin P_u \cap Q_u$ and $pq_{\min} \notin P_l \cap Q_l$.
Cases i) and ii) can be tested in $O(\log n)$ time using Shamos' algorithm [13] for the point-in-convex-polygon problem (no preprocessing necessary). Whether to test case iii) or iv) we can determine from the relative position of the line segments $\overline{p_{\min}p_{\max}}$ and $\overline{q_{\min}q_{\max}}$.
The intersection of an upper and a lower chain is handled similar to [6]. However, we have to choose the correct intersection point if two of them exist. (This can easily be done by examining the slopes of the edges defining the intersection point.) At each step half of the edges of at least one of the chains are eliminated, hence the algorithm runs in $O(\log n)$ time.

To determine the x-minimum of the common intersection of two polyhedra, we can modify the algorithm of [6] for three-dimensional intersections in a similar way (details can be found in [11]). We only state that it is possible in $O(\log^2 n)$ time.

*

Finding a common intersection of k convex polyhedra ($k > 2$) can be done using the following observation: The vertex $v_{\min}$ with minimal x-coordinate of the common intersection is either the x-minimal vertex of one of the polyhedra or the x-minimal vertex of an intersection of two polyhedra. Each of these $O(k^2)$ points can be determined in $O(\log^2 n)$ time. Obviously only their x-maximum is a candidate for $v_{\min}$. We have to test whether this maximal point lies inside all polyhedra, which can be done in $O(k \log n)$ time. Hence, we can determine $v_{\min}$ in $O(k^2 \log^2 n)$ time. It is especially possible in $O(\log^2 n)$ time to decide whether three convex polyhedra have a point in common.

An other approach is the transformation to a three-variable linear program: Each polyhedron is the common intersection of $O(n)$ halfspaces. Therefore our problem is equivalent to finding a feasible point of a linear program in $\mathbf{R}^3$ with $O(kn)$ restrictions. Using the algorithms of Dyer [7] and Megiddo [10] this can be solved in $O(kn)$ time.

3 The algorithm

Our algorithms consists of two parts. Initially for each polygon we determine the drum which must contain the common intersection I if I is not empty. It is possible that a common intersection is already detected in this step. Afterwards we have to search for a common intersection of k drums each containing at most n vertices.

During the algorithm it is necessary to intersect a drum and a plane and to use the resulting polygon as input for a two-dimensional algorithm. The polygon must not be determined explicitly (this would require linear time), but it is considered as an implicitly specified object. Only a constant number of operations is necessary to obtain a special vertex (see [6]).

In the first part of the algorithm we establish upper and lower bounds for the common intersection I. Each bound is a plane parallel to the xy-plane passing through one of the $k \cdot n$ vertices of the polyhedra. Hence it is a plane containing the top or bottom face of a drum. The bounds are updated by eliminating a constant part of the vertices (resp. drums). After $O(\log kn)$ steps we have either proven the existence or non-existence of a common intersection, or between the bounds each polyhedron consists of a single drum.

Initially we determine for each polyhedron the vertices with minimal and maximal z-coordinate. As the first upper bound we take the plane through the maxima's minimum and as the first lower bound the plane through the minima's maximum. (The common intersection I must lie between these two planes.)

Furthermore we have to decide whether a plane P parallel to the xy-plane intersects I or whether I lies above or below P. (Due to convexity one of these three cases must hold.) Using binary search we determine for each polyhedron its drum intersecting P. The intersection is a convex polygon with the vertices implicitly specified. With these k polygons we start an algorithm for detecting a common intersection of k convex polygons [12]. If there exists a common point, we are ready: This point is an element of I. Otherwise there must exist a triple of non-intersecting polygons (this is an immediate consequence of Helly's theorem [8]). The two-dimensional algorithm of [12] can easily be extended to obtain such a triple (see [11]). We consider the three polyhedra P_1, P_2, P_3 which defined the triple of polygons. Obviously the common intersection I of all polyhedra must be contained in the common intersection I_3 of P_1, P_2, P_3. Therefore I lies on the same side of P as I_3. As denoted in chapter 2, we can determine in $O(\log^2 n)$

time a point common to three polyhedra. Hence it is clear on which side of P we have to search for I. If I_3 is empty, of course I is empty, too.

It remains to show how a constant part of the drums can be eliminated. Here we can use the algorithm of chapter 2 to find an approximative median m of the z-coordinates of the vertices. (The vertices in each polyhedron are sorted with respect to their z-coordinates.) At least $\frac{1}{16}$ of all drums lies above resp. below the plane P passing through m. Therefore the number of drums not on the same side of P as the common intersection I is at least $\frac{1}{16}$; these drums can be excluded from further consideration.

Turning to the complexity of this part of the algorithm, we can state the following: Establishing the first bounds takes $O(k)$ time. Testing with the plane P needs $O(k \log n)$ time to find the relevant drum; the algorithm from [12] to detect a common intersection of k convex polygons takes $O(k \log^2 n)$ time. The test with I_3 needs $O(\log^2 n)$ time. $O(\log kn)$ steps are necessary to reduce each polyhedron to a single drum. Therefore in $O(k \log^2 n \log kn)$ time we have found the k drums which must contain I. (Of course the algorithm may stop if one of the planes has intersected I).

*

What we need now is an algorithm detecting a common intersection of k drums. Each of them has at most n vertices; the top and bottom faces are coplanar and have no common intersection.

We proceed as follows: At first we determine for each drum D_i the edges joining the vertices of top and bottom faces with maximal resp. minimal x-coordinate. These edges divide D_i's lateral sides — the belt — into the anterior and the posterior part with the edge sets B_i^a and B_i^p respectively. We now take two planes P_i^l and P_i^r parallel to the y-axis which support the x-extremal edges respectively. Hence, D_i is enclosed by P_i^l and P_i^r. The common intersection of the k drums must lie between an arbitrary pair of planes P_i^l, P_j^r. Such a pair can serve for the first left and right bounds.

In the following we determine a plane parallel to the y-axis, i. e. perpendicular to the xz-plane, in a way that at each side of it lies a constant part of the edges of all k drums. Here we use again the technique of chapter 2 for finding an approximative median. However, in this algorithm we have to find a median of $O(k)$ planes which are not parallel. A key for the solution is Megiddo's technique described in chapter 2. Instead of looking for a zero of the function f we search for the common intersection by applying the two-dimensional algorithm of [12]. Thereby the bounds concerning the z-coordinate (top and bottom faces of the drum) are updated. After determining the side of the median plane on which the common intersection of the drums must lie we can eliminate a constant part of the edges. We repeat this process until only $O(k)$ edges remain.

In more detail the algorithm is as follows:

Step 1. The drums' belts are divided into anterior and posterior parts. Only the z edges falling partially between left and right bound are relevant. Let $n_i := |B_i^a|$ for $i = 1, \ldots, k$, $n_i := |B_{i-k}^p|$ for $i = k+1, \ldots, 2k$; hence $z = \sum_{i=1}^{2k} n_i$. The edges of each half-belt are sorted, therefore the algorithm of chapter 2 can be applied: Let $d := \lceil z/(4k) \rceil$. We divide each set of edges with $n_i > d$ into $\lfloor n_i/d \rfloor$ parts of size between d and $2d-1$ and determine the median of each part. Then we take a plane parallel to the y-axis through this median edge. (Such a plane does not intersect any other edge of the corresponding half-belt.)

Step 2. We have to determine the median of these $O(k)$ planes. Because the planes can intersect, we compute the median in parallel, thereby updating the drums' z-coordinates. (It suffices to consider only the intersections between the planes and the xz-plane.) We apply Megiddo's technique using the parallel median algorithm of Cole and Yap.

Step 3. The planes now do not intersect and their median M is determined. We decide on which side of M the common intersection of the drums must lie, using the algorithm of [12] for detecting a common intersection of k convex polygons. M replaces one of the bounds.

Step 4. At last we eliminate all drum edges falling between M and the bound replaced by M. These are at least half of the edges of at least half of the parts (inside a part the edges are separated; the medians are separated because of step 2). Therefore at least $z/16$ edges can be eliminated.

We repeat this procedure with the reduced drums until $z \leq 4k$.
Between the bounds per drum there lie only two lateral faces more than lateral edges. With these $O(k)$ faces we search for a feasible point of the related 3-dimensional linear program, using the algorithms of Dyer [7] or Megiddo [10].

The complexity of the algorithm is as follows:
Finding the first bounds takes $O(k \log n)$ time. The steps 1 through 4 have to be executed $O(\log n)$ times. Step 1 requires linear time. Step 2 can be done in $O(k \log^2 n \log k (\log\log k)^2)$ time (m in the formula of chapter 2 has to be replaced by $k \cdot \log^2 n$, the complexity of the algorithm in [12]). Step 3 needs $O(k \log^2 n)$ time, and step 4 can be executed in time linear in k. The linear program is solvable in linear time, too. Hence, the total complexity is $O(k \log k (\log\log k)^2 \log^3 n)$.

Applying Cole's modification the complexity of step 2 becomes $O(k \log^2 n \log k)$. Therefore the total complexity reduces to $O(k \log k \log^3 n)$. However, Cole's modification implies a large constant factor [4].

*

If the k polyhedra have different numbers of vertices n_i, $i = 1, \ldots, k$, the algorithm can applied as well. Let $z := \sum_{i=1}^{k} n_i$ be the whole number of vertices. Then the complexity of the algorithm is

$$O\left(\log^2 \frac{z}{k} \cdot \log k \cdot (\log\log k)^2 \cdot \log \prod_{i=1}^{k} n_i\right)$$

and with Cole's modification

$$O\left(\log^2 \frac{z}{k} \cdot \log k \cdot \log \prod_{i=1}^{k} n_i\right).$$

Finally, we can summarize the results:

Theorem: Given k convex polyhedra P_i, $i = 1, \ldots, k$, with n_i vertices respectively and let $z := \sum_{i=1}^{k} n_i$. It can be decided in $O\left(\log^2 \frac{z}{k} \cdot \log k \cdot \log \prod_{i=1}^{k} n_i\right)$ time whether they have a point in common. If $n_i = n$ for all $i = 1, \ldots, k$, the complexity is $O(k \log k \log^3 n)$.

*

References

[1] AJTAI, M.; KOMLOS, J; SZEMEREDI, E.: *An $O(n \log n)$ Sorting Network,* Proc. 15th Annual ACM Symposium on Theory of Computing 1983, 1–9.

[2] CHAZELLE, BERNARD; DOBKIN, DAVID P.: *Detection is easier than computation,* Proc. 12th Annual ACM Symposium on Theory of Computing 1980, 146–153.

[3] CHAZELLE, BERNARD; DOBKIN, DAVID P.: *Intersection of Convex Objects in Two and Three Dimensions,* Journal of the ACM **34** (1987), 1–27.

[4] COLE, RICHARD: *Slowing Down Sorting Networks to Obtain Faster Sorting Algorithms,* Journal of the ACM **34** (1987), 200–208.

[5] COLE, RICHARD; YAP, CHEE K.: *A parallel median algorithm,* Information Processing Letters **20** (1985), 137–139.

[6] DOBKIN, DAVID P.; KIRKPATRICK, DAVID G.: *Fast detection of polyhedral intersection,* Theoretical Computer Science **27** (1983), 241–253.

[7] DYER, M. E.: *Linear time algorithms for two- and three-variable linear programs,* SIAM Journal on Computing **13** (1984), 31–45.

[8] HELLY, ED.: *Über Mengen konvexer Körper mit gemeinschaftlichen Punkten,* Jahresbericht der Deutschen Mathematiker-Vereinigung **32** (1923), 175–176.

[9] MEGIDDO, NIMROD: *Applying Parallel Computation Algorithms in the Design of Serial Algorithms,* Journal of the ACM **30** (1983), 852–865.

[10] MEGIDDO, NIMROD: *Linear-time algorithms for linear programming in* $\mathbf{R}^3$ *and related problems,* SIAM Journal on Computing **12** (1983), 759–776.

[11] REICHLING, MATTHIAS: *Entdeckung eines gemeinsamen Schnitts von k konvexen Objekten (On the detection of a common intersection of k convex objects),* PhD-thesis, Würzburg 1987 (in German).

[12] REICHLING, MATTHIAS: *On the detection of a common intersection of k convex objects in the plane,* Information Processing Letters, to appear.

[13] SHAMOS, MICHAEL IAN: *Geometric Complexity,* Proc. 7th Annual ACM Symposium on Theory of Computing 1975, 224–233.

Time Coherence in Computer Animation by Ray Tracing[1]
- Extended Abstract -

Heinrich Müller
Department of Computer Science, University of Karlsruhe
7500 Karlsruhe, Germany

Abstract

Raytracing is a powerful, but relatively time consuming technique for realistic image synthesis in computer graphics. Two algorithms are presented accelerating raytracing of animations by considering coherence over time. The look-ahead algorithm avoids repeatingly tracing the same ray for several frames. More precisely, if a ray remains constant over l frames, the look-ahead algorithm does not need more than $O(\log l)$ queries into the scene, compared to l queries for the frame-by-frame approach. This type of coherence particularily occurs for a fixed camera, and scenes only partly changing over time. The preprocessing time is increased by a factor 2 compared to the frame-by-frame approach, while space requirements grow by a factor of $\log f$, f the number of frames to be calculated. The frame interpolation algorithm is based on image interpolation using knowledge about the scene. The central algorithmic task is the comparison of two transformed pixel grids which is solved by a modified version of the plane sweep algorithm for line segment intersection reporting. The frame interpolation algorithm is not restricted to fixed cameras, but may lose information against the frame-by-frame approach.

1. The problem

In the last few years, research in generation of realistic raster images with multiple reflection and refraction has been concentrated on raytracing (Whitted, 1980). Raytracing is a method deduced from the optics of rays in physics. In its simplest form, for each pixel of the image a ray is traced from the viewpoint into the 3-d scene to calculate the first intersection of the ray with a surface (figure 1). If the surface is reflecting or refracting, an appropriate ray is determined by the law of reflection or refraction. These new rays are traced analogously. To calculate shadows, the ray-surface intersection points are connected by line segments to the point light sources illuminating the scene. If there is a surface intersecting the line segment, the intersection point lies in the shadow of this light source, and its intensity is not taken into account for intensity calculations.

Finding the first intersection points with the surfaces is time consuming, even if it is supported by appropriate data structures. This is due to the large number of rays. The question arises if indeed all rays are necessary, or if a small number of sampling rays suffices to deduce the necessary information for the others. We investigate this question in case of animations. Animation

[1] Work was partially supported by DFG (Mu744/1-1).

means that a sequence of pictures is generated from a scene of surfaces changing their shape or location over time. When considering consecutive frames one realizes that visibility changes only slightly from frame to frame.

Raytracing algorithms for large scenes start by preprocessing the scene into a data structure which allows to answer arbitrary ray queries quickly. In an animated scene, only a few objects may change from frame to frame. So it may make sense to lay out the data structure dynamically, i.e. that it allows fast insertions and deletions of surfaces. However, experience shows that preprocessing time is only a small portion of the total time of image synthesis, at least if the number of surfaces is small compared to the number of rays. Better profit can be expected by reducing the number of ray queries. A typical situation when it may be unnecessary to perform several ray queries for consecutive frames arises if the ray remains geomatically fixed and a static surface is hit. This occurs in animations if the camera (i.e. the viewpoint and the image plane) is fixed, and there are only a few kinetic surfaces of reasonable size. In the first part of this paper, an algorithm is presented which makes use of this special situation. This so-called look-ahead algorithm alternates between phases of preprocessing, raytracing, and reporting. The information necessary to use coherence is gained in the preprocessing phase. Under the condition that the number of surfaces remains the same during the movie, the total time for preprocessing is not more than twice of that of the frame-by-frame approach. The storage requirements are a factor $\log f$ higher, with f the number of frames of the movie. However, the most important property of the algorithm is that for a ray query whose answer holds for l consecutive frames, at most $O(\log l)$ queries must be carried out (instead of l for the frame-by-frame approach).

If the camera is moving, the visible portions of a scene are translated, resized, or deformed from frame to frame. The idea of the algorithm presented in the second part of this paper is to interpolate the intermediate frames between two key frames on picture level. Regions of the keyframe images are identified which represent the same parts of the surfaces. For this purpose, knowledge about the 3-d scene is used. By interpolating the shape of the regions as well as their colors, the shapes for the frames in-between are derived.

The problem to be solved here can be seen as the inverse of the object recognition problem in image sequence analysis. In image sequence analysis, a sequence of images of a three-dimensional scene is given. The problem is to identify moving objects in the scene and possibly reconstruct their three-dimensional shapes. One method is to calculate displacement vector fields from the distribution of the pixel intensities in order to identify the same features in consecutive images (Nagel, 1985). In contrast to this, we know the displacement vector field (we can calculate it from the scene) and have to determine the pixel intensities.

2. The look-ahead algorithm

2.1 Ray queries in animated scenes

For our movie synthesis algorithm we presume a data structure for answering ray queries on a static scene, for example one of those surveyed by Schmitt, Müller, Leister (1988). The details of this structure are irrelevant. We only require that the kinetic hulls of the surfaces making up the scene can be preprocessed into this data structure. A kinetic surface S is a family of surfaces indexed by the numbers in a real interval $[s(S), t(S))$, i.e. $S = \{S_t : t \in [s(S), t(S))\}$.

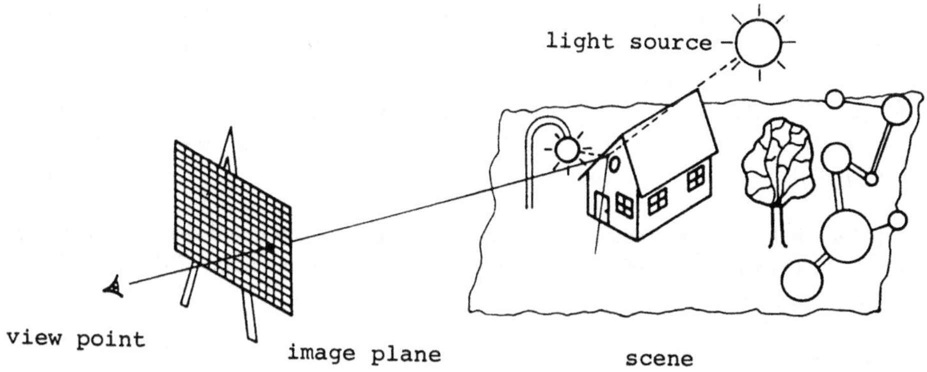

Figure 1. Calculating raster images by ray tracing.

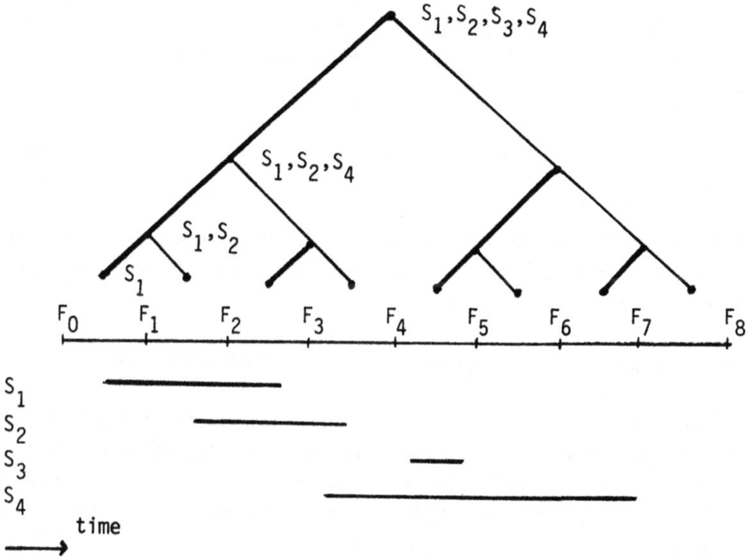

Figure 2. The hierarchy of scenes for look-ahead.

The interval $[s(S), t(S))$ defines the time interval during which S exists. S_t is the state of S at time t. The kinetic hull $H(S, s, t)$ is the set of all points which are element of S_t for some $t \in [s(S), t(S))$, i.e. $H(S, s, t) := \{p \in S_t : t \in [s, t) \cap [s(S), t(S))\}$. A kinetic surface S is called static in the interval $[s, t)$ if $s(S) \leq s < t \leq t(S)$ and $H(S, s, t) = S_v$ for $v \in [s, t)$. The kinetic hull is a static spatial object. The first intersection point of a ray $p + \lambda v$, p a point in space, v a 3-d vector, with a kinetic hull is $p + \lambda_0 v$ with $\lambda_0 := \inf\{\lambda \geq 0 : p + \lambda v \in S_t$ for some $t \in [s(S), t(S))\}$.

The queries are defined by interval rays. An interval ray r is a ray in 3-d space attributed by a real interval $[s(r), t(r))$ of its life time. Now we come to the formulation of the problem to be solved in this section.

PROBLEM "Interval Ray Query"

Input. A scene **S** of n kinetic surfaces, a set R of m interval rays, a sequence F of $f + 1$ sample points $F_0, ..., F_{f-1}, F_f$.

Output: A sequence $I_0, ..., I_{f-1}$, defined by

$$I_j := \{(r, S) : r \in R,\ S \in \mathbf{S},\ [s(r), t(r)) \cap [F_j, F_{j+1}) \neq \emptyset,\ [s(S), t(S)) \cap [F_j, F_{j+1}) \neq \emptyset,$$
$$\text{the first intersection point of } r \text{ on } H(S, F_j, F_{j+1}) \text{ also is a first}$$
$$\text{intersection point of } r \text{ with the scene } H(S, F_j, F_{j+1}) := \{H(S, F_j, F_{j+1}) : S \in \mathbf{S}\}\}.$$

If several surfaces S satisfy the condition for a ray r, one of them is chosen arbitrarily.

This means that the kinetic surfaces are not sampled at a fixed time t, but their kinetic hulls over a sample interval are sampled instead. The sequence I_j gives the first hit kinetic hull for all rays valid in this interval.

For the following we suppose f to be a power of 2. This simplifies the presentation.

Preprocessing of the interval rays $r \in R$ goes sequential in time according to their times of birth $s(r)$. For every ray, an initial subinterval, as long as possible, of its life time interval is processed, i.e. possibly more than one sample interval at once. The remainder of the ray is re-inserted into R, and not earlier processed than when it becomes next in time sequence. For this purpose, R is organized in a priority queue RE, with birth time as key. The result is a first intersected surface and its interval of validity. It is stored in a list L. If all rays in RE with inital point in the current sample interval $[F_j, F_{j+1})$ are processed, I_j is constructed from L. Further, all results are deleted from L which do not hold beyond F_{j+1}.

There remains to find a first intersection point and its interval of validity. For every sample interval $[F_j, F_{j+1})$, a family of scenes $\{S_j(i) : i = 1, ..., k_j\}$ is built up. Since only one sample interval is processed at once, only those scenes for the current sampling interval are required. The scenes of a family belong to a family of intervals of exponentially increasing length. Each of these scenes contains the kinetic hulls of those surfaces whose time intervals intersect the interval. The kinetic hull is taken over this intersection. If the union of the leaf intervals in any subtree of the binary tree in figure 2 is assigned to its root, the families of the scene intervals just correspond to the paths drawn fat in the figure. More exactly,

$$\mathbf{S}_j(i) := \{H(S, F_j, F_{j+2^i}) : S \in \mathbf{S},$$
$$[s(S), t(S)) \cap [F_j, F_{j+2^i}) \neq \emptyset\}, \quad i = 0, ..., k_j,$$
$$k_j := \max\{i : 2^i \text{ divides } j, \quad j \leq \log f\}, \quad j = 0, ..., f-1.$$

Now let be r an interval ray in RE to be processed next for the interval $[F_j, F_{j+1}), 0 \leq j < f-1$. Hence $s(r) < F_{j+1}$. The rays in RE are attributed by "up" or "down". The search strategy depends on these attributes as follows.

ALGORITHM SearchMaxInterval

Input: The index j, $0 \leq j < f-1$, of the current sample interval, a ray $r \in R$ with $s(r) < F_{j+1}$ and $t(r) > F_j$.

Global Data Structures: $\mathbf{S}_j(i), i = 0, ..., k_j$, RE, L, see above.

```
BEGIN
    k := k_j
    S := first surface in S_j(0) hit by r;
    IF S not static in [F_j, F_{j+1}) THEN BEGIN
        insert (r, S, j + 1) in L;
        insert (r, j + 1,up) in RE
    END ELSE BEGIN {S static}
        S := first surface hit in S_j(k);
        IF S static in [F_j, F_{j+2^k}) and t(r) > F_{j+2^k-1} THEN BEGIN
            insert (r, S, j + 2^k) into L;
            insert (r, j + 2^k,up) into RE;
        END ELSE
        IF attribute of r is "up" THEN BEGIN
            find, starting with S_j(1), the largest index
            i' so that the first surface S' in S_j(i') hit by r
            is static in [F_j, F_{j+2^i'}) and t(r) > F_{j+2^i'-1};
            insert (r, S', j + 2^i') in L;
            insert (r, j + 2^i',down) in RE
        END ELSE
        BEGIN {attribute of r is "down"}
            find, starting with S_j(k - 1), the largest index
            i', so that the first surface S' in
            S_j(i') is static in [F_j, F_{j+2^i'}) and t(r) > F_{j+2^i'-1};
            insert (r, S', j + 2^i') in L;
            insert (r, j + 2^i',down) in RE
        END
    END {S static}
END.
```

There is the special case that none of the surfaces is intersected. Then S is set to "background" and is treated like a static object.

With the subalgorithm *SearchMaxInterval* as central subroutine, the look-ahead algorithm solving the "Interval Ray Query" problem works as follows.

ALGORITHM Look-Ahead

Input, Output cf. "Interval Ray Query" problem

Data Structures:

RE: a priority queue for interval rays with their birth times as keys, implemented as an array of lists with indices in $\{0, ..., f-1\}$;

SE: sorted list of birth and death times of scene surfaces;

L: list of intermediate results, built up by tuples (r, S, t), with S the first surface intersected by r, up to time t;

S: ARRAY[1..log f] of scenes preprocessed into a data structure for efficient ray queries on static surfaces;

F: sorted list of sample times.

SUBALGORITHM Preprocessing(M,S);
{builds up the data structure **S** from a set M of kinetic hulls}

SUBALGORITHM SearchMaxInterval(j, r);
{see above.}

BEGIN
    initialize RE with R, attributing the rays with "up";
    initialize SE with **S**;
    FOR $j := 0$ TO $f - 1$ DO BEGIN
        remove the surfaces S from SE with death time $t(S) \leq F_j$;
        remove the rays r from RE with death time $t(r) \leq F_j$;

        { preprocessing }

        k := $\max\{i : 2^i \text{ divides } j, \quad i \leq \log f\}$;
        FOR $i := 0$ TO k DO BEGIN
            M:=$\{H(S, F_j, F_{j+2^i}) : S$ an surface in SE with $s(S) < F_{j+2^i}\}$;
            Preprocessing(M,**S**[i])
        END;

```
        { raytracing }

        r:= first ray in RE;
        WHILE s(r) < F_{j+1} DO BEGIN
            remove r from RE;
            SearchMaxInterval(j, r);
            r := first ray in RE
        END;

        { reporting }

        I_j := ∅;
        FOR (r, S, t) ∈ L DO BEGIN
            I_j := I_j ∪ {(r, S)};
            IF t ≤ F_{j+1} THEN remove (r, S, t) from L
        END
    END {FOR j := 0 TO f − 1 DO ...}
END.
```

The complexity of the look-ahead-algorithm is as follows.

Theorem
Suppose given a scene of n kinetic surfaces, and an algorithm to preprocess the kinetic hulls of n' of these surfaces within time $p(n')$ into a data structure of size $s(n')$ so that the first intersected kinetic hull can be found within $q(n')$ time. Further, let be R a set of m interval rays, and f the number of sample intervals, w.l.o.g. of length 1. Then the space and time requirements of the look-ahead algorithm is bounded by

$$S(n,m,f) \leq s(n) \cdot \log f + O(n+m+f)$$

$$T(n,m,f) \leq 2 \cdot f \cdot p(n) + O(n \log n + m \log m) + \sum_{r \in R} O(t(r) - s(r))$$

$$+ \sum_{r \in R} \Big(\sum_{\substack{r' \text{ subinterval of } r \\ \text{with same answer}}} (q(n) + O(1)) \cdot O(\log(\lceil t(r') \rceil - \lfloor s(r') \rfloor)).$$

Proof. The bound of space requirement $S(n,m,f)$ is immediate, since at any time at most $\log f$ preprocessed scenes are stored. The first term of the bound for $T(n,m,f)$ is the time to preprocess these scenes. The binary tree of scenes does not have more than $2f$ nodes, and for every node at most one scene is preprocessed. The second term is composed by the time necessary to sort the end points of the life time intervals of the surfaces in SE ($O(n \log n)$) resp. ray intervals in RE ($O(m \log m)$). $\sum_{r \in R} O(t(r) - s(r))$ is the total time to build up the I_j from L. Finally, there remains the time to construct the intermediate results inserted into L during calculation, at which we will focus now.

The look-ahead algorithm processes scenes belonging to maximal subintervals with the same answer, i.e. for the whole interval the first intersected surface remains the same. These maximum subintervals are intervals of nodes in the basic binary tree of figure 3. These nodes are sons of nodes on two paths related by a common node. In the following, these paths are called left resp. right path. For every call of *SearchMaxInterval* one of the nodes drawn fat in figure 3 will be found. *SearchMaxInterval* always starts with a most left leaf of the subtree induced by this node. The scene found by *SearchMaxInterval* lies on the straight path from this leaf to the root.

The algorithm is divided in an "up"- and in a "down"-phase, cf. figure 3. In both phases, the last scene on the path is tested first. If the first intersected surface is static over the whole interval, search continues in the "up"-phase at the next leaf of interest. Otherwise, in the "up"-phase, the first marked scene is searched, starting at the leaf. Further, "up" is switched to "down". In the "down"-phase, the scene before the last is tested. If no static intersected surface is found, a first scene with a static intersected surface is searched in direction to the leftmost leaf.

Every interval found in the up phase costs a query into a preprocessed scene. The number of queries is bounded by $\log l$, l the length of the interval with constant answer. This comes from the fact that the lengths of node intervals increase monotonously, at least by a factor 2 on the left path. In the "down"-phase, for every node of the right path, at most one query is carried out, i.e. at most $\log l$ queries by an analogous argument. Finally, finding the marked node does not cost more queries than the length of the left path, i.e. $\leq \log l$ queries. Altogether, we have $O(\log l)$ queries with time requirement $\leq q(n)$. For every query, the remainder of the ray must be inserted into RE which is possible in $O(1)$ time. With $l = \lceil t(r')\rceil - \lfloor s(r')\rfloor$ we get the last term of the estimation of $T(n, m, f)$.

For comparison, the complextiy of the frame-by-frame approach is

$$S(n, m, f) = s(n)$$

$$T(n, m, f) = f * p(n) + \sum_{r \in R} q(n) * (\lceil t(r)\rceil - \lfloor s(r)\rfloor).$$

2.2 Animation synthesis by the look-ahead algorithm

It is easy to extend the look-ahead algorithm from previous section to a movie synthesis algorithm. Suppose the camera is fixed over the whole movie. This means that all rays of view remain the same for the whole time. RE is initialized with the corresponding interval rays. With every ray, its pixel address is stored. During movie synthesis, these rays will induce other rays, i.e. rays of reflection and refraction, and rays going to the light sources. For any ray to the light source, the surface it is coming from is stored. The list L becomes an array of the same dimensions as the frames. For every pixel, it stores several intensities, as well as the time when they expire. These intensities are those coming in from the different rays of reflection and refraction induced by the pixel. I_j is a raster image, i.e. the current frame, and is updated from the intensities in L, for example by summing up all intensities in L for one pixel. The modified look-ahead algorithm processes the sample intervals $[F_j, F_{j+1})$ as before. A ray of view, or reflection, or refraction, in RE with initial point in the current sample interval, is

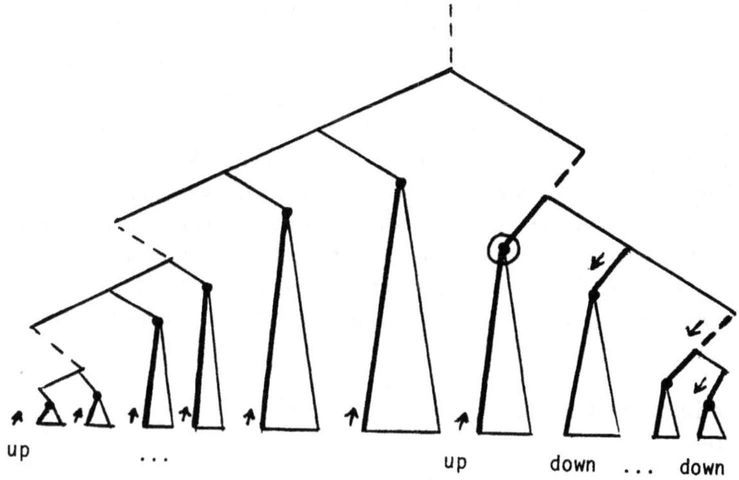

Figure 3. The traversal of the tree of scenes for a ray interval.

$$p = \lambda p_1 + \mu p_2 + \nu p_3, \quad \lambda+\mu+\nu=1 \quad \rightsquigarrow \quad (u,v) = \lambda(u_1,v_1) + \mu(u_2,v_2) + \nu(u_3,v_3)$$

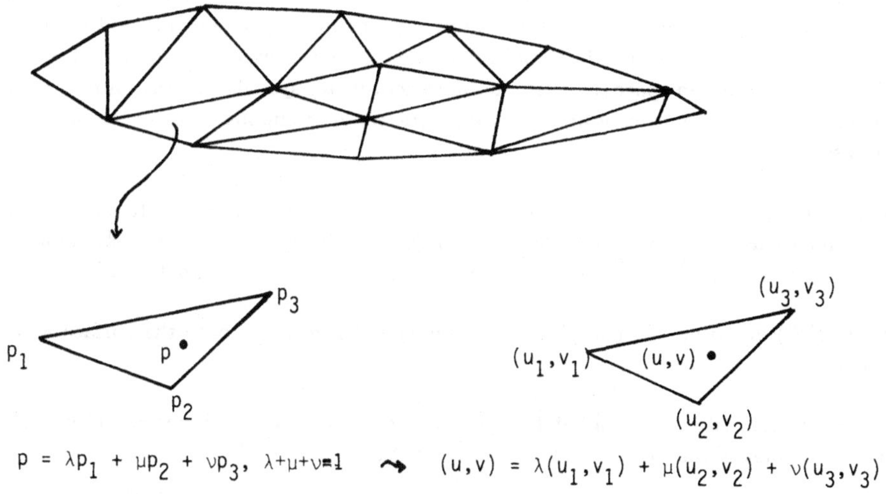

Figure 4. Defining an inner coordinate system for a surface of triangles.

treated as before. Instead of inserting the result (r, S, t) into L, for every light source a ray r' with time interval $[F_j, t)$ is inserted into RE. If the surface intersected by r is reflecting or refracting, the rays of reflection and refraction are evaluated and inserted into RE. The respective time interval is $[F_j, t)$. (Rays of reflection and refraction usually are calculated only up to a certain limited depth. Higher order of reflection and refraction is neglected then.) For the second sort of rays, i.e. the light source rays, a blocking surface between the light source is searched. This works analogously to intersection searching, including the update of RE. If no blocking surface is found, the intensity induced by the light source on the surface stored with the ray is evaluated. This intensity, as well as the time when it expires, is inserted into L at the appropriate pixel. If all rays, the old and the newly inserted ones, with inital time in $[F_j, F_{j+1})$, are processed, I_j is calculated from L and reported as the next frame.

3. The image interpolation algorithm

3.1 The strategy

In addition to the previous section, we assume surfaces of the given scene supplied with an inner coordinate system. For example, the parameter values of a surface in parameter representation, $F : R^2 \rightarrow R^3$, $F : (u, v) \rightarrow F(u, v)$, $(u, v) \in I$, I an interval in the plane, can be used as the inner coordinates. For polyhedral surfaces consisting of triangles, inner coordinates (u, v) can be assigned to every vertex p. The coordinates for the inner points of triangles can be obtained by interpolation of the vertex coordinates (figure 4). Important for the method is that large parts of the scene can be uniformly parameterized. Surfaces are animated by moving or transforming them in space, so that the inner coordinate system remains untouched, i.e. a point with the same inner coordinates at different times is supposed to be the same, up to deformation.

Now let S be a surface of a given scene. Again S_t is the surface at time t. In this section, we restrict the discussion to visibility calculation only, i.e. only the rays immediately coming from the view point are considered. Further generations of rays are treated analogously.

The set $R(S_t)$ describes the mapping of the rastered image plane onto the surface at time t (figure 5),

$R(S_t) := \{(i, j, u, v) | S_t$ is visible at the pixel corner (i, j), and the inner coordinates of the intersection point of the ray through this corner with S_t are $(u, v)\}$.

Now we consider the frames at time t_s and time t_e. The parts of S visible in both frames are approximated by the set $I(S_{t_s}, S_{t_e})$ of all tuples in $R(S_{t_s})$ whose pixel is totally covered by pixels in $R(S_{t_e})$,

$I(S_{t_s}, S_{t_e}) := \{(i, j, u, v) | (i, j, u, v) \in R(S_{t_s})$, the pixel of (i, j, u, v) is covered by pixels in $R(S_{t_e})\}$.

The pixel of a pixel corner (i, j) is that defined by the corners $(i, j), (i+1, j), (i, j+1), (i+1, j+1)$. The pixels in a set $R(S_t)$ are those with all four corners in the set.

The first step of interpolation is to calculate $I(S_{t_s}, S_{t_e})$. In addition, the images of the pixels

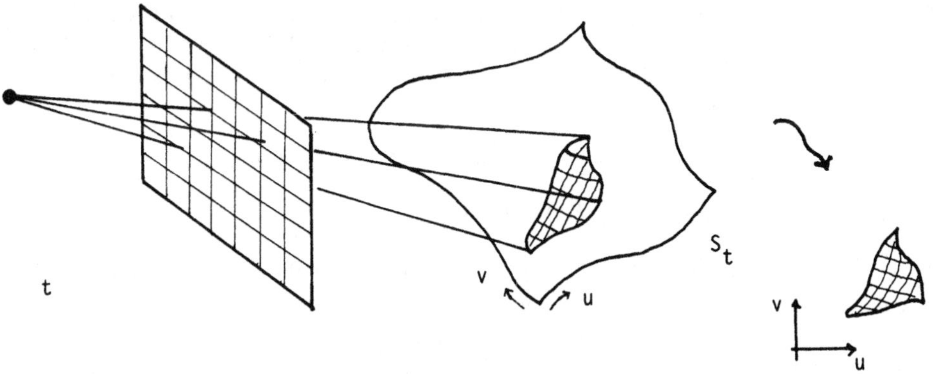

Figure 5. The set $R(S_t)$ of pixels visible on surface S at time t.

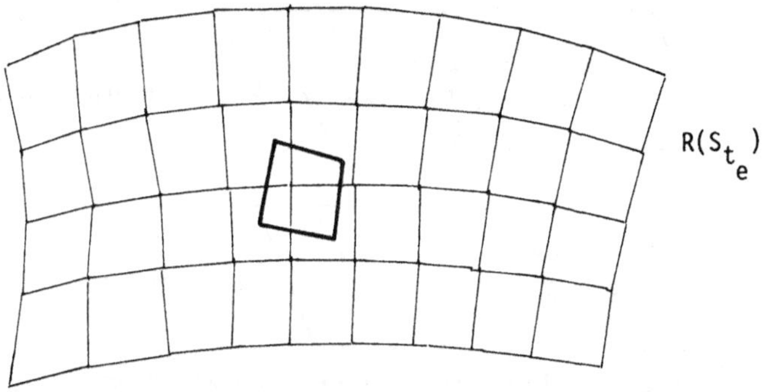

Figure 6. The image of a pixel of one raster in the other raster.

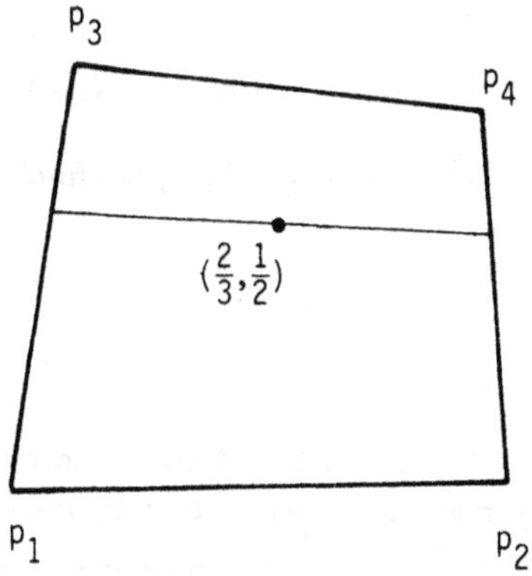

Figure 7. The blending coordinates of a point in a pixel.

in $R(S_{t_s})$ relative to the raster induced by the pixels in $R(S_{t_e})$ is calculated (figure 6). For this purpose we have to define local coordinates in the quadrangles induced by pixels (figure 7). The blending coordinates of a point q in a pixel with corners p_1, p_2, p_3, p_4 are obtained as the solution of the equation

$$(1-\mu)\cdot((1-\lambda)\cdot p_1 + \lambda\cdot p_3) + \mu\cdot((1-\lambda)\cdot p_2 + \lambda\cdot p_4).$$

The result of this step is a set $P(S_{t_s}, S_{t_e})$ of tuples,

$P(S_{t_s}, S_{t_e}) := \{(i, j, i', j', \lambda, \mu) | (i, j, ., .) \in I(S_{t_s}, S_{t_e}),\ (i', j', ., .)$ the pixel into which $(i, j, ., .)$ falls, λ, μ the pixel coordinates of $(i, j, ., .)$ w.r.t. pixel $(i', j', ., .)\}$.

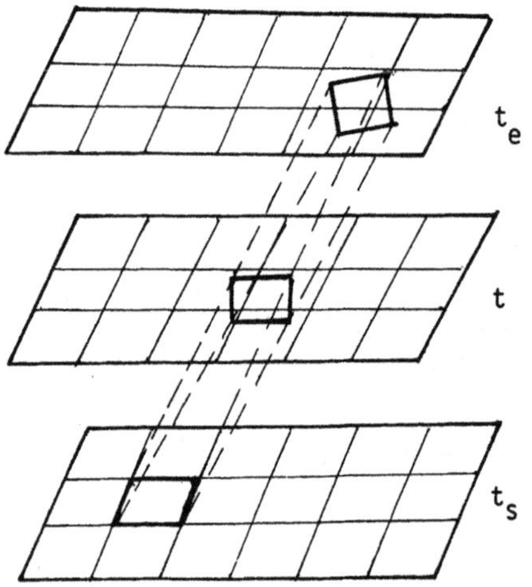

Figure 8. Interpolation between two key frames.

The second step is interpolation (figure 8). The coordinates of the intermediate pixels are obtained from

$$(i, j) + \frac{t - t_s}{t_e - t_s}\cdot(i' - i + \lambda, j' - j + \mu),\ (i, j, i', j', \lambda, \mu) \in P(S_{t_s}, S_{t_e}).$$

Rounding yields the respective pixel in the image plane, the fractional remainder is the location within this pixel.

3.2 The algorithmic details

The key task is to calculate the mutial location of two quadrangular planar decompositions, obtained from $R(S_{t_s})$ and $R(S_{t_e})$ (figure 9). Their intersections can be found by the algorithm of Mairson and Stolfi (1988) for intersecting two sets of disjoint line segments within $O(m \log m + k)$ time, m the number of polygon edes, k the number of intersecting edges. Guibas and Seidel (1986) gave an algorithm for intersecting two convex partitions of the plane in time $O(m + k)$. The second result is not immediately applicable to our problem, while the first one is too general.

The pixel grids can be supposed to differ only slightly in location and size of pixels. This allows to design a simpler algorithm. Our algorithm follows the line-segment intersection algorithm

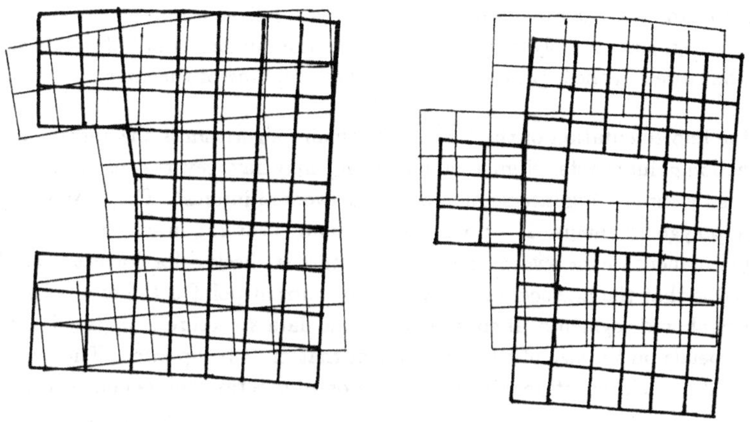

Figure 9. Matching of grids.

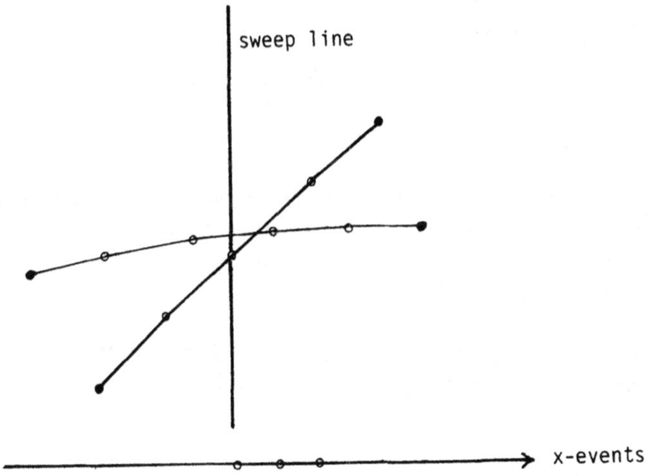

Figure 10. Polygonal chains as segments.

by Bentley, Ottmann (1979). The line segments are replaced by x-monotonous chains of short line segments. We obtain the polygonal chains by fixing the i resp. j (i', j' respectively) pixel index and look for consecutive intervals of the other index (figure 10). The resulting chain can be expected monotonous in x-direction. Otherwise it is subdivided into monotonous subchains. Going from line segments to polygonal chains reduces the number of initial points significantly that have to be stored in the x-priority queue at the beginning of the sweep.

Crucial for the implementation of the Bentley/Ottmann algorithm is the correct arrangement of the intersection point events. Numerical inaccuracy may have severe consequences. We avoid these problems by eliminating intersection points as event of interest. They are replaced by the points on the polygonal chains. If a point on the chain is coming up as the next event of the sweep, its environment in the sweep line data structure is checked for correctness. What may have happened is that the line segment of which the current point is the end point could have intersected neighbored segments. If so, the sweep line data structure is updated. The time for this update depends on the size of the environment that must be updated. The overall number of update operations is proportional in the number of intersection points between the two grids.

A further advantage of this approach is that arithmetical operations are saved. Our algorithm only requires to compare the y-coordinates of intersection points of line segments with the sweep line against a point r. This can be performed by

$$(r_y - p_y) \cdot (q_x - p_x) > (r_x - p_x) \cdot (q_y - p_y)$$

at a cost of 2 $*$ and 4 $+-$. For comparison, the calculation of the x-coordinate i_x of the intersection point of two line segment pq and rs is obtained from

$$i_x = p_x + \frac{|r - p\ q - p|}{|q - p\ r - s|} \cdot (q_x - p_x)$$

at a cost of 5 $*$, 1 $/$, and 11 $+-$.

The sweep line data structure consists of elements of two types. The two types of elements correspond to the active line segments of the two grids. An element refers to its neighbors in the other grid, as well as to the neighbors in its own grid. A dynamic balanced search tree is used for every type to insert or delete polygonal chains.

The overall complexity of this algorithm is $O(m' \log m' + m + k)$, for m' polygonal chains, m line segments, and k intersection points between the grids.

3.3 Full Raytracing

In the previous section, the exact location of pixels was calculated in order to interpolate their intensities exactly by weighted sums of the intensities of the covering pixels. The discussion was restricted to the rays of view. Extension to reflecting and refracting rays goes by treating every generation of rays in an analogous manner. A generation of rays consists of all those rays corresponding to the same node in the raytrace tree, for example the first generation of reflected rays, the first generation of refracted rays, etc..

After interpolation, their remain parts in the interpolated images that could not be calculated by this method. These rays are treated by the usual raytracing algorithm. Parts that cause these problems are at the silhouette of objects, and small objects for which interpolation is too inefficient. A typical example is a rotating sphere, where parts previously hidden are coming up at the silhouette. The size of this part depends on the rotational speed of the sphere.

4. Remarks

The look-ahead algorithm asymtotically speeds up the computation of realistic animations by reducing the number of ray queries. For its application, some details must be inspected closer, which were left open here. For example, the calculation of the kinetic hulls can be a nontrivial task, depending on the type of surfaces and animations. Further, preprocessing the kinetic hulls for efficient static raytracing becomes more complicated, due to the more complex geometry of the hulls. In practice, the exact hulls will be replaced by approximations of simpler type, like parallelepipeds, spheres, or ellipsoides.

The image interpolation is less resticted than the look-ahead algorithm with respect to situations favorable for a speed-up. However, it requires a visual check how far interpolation reduces the quality of the resulting pictures. It is known from the key frame technique of classical 2-d animation that interpolation may cause artifacts. Further, the storage requirements of the sets $R(S_{t_s})$, $R(S_{t_e})$ and those arising from them may be considerable. This causes no fundamental problems, however, they should be stored so that they can mainly processed sequentially. Sorting or random access should be avoided. And finally, it must be checked whether the pixel matching algorithm is fast enough in order to save time significantly.

References

Bentley, J.L., Ottmann, T. (1979) Algorithms for reporting and counting geometric intersections, IEEE Transactions on Computers C28, 643-647

Guibas, L.J., Seidel, R. (1986) Computing convolutions by reciprocal search, Proc. 2nd ACM Symp. on Computational Geometry, 90-99

Mairson, H.G., Stolfi, J. (1988) Reporting and Counting Intersections between two sets of line segments, in: NATO ASI Series, Vol. F40, Theoretical Foundations of Computer Graphics and CAD, R.A. Earnshaw, ed., Springer-Verlag, Berlin, 1988

Nagel, H.-H. (1985) Analyse und Interpretation von Bildfolgen, Informatik-Spektrum 6, 178-200

Schmitt., A., Müller, H., Leister, W. (1987) Ray Tracing Algoithms - Theory and Practice, in: NATO ASI Series, Vol. F40, Theoretical Foundations of Computer Graphics and CAD, R.A. Earnshaw, ed., Springer-Verlag, Berlin, 1988

Whitted, T. (1980), An Improved Illumination Model for Shaded Display, CACM 23, 343-349

USING GALE TRANSFORMS IN COMPUTATIONAL GEOMETRY

by

Franz Aurenhammer[1)]

Abstract

Let P denote a set of $n > d+1$ points in d-space R^d. A Gale transform of P assigns to each point in P a vector in space R^{n-d-1} such that the resulting n-tuple of vectors reflects all affinely invariant properties of P. First utilized by Gale in the 1950s, Gale transforms have been recognized as a powerful tool in combinatorial geometry.

This paper introduces Gale transforms to computational geometry. It offers a direct algorithm for their construction and sketches applications to convex hull and visibility problems. An application to scene analysis is worked out in some more detail.

---

1) Institutes for Information Processing,
Technical University of Graz and Austrian Computer Society,
Schiesstattgasse 4a, A-8010 Graz, Austria.

1. Introduction

The young discipline of computational geometry - concerned with algorithmic aspects of geometric questions - is a continuing source of problems and scientific interest. Various powerful solution techniques have been developed. Among them is one that, in some sense, exploits the fact that the problems are geometrical in nature the most: geometric transformation. Its first systematic use dates back to Brown [3]. The important role of transforming geometric problems (into easier solved ones) is best documented in the recent book by Edelsbrunner [4].

This paper introduces a technique called Gale transform to computational geometry. The concept has been invented by Gale [5] and has been recognized as a very useful tool in combinatorial geometry during the last few decades. Most of its applications concern the combinatorial structure of convex polytopes. The interested reader is referred to [6] for a catalogue of properties of Gale transforms.

To the author's knowledge, Gale transforms (GTs, for short) did not receive attention in computational geometry yet. Here we present a first algorithmic treatment of GTs. Two features of GTs suggest their relevance to algorithmic applications: They can be computed by a simple and efficient algorithm, and they (usually) drastically change the dimension of the underlying problem. The following section is devoted to a definition and to the construction of Gale transforms. In Sections 3 through 5, applications to convex hull problems, visibility problems, and to scene analysis are addressed.

2. Constructing Gale transforms

We start with an analytic definition of Gale transforms following [6]. Let $P=\{p_1,\dots,p_n\}$ be a set of points in Euclidean d-space R^d. P *affinely spans* R^d if the affine hull of P is R^d. The *affine hull* of P is defined as the set

$$\text{aff } P = \{x=\sum_{1\leq i\leq n}\lambda_i p_i \mid \sum_{1\leq i\leq n}\lambda_i=1\}.$$

So aff P is the affine subspace of R^d of lowest dimension that contains P. Throughout, we assume that P affinely spans R^d which particularly implies $n \geq d+1$. This is no loss of generality since the discussion can be carried out in R^k instead of R^d, otherwise, for k the dimension of aff P. An *affine dependence* of P is a vector $t=(\tau_1,\ldots,\tau_n) \in R^n$ with

$$\sum_{1 \leq i \leq n} \tau_i p_i = 0 \quad \text{and} \quad \sum_{1 \leq i \leq n} \tau_i = 0.$$

It is easily verfied that the set of all affine dependences of P forms an m-dimensional linear subspace of R^n, for $m=n-d-1$. Let $t_1,\ldots,t_m$ be a basis of this vector space, arranged in the m by n matrix

$$\begin{pmatrix} t_1 \\ \vdots \\ t_m \end{pmatrix} = (\bar{p}_1 \ldots \bar{p}_n).$$

The result of this construction is the assignment of a vector $\bar{p}_i \in R^m$ to each point $p_i \in P$. The n-tuple $\bar{P}=(\bar{p}_1,\ldots,\bar{p}_n)$ is called a *Gale transform* of P.

Observe that $\bar{P}$, which linearly spans R^m, does not necessarily consist of n different vectors. Clearly $\bar{P}$ is not unique since it depends on the choice of the basis $t_1,\ldots,t_m$. Nevertheless, any GT of P reflects all properties of P that are invariant under affine transformations. We shall see several such properties and their counterparts in the terminology of GTs, and exploit these correspondences for efficient algorithmic solutions in Sections 3 through 5. The remainder of this section deals with the algorithmic construction of GTs.

The definition of a GT $\bar{P}$ of P already suggests a method for its computation. Obviously, any affine dependence t of P lies in the solution space D(P) of the linear homogenous system

$$\begin{pmatrix} p_1 & \ldots & p_n \\ 1 & \ldots & 1 \end{pmatrix} t = 0.$$

So we are left with the problem of finding a basis of D(P). As is well known from elementary linear algebra, this can be accomplished by means of Gaussian elimination (see e.g. [9]) as follows. Assume that, without loss of generality, $\{p_1,\dots,p_{d+1}\}$ affinely spans R^d. Then the d+1 by n matrix above can be brought, by application of elementary row operations, into the form

$$\begin{pmatrix} -1 & . & . & 0 & & & \\ . & & & . & c_1 & \dots & c_m \\ 0 & . & . & -1 & & & \end{pmatrix}$$

without effecting the solution space D(P) of the corresponding system. The submatrix $U = (c_1 \dots c_m)$ contains $m=n-d-1$ columns and d+1 rows. We denote the latter by $r_1,\dots,r_{d+1}$ and extend U to an n by m matrix U' by adding m rows $r_i = e_{i-d-1}$, for $d+2 \leq i \leq n$, with e_k denoting the k-th standard basis vector of R^m. Then the columns of U' obviously constitute a basis of D(P) so that, by definition, the rows of U' represent a Gale transform $\bar{P}$ of P. To be more precise, $\bar{P}=(\bar{p}_1,\dots,\bar{p}_n)$ has the form

$$\bar{p}_i = r_i, \quad \text{for } 1 \leq i \leq d+1,$$

$$\bar{p}_i = e_{i-d-1}, \quad \text{for } d+2 \leq i \leq n.$$

What is the computational complexity of this construction? We only need to compute the submatrix U whose rows are the vectors $\bar{p}_1,\dots,\bar{p}_{d+1}$ since $\bar{p}_{d+2},\dots,\bar{p}_n$ are the m basis vectors of R^m. U is obtained by eliminating d(d+1) entries and normalizing d+1 entries of the initial matrix using elementary row operations. Clearly $O(d^2n)$ time and O(dn) space suffice for this task that, as a byproduct, also recognizes when $\{p_1,\dots,p_{d+1}\}$ does not affinely span R^d (as is required in the definition of U.) If this happens, we exchange (that is, relabel) appropriate rows in the current matrix in order to meet the requirement. This is always possible (because aff P = R^d), does not affect the asymptotic runtime, and amounts to relabelling the vectors in $\bar{P}$. As an additional property, $\bar{P}$ realizes a total of at most (d+1)(n-d-1) + n-d-1 = O(d(n-d)) non-zero entries (for $\bar{p}_1,\dots,\bar{p}_{d+1}$ and $\bar{p}_{d+2},\dots,\bar{p}_n$, respectively). We conclude:

Lemma 1. Let P be a set of n points that affinely spans R^d. A Gale transform $\bar{P}$ of P can be computed in $O(d^2n)$ time and optimal O(dn) space.

If d is considered as a constant then an $O(n)$ time and space construction of GTs results which clearly is optimal. It is interesting to note that an essentially distinct method for constructing GTs results from their relationship to so-called Voronoi diagrams established in [2]. Though being more complicated than ours, that method achieves $O(n)$ space complexity for $d=2$, too.

3. Extreme points

If we recall the definition of a Gale transform we recognize that sets which consist of few high-dimensional points have low-dimensional GTs. This fact suggests the application of GTs to problems involving a relatively small number of high-dimensional points. Two such problems are addressed here and in Section 4, respectively.

As usual, let the point-set $P=\{p_1,\ldots,p_n\}$ affinely span R^d. For $X \subseteq P$ let conv X denote the convex hull of X. We say that $p_i \in P$ is an *extreme point* of P if

$$\text{conv } P \neq \text{conv}(P-\{p_i\}).$$

Thus the set of extreme points of P is the minimal subset X of P with conv X = conv P. X is usually called the set of *vertices* of conv P. Extremality of a point can be expressed in terms of GTs as below.

Property 1. [6]. p_i is an extreme point of P if and only if $0 \in \text{conv}(\bar{P}-\{\bar{p}_i\})$, for $\bar{P}=(\bar{p}_1,\ldots,\bar{p}_n)$ a Gale transform of P.

Property 1 enables us to solve the following problem efficiently: Given a set P of $n=d+1+k$ points in (the potentially high-dimensional space) R^d, for some constant k, enumerate those points that are vertices of the convex hull of P. As a first step, we compute a Gale transform $\bar{P}$ of P in $O(d^2(d+k+1)) = O(d^3)$ time and $O(d(d+k+1)) = O(d^2)$ space; compare Lemma 1. This leaves us with the problem of deciding, for each $p_i \in P$, whether the origin falls into the convex hull $\text{conv}(\bar{P}-\{\bar{p}_i\})$, where $\bar{P}-\{\bar{p}_i\}$ consists of $d+k$ vectors in $R^{d+k+1-d-1} = R^k$. It is well known that this problem can be dualized into a linear programming problem that involves d+k constraints in k variables. Consult [4] for a description of both the dual transform and an algorithm (by Megiddo) that solves this linear program in O(d+k) = O(d) time provided the dimension k is fixed. Since one linear program arises from each of the O(d) points of P, we obtain the following overall complexity.

Theorem 1. From a set of d+k+1 points in R^d, those being convex hull vertices can be singled out in $O(d^3)$ time and $O(d^2)$ space provided k=O(1).

Note that the space requirement is optimal since the given set needs $O(d^2)$ space itself. Interestingly, the time for computing the Gale transform dominates the time spent for the linear programs (at least asymptotically). In other words, reducing the dimension of the problem is more costy than solving the low-dimensional problem. Let us finally analyze the complexity of the problem when attacked in its original, that is, if $p_i \in$ conv P is tested directly for each $p_i \in P$. Now O(d) linear programs, each with O(d) constraints in d variables, have to be treated. Megiddo's method does not apply because d is not constant, so we need to rely on (eventually costy) algorithms like the simplex method that are not even guaranteed to run im $O(d^2)$ time on a single linear program.

4. Star watcher

Here we show that the method of GTs efficiently solves a visibility problem best defined intuitively in Marcus [8] as follows. Suppose that from each point on a very small planet, you have an unobstructed view of a hemisphere of the sky. How should a (minimal) set of stars be arranged to guarantee that at least k stars are visible from any point on the planet?

The problem can be reformulated in terms of spanning sets of vectors when the center of the planet is identified with the origin. Let $Q=\{q_1,\ldots,q_n\}$ be a set of vectors in R^d. The *positive hull* of Q is defined as

$$\text{pos } Q = \{x=\sum_{1 \leq i \leq n} \lambda_i p_i \mid \lambda_i \geq 0\}.$$

Q is called a *positive spanning set* for R^d provided pos Q = R^d. More generally, Q is a *positive k-spanning set* for R^d if pos(Q−X) = R^d, for all $X \subsetneq Q$ with $|X|=k-1$. Equivalently, $|Q \cap h| \geq k$ holds for each open halfspace h of R^d with $0 \in h$. In our context, this means that each hemisphere includes at least k points of Q. So what here is essentially asked for is an efficient method for computing a minimal positive k-spanning set. A result in [8] draws the connection to Gale transforms. It is the basis of our algorithm and can be stated as follows. Let B be any set of n real numbers, and define the set P(n,m) of n points in R^m as

$$P(n,m) = \{p(\pi)=(\pi,\pi^2,\ldots,\pi^m) \mid \pi \in B\}.$$

Property 2. [8]. Let n=2k+d−1 and m=2k−2. Any Gale transforn of P(n,m) is a positive k-spanning set for R^d.

The strategy for finding stars is now evident. First, the set P(n,m) is calculated in a straightforward manner in O(nm) = O(k(k+d)) time. Computing a GT of P(n,m) takes $O(m^2n) = O(k^2(k+d))$ time and O(mn) = O(k(k+d)) space according to Lemma 1. Finally, the vectors of the GT are translated by the vector − c, for c the center of the planet, in additional O(k(k+d)) time.

The obvious question arises if a set of less than 2k+d−1 points can positively k-span R^d. This illusion is destroyed by a simple counterexample. Choose some hyperplane g in R^d that passes through the origin and through d−1 points of the set. Then k additional points must lie in either of the two open halfspaces bounded by g which implies a lower bound of 2k+d−1 on the cardinality of the set.

From the above discussion we conclude:

Theorem 2. $O(k^2(k+d))$ time and O(k(k+d)) space suffice to arrange a minimal set of stars on the d-dimensional sky so that any hemisphere includes at least k stars.

5. Simplicial scenes

The branch of scene analysis is concerned with the reconstruction of a 3-dimensional scene from its planar projection. Particular attention has been paid to the analysis of scenes which are polyhedral, that is, where the boundaries of the objects in the scene consist of planar faces. Work in this area has been done, among others, by Sugihara [11], Imai [7], Whiteley [12], and Aurenhammer [1]. Here we are interested in a scene formed by the boundary of a single convex polyhedron. Let us first formulate the problem more precisely. We then will show that Gale transforms provide an efficient solution.

A *d-polytope* S is the convex hull conv X of a finite set X in R^d. The boundary of S consists of finitely many j-polytopes, for $0 \leq j \leq d-1$, which are called *vertices* for j=0 and *facets* for j=d-1. The vertex set $V=\{v_1,\ldots,v_n\}$ of S defines the position of S in R^d. The combinatorial structure of S can be described by the set

$$F = \{I(f) \mid f \text{ is facet of } S\},$$
$$\text{with } I(f) = \{i \mid v_i \text{ is vertex of } f\}.$$

So S is uniquely determined by the pair (V,F). A d-polytope S=(V,F) is called *simple* if each index i, $1 \leq i \leq n$, appears in exactly d sets in F. That is to say, each vertex of S is a vertex of exactly d facets. S is called *simplicial* if each set in F contains exactly d indices, which means that each facet of S is the convex hull of exactly d vertices, that is, a *(d-1)-simplex.*

A *d-picture* is a pair (P,φ), where $P=\{p_1,\ldots,p_n\}$ is a subset of R^d and φ is a subset of $2^{\{1,\ldots,n\}}$. We shall call a d-picture $C=(P,\varphi)$ *polytopical* if there exists some (d+1)-polytope S=(V,F) so that $F = \varphi$ and so that V projects vertically (i.e., parallel to the (d+1)-st coordinate axis) to P in R^d. C will also be referred to as a *projection* of S in this case. Note that the attributes of "simple" and "simplicial" carry over from (d+1)-polytopes to d-pictures in an obvious manner.

Under the introduced notation, our scene analysis problem reads:

Decide whether or not a given d-picture C is polytopical and, in the affirmative case, compute a (d+1)-polytope S whose projection is C.

For simple d-pictures, the problem is settled completely in [1]. The strategy used there, however, does not carry over to non-simple pictures. For general pictures in the plane, the decision part of the problem is attacked sucessfully by a method proposed in [11] and refined in [7] and [12]. The aim of this section is to provide a solution for the other extreme, that is, for simplicial d-pictures.

5.1. The overall algorithm

Our approach is based on the technique of Gale transforms and, in particular, on the following result by Sturmfels [10]. Let rpos X stand for the relative interior of the positive hull of the point-set X.

Property 3. [10].

(i) A d-picture $C=(P,\varphi)$ is polytopical if and only if

$$\bigcap_{I\in\varphi} \operatorname{rpos}(\bar{P}-\bar{P}(I)) \cup -\operatorname{rpos}(\bar{P}-\bar{P}(I))$$

is non-empty, for a Gale transform $\bar{P}=(\bar{p}_1,\ldots,\bar{p}_n)$ of P and for $\bar{P}(I)=\{\bar{p}_i \mid i\in I\}$.

(ii) Let $C=(P,\varphi)$ be the projection of some (d+1)-polytope (V,φ) such that a point $\bar{z}$ in the intersection above exists. Then $(\bar{p}_1,\ldots,\bar{p}_n,\bar{z})$ is a Gale transform of $V\cup\{z\}$, for z the point on the vertical axis of $\mathbb{R}^{d+1}$ at plus infinity.

In the light of Property 3, the following overall structure of an algorithm for analyzing a d-picture (P,φ) suggests itself.

Step 1. Compute a GT $\bar{P}$ of P.

Step 2. Decide the existence of a point $\bar{z}$ for $\bar{P}$ and φ. If the answer is positive then compute $\bar{z}$ else report that (P,φ) is non-polytopical and terminate.

Step 3. Derive a GT $\bar{V}$ of V from the GT of $V\cup\{z\}$ that is specified by $\bar{P}$ and $\bar{z}$.

Step 4. Reconstruct the (d+1)-st coordinates for V's vertices from its GT $\bar{V}$ and its projection P.

We shall develope and analyze this algorithm for recognizing polytopical d-pictures and reconstructing projection polytopes in some detail and under the (restrictive) assumption that the underlying d-pictures are simplicial. Step 1 has been solved in Section 2 so we focus on Step 2 and will come back to Steps 3 and 4 later.

5.2. Detailing Step 2

Let $S=(V,F)$ be a simplicial (d+1)-polytope. Recall that $V=\{v_1,\dots,v_n\}$ is the vertex-set of S, and that $I(f)\in F$ holds for each facet f of S, where $I(f)=\{i \mid v_i$ is vertex of $f\}$. By simpliciality of S, we have $|I|=d+1$ for each $I\in F$. Consider a Gale transform $\bar{V}=\{\bar{v}_1,\dots,\bar{v}_n\}$ of V. For each $I\in F$, define $\bar{V}(I)=\{\bar{v}_i \mid i\in I\}$. The following property of GTs is well-known.

Property 4. [6].

$\mathrm{conv}(\bar{V}-\bar{V}(I))$ is an (m-1)-simplex in R^{m-1}, for $m=n-d-1$.

Now let the d-picture (P,φ) be a projection of the above polytope S. Clearly, $\varphi = F$ holds such that (P,φ) is simplicial, too. For each $I\in\varphi$, consider the set $\bar{P}-\bar{P}(I)$ as defined in Property 3(i). This set consists of m points in R^m and (as can be seen by combining Property 3(ii) and Property 5) projects to $\bar{V}-\bar{V}(I)$. Hence, by Property 4, $\mathrm{conv}(\bar{P}-\bar{P}(I))$ is an (m-1)-simplex in R^m. This implies that $\mathrm{rpos}(\bar{P}-\bar{P}(I))$ is a *simplicial cone* of dimension m and with apex 0, expressible as the intersection of m open halfspaces of R^m, each bounded by a hyperplane that passes through 0 and through $m-1$ points in $\bar{P}-\bar{P}(I)$.

Our goal is to find a point $\bar{z}$ in the intersection of the double-cones $D(I) = \mathrm{rpos}(\bar{P}-\bar{P}(I)) \cup -\mathrm{rpos}(\bar{P}-\bar{P}(I))$ for all $I \in \varphi$. Clearly, if $\bar{z}$ is a solution then so does $-\bar{z}$, so only one of the cones, either $\mathrm{rpos}(\bar{P}-\bar{P}(I))$ or $-\mathrm{rpos}(\bar{P}-\bar{P}(I))$, need to be considered for intersection, for each particular $D(I)$. To single out the relevant cone we exploit the fact that the double-cones have several bounding hyperplanes in common. Let $I, I' \in \varphi$ such that the corresponding facets f and f' of S are *adjacent*, i.e., the relative boundaries of f and f' intersect in a $(d-1)$-simplex. Then $|I \cap I'| = d$ which implies that the sets $\bar{P}-\bar{P}(I)$ and $\bar{P}-\bar{P}(I')$ have $n-d-2 = m-1$ points in common. That is to say, $D(I)$ and $D(I')$ share a bounding hyperplane $h(I,I')$ of R^m. Now observe that the cones relevant for $D(I)$ and $D(I')$, respectively, have to lie in the same halfspace bounded by $h(I,I')$ since the whole intersection were trivially empty, otherwise. Each $D(I)$ shares a bounding hyperplane with some other double-cone (because each facet is adjacent to some other facet), so the relevant cone (termed cone(I)) for each double-cone $D(I)$ is determined in this way.

To find a desired point $\bar{z}$ with

$$\bar{z} \in \bigcap_{I \in \varphi} \mathrm{cone}(I)$$

we proceed as follows. Since cone(I) is the intersection of m halfspaces of R^m, $\bar{z}$ has to lie in the intersection of $m|\varphi|$ halfspaces (some of which are counted twice due to the discussion above). We first calculate the describing inequality for each halfspace and then solve the resulting linear program that consists of less than $m|\varphi|$ constraints in m variables. It will turn out, however, that not m but only $d+1$ variables are involved in each constraint.

To ease the description of the following algorithm it is assumed that each $I \in \varphi$ is associated, via pointers, with the $d+1$ sets $I' \in \varphi$ with $|I \cap I'| = d$. Intuitively speaking, the adjacencies among the facets of our virtual $(d+1)$-polytope $S=(V,\varphi)$ that should project to the d-picture (P,φ) are reflected by pointers among the respective index sets in φ. We will call the sets themselves adjacent in the sequel. In most applications, these adjacencies naturally will be part of the input picture.

As a first step, an arbitrary $I\in\varphi$ is taken and the m hyperplanes bounding cone(I) are calculated as below. Each of them passes through the origin and through m-1 points of $\bar{P}=(\bar{p}_1,\ldots,\bar{p}_n)\subseteq R^m$. Thus we need the hyperplanes

$$h_j = \text{aff}(\{\bar{p}_k \mid k\in I-\{j\}\}\cup\{0\}), \text{ for all } j\in I.$$

The normal vector v of h_j obviously satisfies

$$\bar{p}_k v = 0, \quad \text{for all } k\in I-\{j\}.$$

Since, by our construction of a Gale transform, $\bar{p}_{d+1},\ldots,\bar{p}_n$ are the m standard basis vectors of R^m, the set $\{\bar{p}_k \mid k\in I-\{j\}\}$ contains at most $n-m = d+1$ non-basis vectors which implies that at least m-1 coefficients of v will vanish. As a consequence, the equation describing h_j in R^{d+1} can be computed by means of Gaussian elimination in $O(d^3)$ time and $O(d^2)$ space. To obtain the corresponding m open halfspaces whose intersection is cone(I) we determine, for each h_j, the side that contains $\bar{p}_j$. This can be done in O(d) time per halfspace.

In order to process the remaining sets in φ correctly, we mark set I as processed and push the d+1 sets in φ that are adjacent to I onto an initially empty stack ST. While ST is non-empty, the following actions are performed.

Remove the first set J from ST. Compute the hyperplanes bounding cone(J) (in a manner similar as described above), except those which also bound cones that already have their halfspaces computed. Observe that the sets corresponding to these cones are both marked and adjacent to J. By the control flow of the algorithm, J is adjacent to at least one marked set J'. So the orientation of at least one halfspace for cone(J), namely the one contributing to cone(J'), is determined. The remaining halfspaces for cone(J) are oriented accordingly. Finally, all unmarked sets that are adjacent to J are pushed onto ST.

The output of this algorithm essentially consists of less than $|\varphi|m$ open halfspaces in R^m, each specified by a strict inequality in at most d+1 variables. By the above discussion, these halfspaces are produced in $O(|\varphi|md^3)$ time and $O(|\varphi|md)$ space. To find a point $\bar{z}$ (if it exists) in their intersection, it remains to solve the corresponding linear program by some suitable method. This matter is not further pursued here.

5.3. Detailing Steps 3 and 4

The existence of $\bar{z}$ leaves us with the problem of reconstructing the (d+1)-polytope $S=(V,\varphi)$ from its projection, the d-picture (P,φ). According to the algorithm sketched at the beginning of this section, this involves two steps.

First, a Gale transform $\bar{V}$ of the vertex-set V of S need to be calculated. By Property 3(ii), a GT of $V\cup\{z\}$, namely the (n+1)-tuple

$$M = (\bar{p}_1,\dots,\bar{p}_n,\bar{z})$$

is available since $\bar{P}$ and $\bar{z}$ are. Observe that M, when considered as an m by n+1 matrix, continues to represent a GT of $V\cup\{z\}$ when the first m-1 entries of $\bar{z}$ are eliminated by elementary row operations. (This easily follows from the definition of a GT and the effect of elementary row operations.) We end up with a tuple $(\bar{q}_1,\dots,\bar{q}_n,\bar{q}_{n+1})$ where $\bar{q}_{n+1}$ is parallel to the m-th coordinate axis of R^m. Now we take advantage of the following basic result on GTs.

Property 5. [6].

Let $(\bar{q}_1,\dots,\bar{q}_{n+1})$ be any Gale transform of $V\cup\{z\}$. For $1\leq i\leq n$, let $\bar{v}_i$ denote the projection parallel to $\bar{q}_{n+1}$ of $\bar{q}_i$ onto the subspace of R^m orthogonal to $\bar{q}_{n+1}$. Then $(\bar{v}_1,\dots,\bar{v}_n)$ is a Gale transform of V.

Because of the special form of $\bar{q}_{n+1}$ above, $\bar{v}_i$ simply consists of the first m-1 entries of $\bar{q}_i$. Elimination is performed in O(dn) time since each row of the initial matrix M contains at most d+3 non-zero entries. This is because $\bar{p}_{d+2},\ldots,\bar{p}_n$ are the m standard basis vectors of R^m which, in addition, forces the matrix $(\bar{v}_1\ldots\bar{v}_n)$ to contain the $m-1$ by $m-1$ unit matrix as a submatrix (formed by the $(d+2)$-nd through $(n-1)$-st column).

We are now ready to reconstruct V from its projection P and its Gale transform $\bar{V}$. Let h_i be the *height* of $v_i \in V$, that is, the $(d+1)$-st coordinate that distinguishes v_i from $p_i \in P$. Then, by definition of a GT,

$$(\bar{v}_1\ldots\bar{v}_n)\begin{pmatrix} h_1 \\ \cdot \\ \cdot \\ \cdot \\ h_n \end{pmatrix} = 0$$

is the only condition that restricts the heights of the vertices in V. The heights thus can be determined by finding a solution of this linear and homogenous system. This, however, is a trivial task if we keep in mind the special form of the matrix $(\bar{v}_1\ldots\bar{v}_n)$ that immediately yields the d+2 basis vectors of the solution space.

If we summarize the results of this section we see that the complexity of analyzing a simplicial d-picture (P,φ) is dominated by the effort of solving a particular linear program. This program consists of less than $(n-d-1)f$ constraints, for $n=|P|$ and $f=|\varphi|$, each of which containing d+1 out of n-d-1 variables. Of particular interest is, of course, the instance d=2, i.e. pictures in the plane. $f \leq 2n-4$ is necessary for a 2-picture to be polytopical [6], so we end up with $O(n^2)$ constraints in n-3 variables, 3 of which are non-zero per constraint. We left open the problem of selecting a method capable of solving the present linear program most efficiently.

References

[1] Aurenhammer, F. Recognising polytopical cell complexes and constructing projection polyhedra. J. Symbolic Computation 3 (1987), 249-255.

[2] Aurenhammer, F. A relationship between Gale transforms and Voronoi diagrams. Report 247, IIG-TU Graz, Austria, 1988.

[3] Brown, K.Q. Geometric transforms for fast geometric algorithms. Ph.D. Thesis, Report CMU-CS-80-101, Carnegie-Mellon Univ., Dept. Comput. Sci., Pittsburgh, PA, 1980.

[4] Edelsbrunner, H. Algorithms in combinatorial geometry. EATCS Monographs Theor. Comput. Sci., Springer, Berlin-Heidelberg, 1987.

[5] Gale, D. Neighboring vertices on a convex polyhedron. In: Linear Inequalities and Related Systems, H.W. Kuhn and A.W. Tucker, eds. (Princeton, 1956), 225-263.

[6] Gruenbaum, B. Convex polytopes. Interscience, New York, 1967.

[7] Imai, H. On combinatorial structures of line drawings of polyhedra. Discrete Applied Math. 10 (1985), 79-92.

[8] Marcus, D. Gale diagrams of convex polytopes and positive spanning sets of vectors. Discrete Applied Math. 9 (1984), 47-67.

[9] Sedgewick, R. Algorithms. Addison-Wesley, 1983.

[10] Sturmfels, B. Central and parallel projections of polytopes. Discrete Math. 62 (1986), 315-318.

[11] Sugihara, K. An algebraic and combinatorial approach to the analysis of line drawings of polyhedra. Discrete Applied Math. 9 (1984), 77-104.

[12] Whiteley, W. Motions and stresses of projected polyhedra. Structural Topology 7 (1982), 13-38.

Geometrical Abstract Automata

Ulrich Huckenbeck
Universität Würzburg
Lehrstuhl für Informatik I
Am Hubland
D-8700 Würzburg, WEST GERMANY

Introduction

One of the fundamental problems of Computational Geometry is that of the underlying machine model. The most well-known of these automata is an extended RAM described in [3], p. 28. This machine has the following capabilities:

1) applying '+','-','*' and '/' to reals,
2) deciding whether $x \geq y$ where $x,y \in \mathbb{R}$,
3) indirectly addressing its memory (integer addresses only).

But this machine has a slight disadvantage: Its primitives are **arithmetical** operations although **geometrical** problems are to be solved. Therefore it seems to be more adequate to create abstract automata with **geometrical** primitives.
This shall be done in this treatise. Our automata will e.g. intersect two circles instead of multiplying two reals; moreover our machines cannot compare reals, but they will be able to decide geometrical relations, e.g. the parallelity of two lines.
The main purpose of this treatise is to give a survey of some typical definitions and problems in connexion with these geometrical machines; some of the definitions are explained and motivated very carefully. On the other hand we omitted the proofs of our results and the formal details of our definitions. They can be found in [1] and in [2]. Consequently, this paper can be understood very easily and informs the reader about a lot of interesting ideas related to Geometrical Automata Theory.

Our treatise is organized as follows: After the basic definitions in **Chapter 1**, we introduce our geometrical automata in **Chapter 2**. The next two paragraphs are on the power of these machines. In **Chapter 3** we apply our automata to very simple problems while in **Chapter 4** the Convex Hull Problem must be solved. Finally, **Chapter 5** is devoted to a practical implementation of our automata.

1. Basic Notations

We start with the following geometrical terms:

$\mathbb{P} := \mathbb{R}^2$ equipped with the Euclidian norm and the cartesian coordinate system,
$\mathbb{G}$:= set of all straight lines in $\mathbb{P}$,
$\mathbb{K}$:= set of all circular lines in $\mathbb{P}$.
All lines $L \in \mathbb{G} \cup \mathbb{K}$ are considered as subsets of $\mathbb{P}$.

G_x is the x-axis of the cartesian coordinate system, G_y is the y-axis. Moreover, H^+ is the closed half plane $\{(x,y) \mid x \geq 0\}$ while $H_+ := \{(x,y) \mid x > 0\}$.
For every $\varrho > 0$, let $\bar{B}(0,\varrho)$ ($B(0,\varrho)$ resp.) be the closed (open resp.) circular disk with radius ϱ around (0,0).

For every points $Q,Q',Q'' \in \mathbb{P}$ with $Q' \neq Q''$ we define
(Q',Q'') := the straight line through Q' and Q'',
$(Q;Q',Q'')$:= the circular line centered in Q with the euclidian distance of Q' and Q'' as its radius.

Finally, we say that a (partial) function $f: \mathbb{R}^n \dashrightarrow \mathbb{R}^m$ ($f: \mathbb{P}^n \dashrightarrow \mathbb{P}^m$) is **rational-integer**, if each real coordinate function of f can be written as the quotient of two polynomials with integer coefficients.

Example: $f: \mathbb{P}^2 \dashrightarrow \mathbb{P}$, $((x_1,y_1),(x_2,y_2)) \longmapsto \left(\dfrac{5x_2y_1^{14} + 9}{47x_2^5y_1^3 - 71y_2} , \dfrac{y_2}{x_1^7 - y_2^3 + 3} \right)$.

2. The Presentation of the Geometrical Automata GCM_0, GCM_1 and their Extensions

In this paragraph we wish to introduce our geometrical automata. We only describe the main features of these machines and leave out the formal details. The exact definitions can be found in [1] on the pages 12 - 18 and 21 - 22; in that work the GCM_0 is called by 'ZLM', and the 'RLM' is the same as the GCM_1.
For the purpose of simplicity, we assume that our geometric automata can deal with all elements of $\mathbb{P} \cup \mathbb{G} \cup \mathbb{K}$ - even if they depend on irrational parameters (e.g. the point $(\pi,\sqrt[5]{7})$). This indeed makes the description of our algorithms easier, because we elude the problems arising from digitalization or approximation of geometrical objects.

2.1. Definition:

A Geometrical Construction Machine of type 0 (= GCM_0) is an automaton simulating the operations with compass and ruler.

For this end the machine has three types of registers:
Points are stored in p0,p1,p2,p3,.... ,
lines " " " g0,g1,g2,g3,.... , and
circles " " " k0,k1,k2,k3,.... .

These registers are **initialized** such that all fundamental geometric objects of the cartesian coordinate system are available:

p0, p3, p6, p9, ... are initialized with the point (0,0),
p1, p4, p7, p10, ... " " " " " (1,0),
p2, p5, p8, p11, ... " " " " " (0,1),
g0, g2, g4, g6, ... " " " " x-axis G_x,
g1, g3, g5, g7, ... " " " " y-axis G_y,
k0, k1, k2, k3, ... " " " " unit circle around (0,0).

Let us make the following convention: We denote the current **contents** of each register by the corresponding indexed capital letter; e.g., the line stored in g76 is called by G_{76}.

Before a GCM_0 begins its work, we have to input a tuple $t = (P_1,\dots,P_\ell\ ,\ G_1,\dots,G_m\ ,\ K_1,\dots,K_n)$ ($\ell,m,n \in \mathbb{N}$). These objects are loaded into the registers p1,...,pℓ , g1,...,gm , k1,...,kn resp. The remaining registers (e.g. p0, g(m+1)) keep the data they received during the initialization.

After that the GCM_0 executes the **statements** of its **program**. Such a program can be seen in Example 3.3. The instructions of the GCM_0 can be structured as follows (where $i,i',i'',j \in \mathbb{N}$):

<1> pj := gi ∩ gi'; ,

<2> pj := gi ∩ ki'; , pj := ki ∩ gi'; , pj := ki ∩ ki';

<3> pj := gi ∩ ki'\ {pi"}; , pj := ki ∩ gi'\ {pi"}; , pj := ki ∩ ki'\ {pi"};

<4> gj := (pi,pi');

<5> kj := (pi ; pi',pi");

<6> pj := pi; , gj := gi; , kj := ki;

<7> write(pj); , write(gj); , write(kj);,

<8> nop; (no operation) ,

<9> end. (last command of every GCM_0-program)

The **meaning** of these statements is almost obvious; but we still have to deal with the following problems:

α) The statements in <2> are to effect the intersection of two lines C',C" where one of them is a circle. Then it "very frequently" happens that C' and C" have two common points. In this case the GCM_0 chooses the point P_j **nondeterministically** within C' ∩ C". If the GCM_0 knows already a point $P_{i''} \in C' \cap C''$, then it can construct the other point of intersection by applying an appropriate instruction of type <3>.
According to this, the statements 'p33 := k1 ∩ g4; p16 := k1 ∩ g4 \ {p33};' can effect the following evaluations of P_{33} and P_{16}:

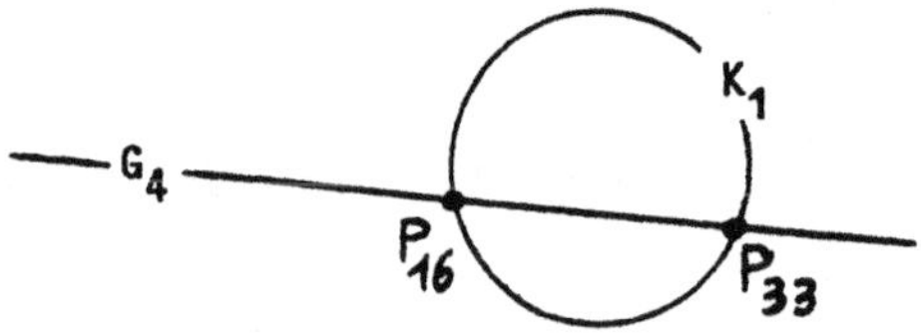

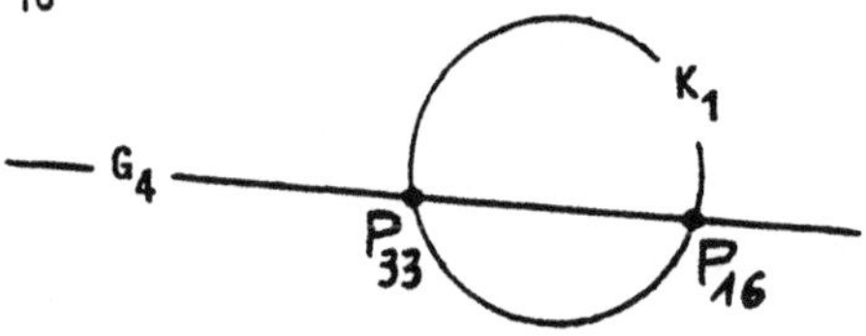

β) A correct execution of the current statement φ is not possible if one of the following **degenerate cases** arises:

- φ is of type <1> and G_i is parallel to $G_{i'}$.
- φ is of type <2> and the lines to be intersected do not have exactly two common points. (E.g.: $G_i \cap K_i = \emptyset$, G_i is a tangent to $K_{i'}$, or $K_i = K_{i'}$).
- φ is of type <3>, where the intersected lines C',C" and the point $P_{i''}$ do not satisfy the following conditions: $|C' \cap C''| = 2$ and $P_{i''} \in C' \cap C''$.
- φ is of type <4> and $P_i = P_{i'}$.
- φ is of type <5> and $P_{i'} = P_{i''}$.

In order to describe the behaviour of a GCM_0 in these cases we define the three simple states N,E and F: During its construction, the GCM_0 is in the <u>n</u>ormal state N. As soon as one of the degenerate cases occurs, the automaton breaks off its work and falls into the <u>e</u>rror state E. When, however, the machine reaches the instruction 'end.', then it takes the <u>f</u>inal state F and terminates its work correctly. ■

Now our definition of the GCM_0 is finished. It is obvious that we can construct similar automata basing on other geometrical tools. In the next definition we shall describe the GCM_1 which is a machine simulating the use of a rectangular ruler. This automaton will have the following advantages: Firstly it is deterministic. Secondly we can use the concept of the GCM_1 to find an automaton basing on **geometric** primitives which has the same power as the extended RAM mentioned above. We shall describe these modifications immediately after Theorem. 3.5.

<u>2.2 Definition:</u>

A <u>G</u>eometrical <u>C</u>onstruction <u>M</u>achine of type 1 (= GCM_1) is an automaton similar to the GCM_0. The only difference is the following: A GCM_1 cannot apply any operation to circles, i.e. the statements <2> , <3> and <5> do not exist. On the other hand, the GCM_1 needs only a single step to drop the perpendicular to a line $G_{i'}$ through a point P_i; this happens when executing the command

<10> gi := l(pi,gi'); . ■

In the next definition we shall describe a simple extension of the GCM_0's and the GCM_1's (see [1], Def. (3.1.1)):

2.3 Definition:

Let $A \subseteq \mathbb{P}^{\ell} \times \mathbb{G}^m \times \mathbb{K}^n$. An A-GCM_0 is a machine which can execute all GCM_0-commands; moreover the following instructions are possible (where $j, \lambda_1,\ldots,\lambda_\ell, \mu_1,\ldots,\mu_m, \nu_1,\ldots,\nu_n \in \mathbb{N}$)

<11> goto j; (a jump to the j-th instruction of the GCM_0-program),

<12> if $(p\lambda_1,\ldots,p\lambda_\ell , g\mu_1,\ldots,g\mu_m, k\nu_1,\ldots,k\nu_n) \in A$ then goto j;

(The machine jumps iff the tuple $(P_{\lambda_1},\ldots,P_{\lambda_\ell} , G_{\mu_1},\ldots,G_{\mu_m}, K_{\nu_1},\ldots,K_{\nu_n}) \in A \subseteq \mathbb{P}^{\ell} \times \mathbb{G}^m \times \mathbb{K}^n$.)

If the program of some A-GCM_0 constists of N lines then every jump address j has to satisfy the condition $j \in \{1,\ldots,N\}$; this means that the j-th program line actually exists.

The A-GCM_1 is defined in an analogous way. ∎

2.4 Remark:

a) The simplest oracles A are the subsets of $\mathbb{P}$, i.e. $\ell = 1$, $m = 0$, $n = 0$. Further typical examples are

$\tilde{A} := \{(G,G') \in \mathbb{G}^2 \mid G \text{ parallel to } G'\}$,

$\hat{A} := \{(P,K) \in \mathbb{P} \times \mathbb{K} \mid P \in K\}$.

The corresponding conditional jumps are 'if $(g\mu_1,g\mu_2) \in \tilde{A}$ then goto j;' and 'if $(p\lambda_1,k\nu_1) \in \hat{A}$ then goto j;' . Consequently the $\tilde{A}$-GCM_0 decides the parallelity of G_{μ_1} and G_{μ_2} while the $\hat{A}$-GCM_0 jumps if P_{λ_1} is situated on K_{ν_1}.

b) It is easy to create 20 or more very simple oracles $A_1,\ldots,A_{20}$, and it is obvious that we can construct extensions of the GCM_0 and the GCM_1 which can use more than only one oracle (e.g. the (A_2,A_5,A_9,A_{11})-GCM_0). This means that the number of possible extensions of our machines is greater or equal to the cardinality of $2^{\{A_1,\ldots,A_{20}\}}$ which is greater than 1 million. Since we cannot deal with such a plenty of geometrical automata, we confine ourselves to a small subclass. This subclass consists of those machines which have only one oracle.

c) The A-GCM_0's are a good model to simualate a real drawer when avoiding nondeterministic GCM_0-operations. For this end, the human constructor implicitely uses control structures to find out the appropriate point of intersection. The same method can be applied by an A-GCM_0 if the oracle A is powerful enough. ∎

3. On the Power of (A-)GCM_0's and (A-)GCM_1's

In this paragraph we want to line out some investigations about the power of our geometrical automata. We begin with considering constructible (= computable) functions. After this we shall deal with the decidability of oracle sets.

3.1 On Functions which can be Constructed by a (A-)GCM_0 or a (A-)GCM_1

3.1 Definition: (see [1], (2.1.3) - (2.1.5))
Given $\ell, m, n, \ell', m', n' \in \mathbb{N}$ and the sets A,D,W where $A \subseteq \mathbb{P}^{\ell} \times \mathbb{G}^{m} \times \mathbb{K}^{n}$, $D \subseteq \mathbb{P}^{\ell'} \times \mathbb{G}^{m'} \times \mathbb{K}^{n'}$.
Let M be a (A-)GCM_0 or a (A-)GCM_1 and let $F : D \longrightarrow W$ (e.g. $F : \mathbb{P} \longrightarrow \mathbb{P}$).
Then the machine M **computes** (or **constructs**) the function F iff for every input $t \in D$ the following is true:

(*) M outputs F(t).

(**) M reaches the final state F. (I.e. M does not run into an infinite loop or into the error state E.)

If M is a (A-)GCM_0 then

(#) the conditions (*) and (**) must be satisfied for **every** nondeterministic path of decisions. ∎

3.2 Remark:
In the normal nondeterministic concepts we only require the **existence** of nondeterministic decisions such that (*) and (**) are true. Therefore, part (#) of Definition 3.1 is rather unusual. Nevertheless there are some reasons why it is sensible:

1) Although condition (#) is seldom mentioned within Automata Theory, it often occurs in Geometry: In a lot of Euclidian constructions we choose an arbitrary auxiliary point P on a circular or straight line L; the **result** of the construction, hovewer, must be independent from P. The same idea occurs in Def. 3.1. The only difference is that P is not chosen among the points of L but among the two points of C' ∩ C" where C' or C" is a circle.
2) Def. 3.1. helps us to create the following extremal situation: The GCM_0 is a very "weak" machine (e.g. without any control structures); on the other hand this machine has to solve the difficult problem of constructing functions under a very sharp condition. Consequently the set $\mathcal{F}$ of the GCM_0-constructible functions F: D ⟶ W is very small. It often occurs that we want to determine the class $\mathcal{F}'$ of those functions which can be constructed by some extended GCM_0 (possibly under a less restrictive condition than (#)). Then it is very helpful to know that $\mathcal{F} \subseteq \mathcal{F}'$.
3) In practice, a nondeterministic computer is only usable if its output does not depend on nondeterministic decisions.

The condition (#) can also be found in [4] and in [5]. In these sources, H. Tietze proved that the constructibility of a point (not of a function) according to (#) is equivalent to its constructibility with a rectangular ruler. This is a further argument for considering this drawing instrument.

3.3 Example:

The following GCM_0 M constructs the function $F\colon D \longrightarrow \mathbb{P}^2$ where $D = \mathbb{P}\setminus\{(0,0)\}$ and $(\forall t = (x,y) \in D)\ \ F(t) = ((x,0),(y,0))$. This means that M decomposes every point $(x,y) \neq (0,0)$ into its coordinates. In the following program the commands of type <3> are marked with a '*'; morover, the value of the underlined variables is the result of the initializations and not of a previous command.

```
  g2  := (p0,p1);

  k3  := (p0;p0,p1);

* p4  := g2 ∩ k3 \ {p1};

  k5  := (p1;p1,p4);

  p10 := g0 ∩ k5;

* p20 := g0 ∩ k5 \ {p10};

  p11 := g1 ∩ k5;

* p21 := g1 ∩ k5 \ {p11};
```

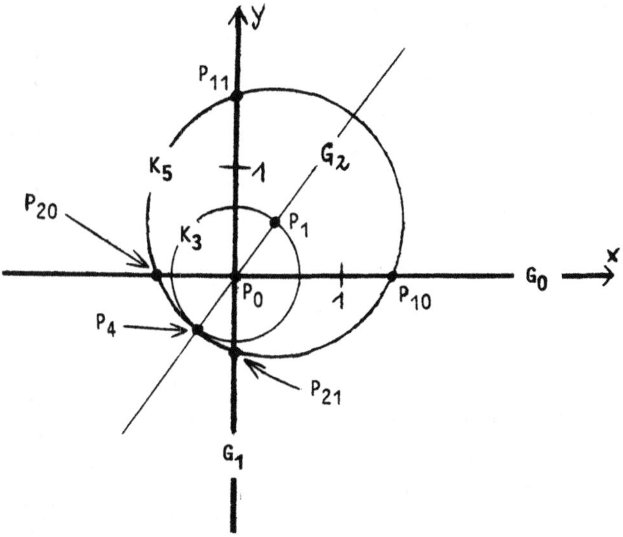


```
  k30 := (p10;p10,p20);

  k40 := (p20;p20,p10);

  p50 := k30 ∩ k40;

* p60 := k30 ∩ k40 \ {p50};

  g70 := (p50,p60);

  p80 := g0 ∩ g70;

  write(p80);
```

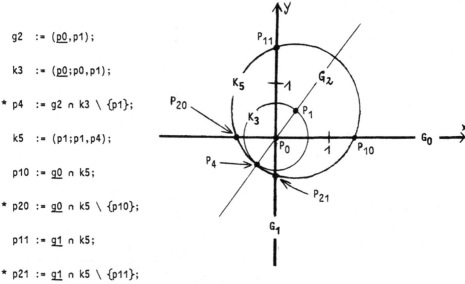

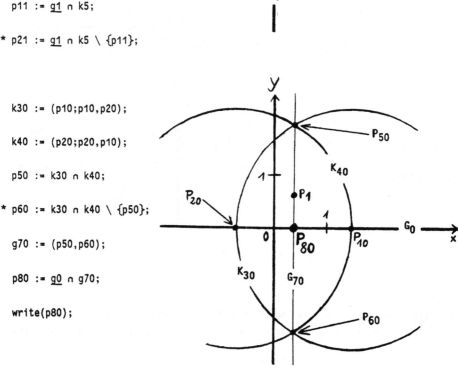

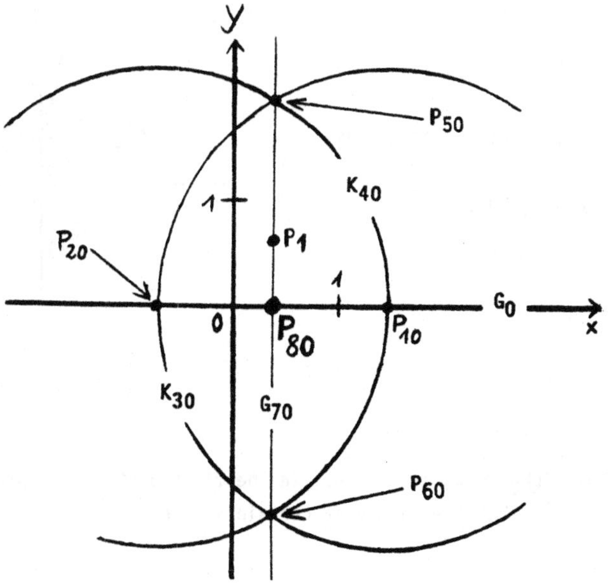

```
  k31 := (p11;p11,p21);

  k41 := (p21;p21,p11);

  p51 := k31 ∩ k41;

* p61 := k31 ∩ k41 \ {p51};

  g71 := (p51,p61);

  p81 := g1 ∩ g71;
```

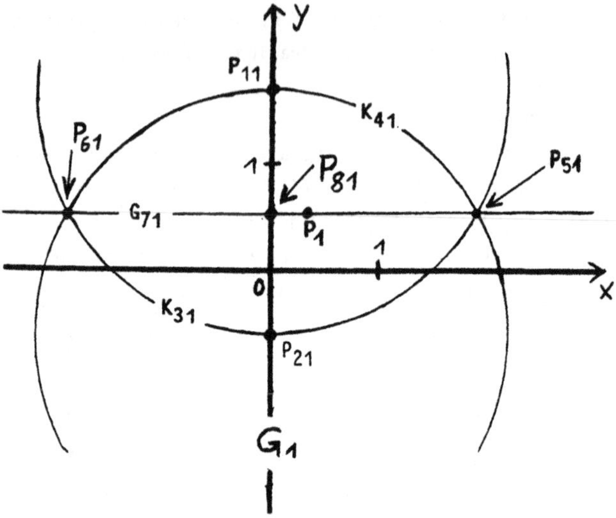


```
  k10 := (p7;p8,p8);

  k11 := (p8;p7,p8);

  p13 := k10 ∩ k11;

* p14 := k10 ∩ k11 \ {p13};

  k13 := (p13;p13,p81);

  k14 := (p14;p14,p81);

* p90 := k13 ∩ k14 \ {p81};

  write(p90);

  end.
```

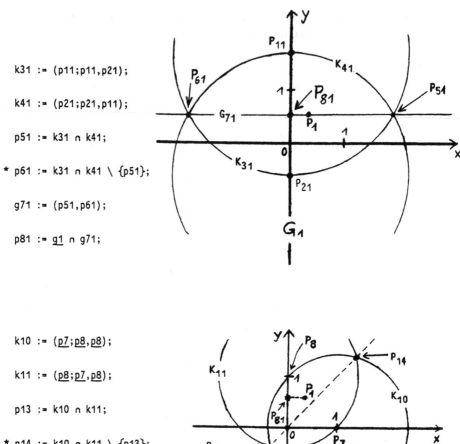

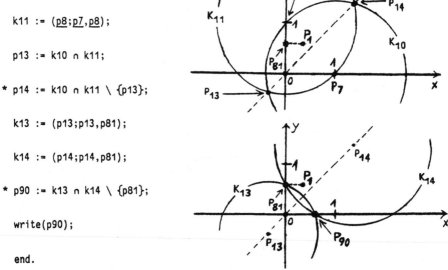

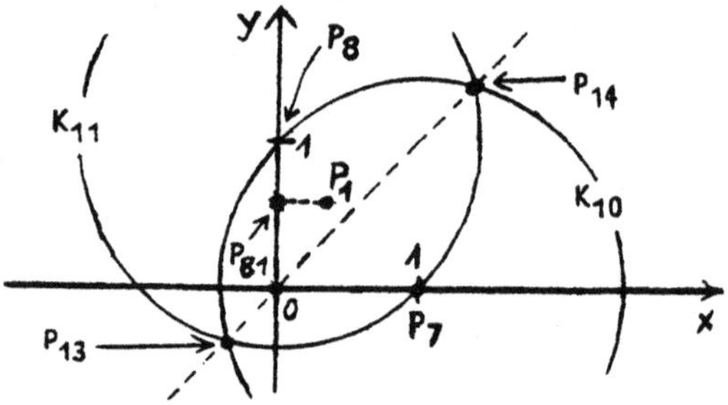

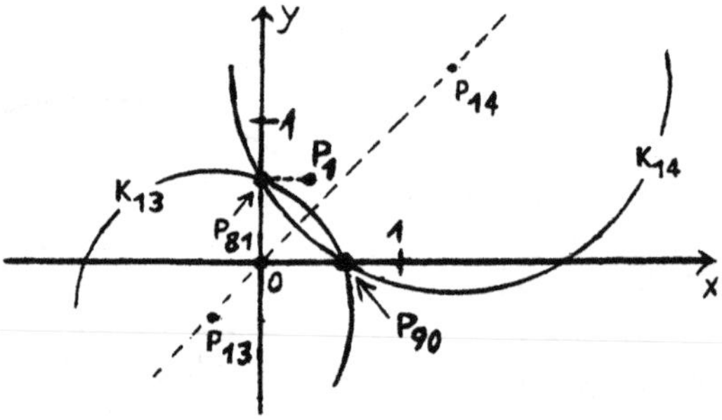

This GCM_0 works as follows: In the first part of its program it constructs the point P_4 such that the circle K_5 certainly intersects the x-axis G_0 (the y-axis G_1, resp.)

in two points P_{50} and P_{60} (P_{51} and P_{61}, resp.). In the second part of the program, M constructs the perpendicular bisector to P_{50} and P_{51}; thus it obtains the point $P_{80} = PR_x(t) = (x,0)$. In an analogous way, the point $P_{81} = PR_y(t) = (0,y)$ is generated in the third part of the program. Finally, the second output $P_{90} = (y,0)$ is the result of reflecting P_{81} in the line $\{(x,x) \mid x \in \mathbb{R}\}$ which contains P_{13} and P_{14}.

It is obvious that M does not work correctly if $t = (0,0)$ is input; already the first instruction effects the state E because of $P_1 = t = (0,0) = P_0$. ∎

After discussing this example we want to investigate the general case. The next theorem is about the GCM_0-constructible functions $F : D \longrightarrow \mathbb{P}$ where $D \subseteq \mathbb{P}$. At the first sight, we expect that every function F consisting of nested square roots is GCM_0-constructible, e.g. $F^{(+)}(x,y) = (x + \sqrt{x^2 + y^4 + 3}\ , x - y)$. The proof of this result seems to be a simple transfer of well-known results from Euclidian Geometry to our GCM_0's. In reality, however, the set of GCM_0-constructible functions is much smaller:

<u>3.4 Theorem:</u> (see [1], (2.3.1) and (2.3.2))

a) The only GCM_0-constructible functions $F: \mathbb{P} \longrightarrow \mathbb{P}$ are the identity on $\mathbb{P}$ and some constant functions.

b) The only GCM_0-constructible functions $F: \mathbb{P}\backslash\{(0,0)\} \longrightarrow \mathbb{P}$ are the rational-integer functions (**without** square roots).

The <u>reasons</u> for these surprising statements are the following:

<u>Fact a)</u> arises from the missing control structures of the GCM_0's. This means that a GCM_0 cannot decide whether some instruction effects the error state E. Let now M be a GCM_0 constructing a function $F \neq$ const, $F \neq Id_{\mathbb{P}}$. Then we can find an input $t \in \mathbb{P}$ such that M runs into the state E. Consequently, M does not satisfy condition (**) of Def. 3.1 which is a contradiction to the assumption that M constructs F.

Our discussion of <u>Fact b)</u> has two aspects. First we compare it to Fact a) and see that the set of constructible functions has become greater. This is because the GCM_0 now "knows" that the input point t is unequal to (0,0). Consequently, the automaton can execute some further operations without running into the state E. One of them is the first instruction of Example 3.3.

On the other hand, the set of constructible functions F remains very small. Although $\mathbb{P}\backslash\{(0,0)\}$ is a smaller domain of definition than that one in Fact a) we do not obtain the GCM_0-computability of nested root functions like $F^{(+)}$. The reason for this is the condition (#) in Def. 3.1. The only functions which are independent from nondeterministic decisions of a GCM_0 are the rational-integer ones. For example, if a GCM_0 tries to construct $F^{(+)}$ then it cannot avoid the wrong result $F^{(-)}(x,y) = (x - \sqrt{x^2 + y^4 + 3}\ , x - y)$ effected by nondeterministic steps of type <2>.
Note that a similar problem was treated by H.Tietze in [4]. ∎

It is clear that a sufficiently strong oracle A can avoid the difficulties just mentioned; the corresponding A-GCM_0 can decide whether the next instruction effects the state E and thus avoid such commands; moreover the oracle A can be used to find the desired point of intersection between a circle and another line (see Remark 2.4 c). Consequently the class of A-GCM_0-constructible functions is often much greater than those classes which we described in the previous theorem.

Let us now deal with the GCM_1-constructible functions. Then we obtain the following simple result:

<u>3.5 Theorem:</u> (see [1], (2.4.1))
Let $D \subseteq \mathbb{P}$. Then $F: D \longrightarrow \mathbb{P}$ is GCM_1-constructible iff F is rational-integer. ∎

This implies that the H^+-GCM_1's with the additional capability of indirect addressing are indeed as powerful as the extended RAM's presented in [3], p.28.

We now wish to finish our discussion about (A-)GCM_0- and (A-)GCM_1-constructible functions. We next investigate the capability of these machines to decide an oracle A'.

3.2 A-GCM_0's and A-GCM_1's as Deciders

In analogy to Def. 2.1 we first have to introduce a fundamental expression:

<u>3.6 Definition:</u> (see [1], (3.1.3))
Let $\ell, m, n, \ell', m', n' \in \mathbb{N}$ and let $D := \mathbb{P}^{\ell} \times \mathbb{G}^m \times \mathbb{K}^n$ and $D' := \mathbb{P}^{\ell'} \times \mathbb{G}^{m'} \times \mathbb{K}^{n'}$.
Given $A \subseteq D$ and $A' \subseteq D'$.
Then we say that A' is GCM_0-decidable by A ($A' \leq A$) iff there are an A-GCM_0 M and two numbers $i_{\in}, i_{\notin}$ such that for every input $d' \in D'$ the following is true:

(*) If $d' \in A'$ then M passes the $i_{\in}$-th instruction and avoids the $i_{\notin}$-th command of its program. If $d' \notin A'$ then M behaves oppositely.
(**) For every $d' \in A'$, M reaches the final state F.

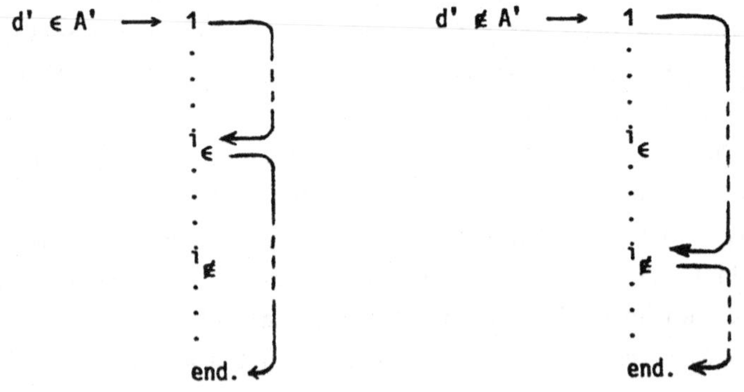

This conditions (*) and (**) must be satisfied for **every** nondeterministic path of decisions.
If even $A' \leq A$ and $A \leq A'$ then the oracles A and A' are called GCM_0-equivalent.

The GCM_1-decidability ($A' \leq\leq A$) and the GCM_1-equivalence are defined in an analogous way.

■

The following remark only deals with the case of A-GCM_0's, but it can also transferred to the case of A-GCM_1's. We shall discuss the relationship between the decidability of A' and the constructibility of the function $1_{A'}: D' \longrightarrow \mathbb{P}$ which is defined as follows:

$$(\forall\, d' \in D') \quad 1_{A'}(d') = \begin{cases} (1,0) & \text{if } d' \in A', \\ (0,0) & \text{if } d' \notin A'. \end{cases}$$

<u>3.7 Remark:</u>
Another version of defining decidability is the following:
A' is GCM_0-decidable by A iff there is an A-GCM_0 M* constructing $1_{A'}$.
But the great advantage of the machine M described in Definition 3.6 is that M immediately helps to simulate conditional jumps "if $a' \in A'$ then goto j;". We only have to insert appropriate jumps into the lines $i_\in$ and $i_\notin$. Consequently, the relation "$A' \leq A$" as defined in 3.6 indeed means that the oracle A' can be replaced by A.

In opposite to this, the A-GCM_0 M* constructing $1_{A'}$ does not immediately yield a simulation of conditional jumps depending on A'. Therefore the question arises whether our suggested modification of Def. 3.6. is equivalent to 3.6 itself, i.e.
$A' \leq A$ (according to Def. 3.6) $\Longleftrightarrow$ $1_{A'}$ is A-GCM_0-constructible.
We now want to discuss this equivalence:

a) The direction " $\Longrightarrow$ " is almost trivial. We only have to modify the machine M mentioned in 3.6 such that the following happens: If M passes its $i_\in$-th program line then it prints (1,0), and the line $i_\notin$ effects the output of (0,0).
b) Also the opposite direction can be proven if $A, A' \subseteq \mathbb{P}$. This is done in [1], (3.3.2) - (3.3.5). If, however, A or A' have a more complicated structure then we do not know whether " $\Longleftarrow$ " is always true.
c) It is very easy to construct an extension of the A-GCM_0 such that the direction " $\Longleftarrow$ " is false. For this end let $D' = \mathbb{P}$. Then we only have enumerably many H^+-GCM_0's. Each of them can only decide finitely many oracles $A' \subseteq \mathbb{P}$ because every A' decidable by some H^+-GCM_0 M must have the form $A' = A_\nu = \{d' \in \mathbb{P} \mid$ M runs through its ν-th program line if d' is input$\}$; consequently we can find a set $B \subseteq \mathbb{P}$ which is not GCM_0-decidable by H^+. Using B we now define the <A-GCM_0> ($A \subseteq \mathbb{P}$); this machine is an A-GCM_0 with the additional command 'write$\big(1_B(p1)\big)$;'. Then it is obvious that 1_B can be constructed by some <H^+-GCM_0> M although B cannot be decided by any <H^+-GCM_0> according to Def. 3.6. Hence, " $\Longleftarrow$ " is not always true for <A-GCM_0>'s, for $A := H^+$ and $A' := B$ are counterexamples.

■

After these principal considerations we now want to deal with more concrete problems. The following theorem presents some results about the decidabilty of very simple oracle sets. The definition of these sets can be found in Chapter 1.

<u>3.8 Theorem:</u>

a) The sets G_x, G_y and $\{(0,0)\}$ are both GCM_0- and GCM_1-equivalent.

b) Also the disks and half planes $B(0,1)$, $\bar{B}(0,1)$, H^+ and H_+ are GCM_0- and GCM_1- -equivalent.

c) Every oracle A' mentioned in a) is weaker than any oracle A described in b), i.e. $A' \leq A$, $A' \leq\leq A$, but none of the inverse relations is true.

d) For every $\varrho > 0$ we have: $\bar{B}(0,\varrho) \leq\leq \bar{B}(0,1) \iff \varrho$ is algebraic.

e) For every $\varrho > 0$ we have: $\bar{B}(0,1) \leq\leq \bar{B}(0,\varrho) \iff$ There are $\xi, \zeta \in \mathbb{Q}$ such that $\varrho^2 = \xi^2 + \zeta^2$.

The proofs of these facts can be found in [1]; see the Theorems and Lemmas (3.5.16), (3.5.9), (3.5.8) and (3.5.1) - (3.5.6). ∎

The results a) - c) are not very surprising, but probably the reader is astonished at the different right sides of the equivalences in d) and e). Consequently it seems to be very difficult to find a general characterization of those pairs (ϱ,ϱ') for which $\bar{B}(0,\varrho')$ is GCM_1-decidable by $\bar{B}(0,\varrho)$. Note that this complicated problem can be **formulated** in a very **elementary** manner: We only need the rectangular ruler and the closed disks $\bar{B}(0,\varrho)$.

We now want to finish our discussion about the power of the (A-)GCM_0's and the (A-)GCM_1's if applied to elementary problems. In our next paragraph a more complex geometrical problem is considered which is typical for Computational Geometry. In order to evade the difficulties arising from the nondeterministic steps of (A-)GCM_0's we shall only deal with (A-)GCM_1's.

4. The A-GCM_1 and the Convex Hull Problem

In our previous problems we always considered inputs with a fixed length. Now, however, we have to treat inputs of the form $(P_1,\ldots,P_n)$ where n is variable. This requires the following extension of the A-GCM_0's: They must also have the capability of indirectly addressing their memory. Then the following fact is true:

<u>4.1 Theorem:</u>

For the H^+-GCM_1, the worst-case time complexity of the Convex Hull Problem is in $\Theta(n\log n)$.

The proof to this theorem is based on the same ideas as the well-known proof in the case of the extended RAM; one of them is the reduction of the Convex Hull Prob-

Our next results are about reduction theory. We assume that an H^+-GCM_1 can execute a particular kind of geometric operations very quickly (e.g. with the help of a parallel processor). Then we want to know the complextity of the Convex Hull Problem if solved by such an improved H^+-GCM_1.
Our first of these theorems is the main result of [2]:

<u>4.2 Theorem:</u>
Let $T: \mathbb{R}_+ \longrightarrow \mathbb{R}_+$ be strictly monotoneous increasing (where $\mathbb{R}_+ := \{x \in \mathbb{R} \mid x > 0\}$).
We assume that the following problem can be solved within $\Theta(T(j))$ - instead of $\Theta(j)$ - time units:

Given j arbitrary points $Q_1,\dots,Q_j$. Find the first of them which is met if we counterclockwise turn a ray R around its starting point Q.

Then for a great class of sublogarithmic functions T the following is true:
The worst-case time for solving the Convex Hull Problem is in $\Theta(n \cdot T(n))$.

Note that there is a simple gift-wrapping method working in $O(n \cdot T(n))$ time:
Let $P_1,\dots,P_n$ be the input points where $P_i = Q$ is that one with the minimal x-coordinate. Moreover let R be the ray starting at Q and directed down. Then our algorithm repeats the following operations until returning to P_i:

```
rotate R counterclockwise until the
  first point Q' ∈ {P1,...,Pn} is met;

let R' be the ray through Q' starting
  from Q;

Q := Q';

R := that part of R' starting at Q';

end.
```

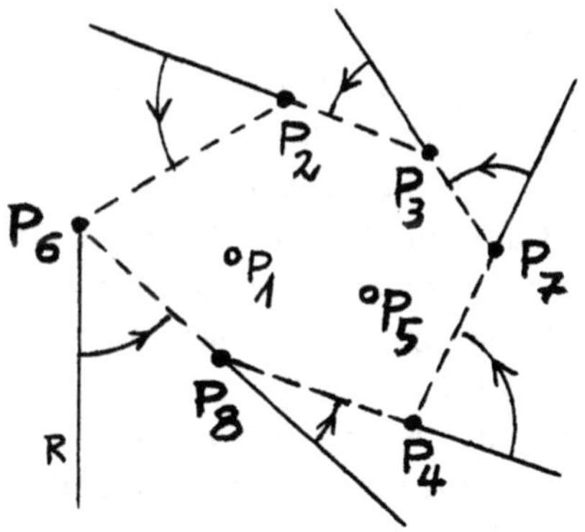

■

A similar result is the following:

<u>4.3 Theorem:</u> (see [1], (5.2.13))
Let $T: \mathbb{R}_+ \longrightarrow \mathbb{R}_+$ be a strictly monotoneous increasing function in $\Omega(\log(n))$.
We assume that every H^+-GCM_1 can execute the following operation within $\Theta(T(n))$ - instead of $\Theta(n)$ - time units:

The machine can nondeterministically create a directed line G and decide which of the input points $P_1,\dots,P_n$ are on the left (on the right, resp.) of G.

Then the complextity of the Convex Hull Problem is in $\Theta(n \cdot \log(T(n)))$. ■

Let us finish this chapter with the following consideration:

4.4 Remark:

Also the proofs to 4.2 and 4.3 are based on the reduction to the classical Sorting Problem. In [1, (5.1.7) - (5.1.8)] and in [2, Def. 2.2, Remark 2.3], a new technique is applied to investigate its complexity. For this end we introduce the term 'sorting-cardinality of $\mathcal{F}$' (= sc($\mathcal{F}$)) where $\mathcal{F}$ is a set of partial functions $f: \mathbb{R}^n$ - -> $\mathbb{R}^n$.

This means for every query tree T symbolizing a sorting algorithm: If only functions $f \in \mathcal{F}$ can be computed along the query paths, then T must have at least sc($\mathcal{F}$) leaves. Consequently sc($\mathcal{F}$) indeed helps to find lower complexity bounds of the Sorting Problem.

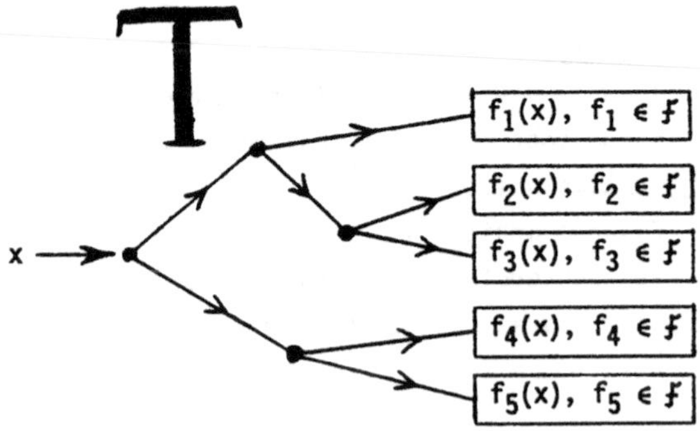

5. An Implementation of the H^+-GCM_0

The author also implemented a program named 'GEO' which simulates arbitrary H^+-GCM_0's; he is very grateful to Mr. Roger Gutbrod for his good ideas to improve this program.

'GEO' works on an 'Olivetti M 24' personal computer with the operating system MS-DOS. Our algorithm is written in an extended TURBO-PASCAL including the instructions of GRAPHIX TOOLBOX.

The simulation of an H^+-GCM_0 M runs as follows:

First the user writes the program of M into a particular text file; this program is followed by three integers ℓ, m, n; they mean that M is to work on inputs of $\mathbb{P}^\ell \times \mathbb{G}^m \times \mathbb{K}^n$.

After that, the user calls 'GEO' itself and inputs the geometrical data. Then 'GEO' simulates the machine M by interpreting its program lines. For this end, the current points, lines and circles in the memory of M are represented by real parameters, so that a lot of H^+-GCM_0-statements can be simulated by applying arithmetical operations to these reals. (By the way, the nondeterministic GCM_0-decisions are replaced by random decisions.) But there is one important situation where 'GEO' makes use of the GRAPHIX TOOLBOX; this happens when interpreting a write-command, e.g. write(g4). Thus the output of the H^+-GCM_0 M finally appears on the screen of the terminal.

The program 'GEO' is used in experimental courses at the University of Würzburg. Our students have to write and to test a program for an H^+-GCM_0 constructing the function $1_{H'}$, where H' := $\{ (x,y) \in \mathbb{P} \mid y \geq x \}$. This problem is much more complicated than it seems to be at the first sight because the error state E must be avoided (see 3.4). Thus the students see the difficulties arising from expressing euclidian construtions in automata theoretical terms.

Concluding Remarks

We gave a survey of some important and typical terms and results related to the geometrical automata (A-)GCM_0 and (A-)GCM_1; moreover we described an interpreter for H^+-GCM_0's. It is obvious that the plenty of possible machines mentioned in 2.4 effects a lot of interesting open questions. We do not want to discuss them, but we wish to present two more general and principal problems:

1) Extend the concept of the GCM_0's such that they are **exactly** as powerful as the normal drawer with compass and ruler. (This also means that the desired automata must not be too powerful.)
2) Give a sensible **formally exact** definition of the term 'abstract geometric automaton'. (In particular, this definition is to comprehend e.g. the (A-)GCM_0's, the (A-)GCM_1's and geometrical RAM's treating points of $\mathbb{R}^d$ ($d \geq 2$).)

References

[1] U.Huckenbeck: Geometrische Maschinenmodelle. Ph.D.thesis, University of Würzburg (1986).

[2] U.Huckenbeck: On the Complexity of Convex Hull Algorithms if Polar Minima can be Found very Fast. To appear in: Zeitschrift für Operations Research (ZOR), Vol. 32, Issue 3 (1988).

[3] F.P.Preparata, M.I.Shamos: Computational Geometry, an Introduction. Springer (1985).

[4] H.Tietze: Über die Konstruierbarkeit mit Lineal und Zirkel. Sitzungsberichte der kaiserlichen Akademie der Wissenschaften. Wien, math.-natw. Klasse 118, Abteilung IIa (1909), p. 735 - 757.

[5] H.Tietze: Über die mit Lineal und Zirkel und die mit dem rechten Zeichenwinkel lösbaren Konstruktionsaufgaben. Math. Zeitschr., Bd. 46 (1940) p. 190 - 203.

Automatizing geometric proofs and constructions

Beat Brüderlin
Institute for Computer Science
Swiss Federal Institute of Technology, ETH
Zürich, Switzerland

Abstract

The theory of Euclidean Geometry is the foundation of almost all Computer-Geometry applications. Also it is one of the first mathematical theories that has been axiomatized systematically by D. Hilbert, in the beginning of this century [HIL 71]. Nevertheless, for most algorithms of "Computational Geometry" the algebraic interpretation of Geometry is of greater importance (see, e.g. [SHA 78]). Already, there exist some very early approaches for proving geometric theorems with Computer-Algebra methods. In this paper a new approach is presented. We use logic programming and directly apply geometric axioms for inferring constructive proofs of elementary geometry theorems. We describe the basic conceptions of the algorithm and its application for interactive modelling of geometric objects.

1) Introduction

Computer Aided Design- (CAD) systems are used for interactively constructing geometric objects. We assume that these objects should fulfill certain given specifications and that the interactive user can chose from a great number of construction tools offered by the system. Let us further assume that the specification can be expressed by geometric constraints, such as, for instance, distances, angles, etc. Therefore, the problem and its solution are of purely geometric nature. If we express the influence of each construction operation in an axiomatic way, the problem of finding the sequence of construction operations for a specified object may be seen as the problem of finding a proof of the constructibility of the object with the given tools. Therefore, a construction problem is a special case of a geometric theorem proving problem.

Most known approaches to this or similar problems first translate from a geometric language to an algebraic formulation, by introducing coordinates. A solution can be found with numerical or algebraic methods. With the approach presented here the geometric constraints (which are geometric relations between points) are represented by predicates. The geometric construction operations are represented by functional expressions. We may express geometric axioms by equations on first order formulas, which allows us to use a rewrite-rule mechanism, and by implications, which may be inferred by a backtracking mechanism, such as implemented in the logic-programming language Prolog. We can prove uniform termination of the inference mechanism using the rewrite-rules, and with a special formulation of the Knuth-Bendix critical pair criterion we may prove completeness (or unique termination).

The algorithm may be applied for supporting the interactive definition of geometric objects by geometric constraints: 1) It may derive a sequence of construction operations automatically from the constraints. 2) It may be used for detecting contradictions between the constraints or redundancies, and for explicitly showing internal degrees of freedom of not yet sufficiently specified objects. Thus, it gives the interactive user a hint, where to add further constraints for uniquely specifying the object. 3) Geometric relations that are implied by the specification by constraints may be inferred. Thus, a certain class of theorems of elementary geometry may be proved automatically by the algorithm.

Since the approach does not require a translation to Algebra it is easy to explain the intermediate or final results in a geometric language. Therefore, the approach is better apt for an integration with an interactive CAD system than other, algebraic approaches.

2) Typical constraint problems

Here we give two typical problems that we might want to solve in an interactive geometric modelling system. Since the problems can be expressed by geometric relations or "constraints" they are referred to as so called "constraint problems" throughout this paper.

Problem 1. Sketching is a natural and convenient way for partially defining geometric objects. Fig. 1.1 shows a typical example for a two-dimensional geometric shape which is approximated by a sketch. The drawing shows that the object consists of points connected by lines and circular arcs. The sketch gives a rough outline of the size relationships, and contains topological information (e.g. that points are to the left, or to the right of line segments, inside or outside areas bounded by line segments, etc.). In addition, some geometric relations (constraints), such as angles and distances, are specified by the dimensioning. Also a congruence relation "$d_1 = d_2$" is specified.

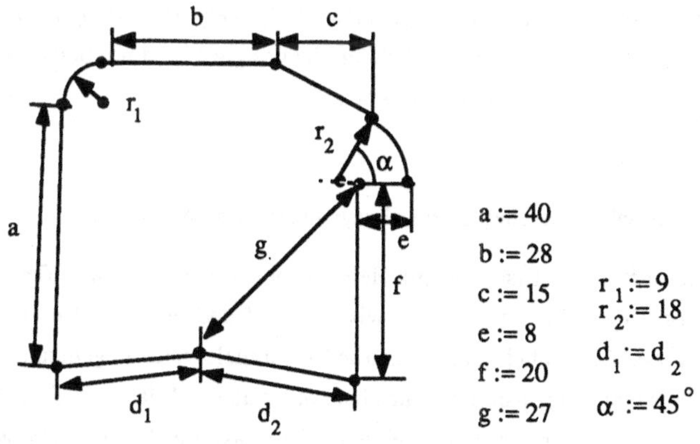

fig. 1.1 A sketch of a two-dimensional geometric object and its dimensioning

A modern CAD system should support the interactive user in solving the "constraint problem". If the constraints are consistent, some algorithm should explicitly find the exact shape of the geometric object meeting these constraints, such that it can be drawn on a graphical screen, or may be manufactured automatically by a NC machine. Such a constraint satisfaction algorithm should also find alternative solutions, if they exist, and it should notify if the description by constraints is insufficient (i.e. an infinite number of solutions exist) or contradictory, in which case finding a consistent solution is impossible. During an interactive design process the dimensions may be changed frequently, therefore, the occurrence of such invalid dimensions is quite likely.

By evaluating the dimensions assigned to the shape in fig. 1.1 we obtain the two solutions shown in fig. 1.2.

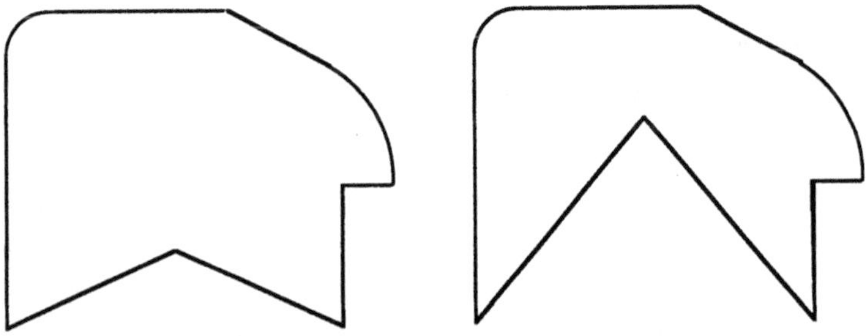

fig. 1.2 The evaluated dimensioning (Two solutions were found)

By splitting the design process in two parts: *sketching*, and second, freely *associating* the *dimensions*, the interactive designer's task is made easier. While sketching, the designer need not bother about exact dimensions and about the question whether a construction is at all possible.

Problem 2. Prove that the diagonals of parallelograms always intersect in the middle.

This question is quite different from the first Problem. Here a geometric object is specified by the fact that the two opposite sides of a quadrilateral are pair-wise parallel and nothing else (fig. 2). This information does not specify a specific parallelogram. It is true for parallelograms in general. We want to derive a certain property which is typical for parallelograms but not for all quadrilaterals. To show that the question is relevant also for constructing a specific parallelogram, let us assume that the position of one of the corner points (point No 1) and the position of the intersection of the diagonals (point No 5) are also given. The opposite point (No 4) may be constructed by adding the relative vector between points No 1 and 5 to the position of point No 5. Before carrying out this operation we need to know that the two vectors are identical, which is a consequence of the property in question.

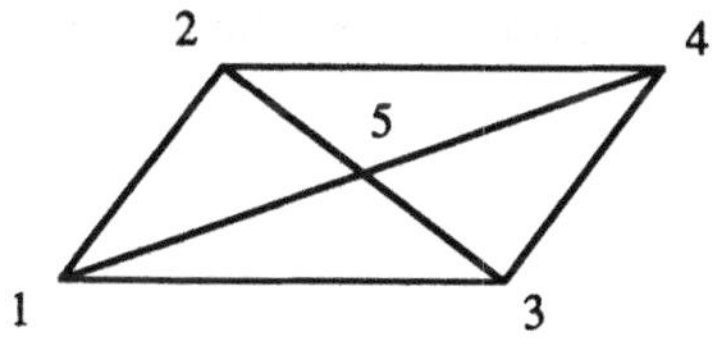

fig. 2

The example shows that it is sometimes necessary to prove that certain properties are the consequence of some other properties, also for constructing simple geometric shapes.

3) Algebraic and numerical methods for solving constraint problems

This section gives an overview on algebraic and numerical methods for solving different problem classes.

Algebraic methods for computing the shape of geometric objects. By introducing a coordinate system where each point P_i is represented by a coordinate tuple $<x_i, y_i, z_i>$, geometric constraint problems may be translated to an algebraic problem. Some examples:

- The fact that the spatial distance between points P_1 and P_2 is equal to the distance between points P_3 and P_4, is expressed by the polynomial equation:

$$(x_2 - x_1)^2 + (y_2 - y_1)^2 + (z_2 - z_1)^2 - (x_4 - x_3)^2 - (y_4 - y_3)^2 - (z_4 - z_3)^2 = 0$$

- The fact that the line P_1P_2 is parallel to the line P_3P_4 in plane, is expressed by the equation:

$$(x_2 - x_1) \cdot (y_4 - y_3) - (y_2 - y_1) \cdot (x_4 - x_3) = 0$$

When solving a system of equations we make a distinction between dependent variables x_j, and independent variables (parameters) u_k. Two methods are usually applied for solving algebraic equations:

- **The numeric approach.** If numeric values are assigned to the parameters u_k, a solution for the variables x_i may be approximated iteratively. Systems like "Sketchpad" by I. Sutherland [SUT 63], "ThingLab" by A. Borning [BOR 81], and "Juno" by G. Nelson [NEL 85], apply numerical approximation methods to solve the algebraic equations (Newton Raphson or relaxation). Starting with a first guess for the point positions, the algorithm finds successive approximations to the object meeting the constraints. The sketched object may serve as a first guess in the iteration process, such that the result is predictable even if there are several possible solutions. On the other hand, it may be difficult to find all alternative solutions with this method. (With the dimensions as assigned in fig. 1.1 we would only find the first solution in fig. 1.2 since it is more similar to the sketch than the second one). A combination of analytical and numerical methods is described in [LIG 82]. By analyzing the equation system it is possible to detect insufficiently constrained and inconsistent object specifications.

- **The symbolic approach.** A symbolic solution to the algebraic problem may be found by solving the system of polynomial equations $p_1(u_1, ..., u_m, x_1, ...x_n) = 0, \; ... \;, p_k(u_1, ..., u_m, x_1, ...x_n) = 0$. The solution is of the form $x_i = f_i(u_1, ..., u_m)$ with some algebraic functions f_i. The approach is more general than a numeric solution, since the parameters are uninstantiated, and arbitrary numeric values may be assigned to them. Computer algebra systems like "MACSYMA", have facilities for solving systems of simultaneous polynomial equations. Encouraged by the success of these systems, some geometric design programs solve the constraint problem by solving the algebraic equations symbolically. Also, the possible alternative solutions for a constraint problem may be found explicitly. Topological information on the other hand would have to be transformed to a system of inequalities. So far useful algorithms are known only for systems of linear inequalities.
  Finding a purely symbolic solution is very time consuming, and therefore specialized algorithms have been invented that are faster, but not always powerful enough for finding a solution in the general case. One approach for algebraic constraint evaluation that has been proposed is described in the Ph.D. Thesis by J. Gosling (see [GOS 83]). An integrated expert system with a "geometry engine" is described by Poppelstone (see [POP 86]).

With the algebraic approaches discussed above it is necessary to transform the geometric problem to an algebraic problem expressed with coordinates. There exist powerful methods in Algebra; the theory is well established. As long as all we want are the coordinates of the object, algebraic methods seem best suited. Problems occur when we try to integrate these methods with an interactive system. If there is an inconsistency in the object specification, it may be difficult to transform the algebraic expressions back to a language with geometric semantics, and therefore to explain the reason for such an inconsistency.

In the following section we discuss algebraic approaches for proving certain geometric theorems. It is possible to prove that geometric relations are a consequence of other relations, or to show that they contradict each other.

Algebraic approaches to theorem proving in elementary geometry. Every theorem in elementary geometry can be translated to the language of algebra by introducing coordinates. It is expressed by a system of polynomial equations and inequalities on the coordinates of the points and some quantifiers for the variables. The first such theorem proving algorithm has been found by A.Tarski. The algorithm can prove or discard theorems of elementary geometry that are expressed algebraically (see [TAR 51]). Unfortunately the algorithm is extremely complicated and intractable even for simple problems. The benefit of Tarski's algorithm is the fact that it shows that problems of elementary geometry are decidable. A newer, and more efficient algorithm for solving the same algebraic problems by quantifier elimination has been developed by Collins [COL 75]. The algorithm is still too inefficient for solving complex problems.

The Chinese mathematician Wu Wen Tsün developed an algorithm for algebraically solving an important class of geometric problems that may be expressed by formulas of the form: $\forall P_1, ..., P_k : A(P_1, ..., P_k)$ $\rightarrow B(P_1, ..., P_k)$, (where A and B are quantifier free formulas on the points P_i). For reference see [WUW 84a], [WUW 84b], and [WUW 86]. When we translate the problem to Algebra we obtain the following polynomial equations.

For the hypotheses:

$$p_1(x_1, ... x_n, u_1, ... u_m) = 0$$

$$\vdots$$

$$p_k(x_1, ... x_n, u_1, ... u_m) = 0$$

For the conclusion:

$$g(x_1, ... x_n, u_1, ... u_m) = 0$$

The method for deciding if the polynomial g follows from polynomials $p_1, ..., p_k$ may be shortly described as follows: The theorem is true if and only if $g \equiv 0 \ mod \ \{p_1, ..., p_k\}$, or in other words, if g is an element of the polynomial ideal spanned by $p_1, ..., p_k$. The algorithm first brings the polynomials to "triangular" form and then applies so called "pseudo-division": The conclusion polynomial g is divided by the polynomial p_k. The remainder r_k is divided by p_{k-1}, and so forth. If the final remainder r_1 is identical to zero the theorem is proved to be true. Sometimes a factorization of the polynomials is necessary. For a description of the implementation of Wu's algorithm see Chou [CHO 84]. The algorithm is very efficient, and many non trivial theorems have been proved with it (see [CHO 86]). An implementation of the algorithm with the computer algebra system MACSYMA is described in [KOH 85].

A similar approach for solving the same algebraic problem uses Buchberger's critical pair algorithm (described in[BUL 82]). First the preliminary polynomials $p_1, ..., p_k$ are brought to normal form, i.e. a so called Gröbner basis is found. The polynomials may be used as reduction rules. The truth of the theorem is proved if g can be reduced to 0 by means of the polynomials of the Gröbner basis of $p_1, ..., p_k$. An application of the Buchberger algorithm to proving theorems in elementary geometry is described in [CHS 86].

The success of the algebraic methods in proving many non trivial problems in elementary geometry comes from the fact that much of the tedious work of theorem proving may be automatized. Nevertheless, there are problems occurring with algebraic methods :

- For deciding which of the variables are the dependent and which are the independent ones, when translating the geometric problem to an algebraic language, the knowledge of a construction sequence for the points in question is required. (The efficiency of the proving mechanism depends largely on a clever choice of the variables too).
- By translating geometric constraints to algebraic equations the geometric meaning is lost. An unambiguous geometrical interpretation of the operations carried out on polynomials is extremely difficult, but still subject to research.

4) A geometric approach to constraint problems

In this section we describe a new approach for solving constraint problems such as mentioned in section 2. Earlier, mostly heuristic approaches can be found, for instance, in [GEL 59] and [COE 86]. A symbolic approach to solving formulas in projective geometry is described in [SCH 88]. The most important difference is that we want to express the problems in a geometric language and do not need to translate to Algebra. First we need to specify the elements of a geometric language for expressing the problems. The elements of this language, the data types, the variables, constants and functions and also the predicates used here, are listed below. Some functions are interpreted in a coordinated model of geometry where the coordinates are the orthogonal projections of points on the coordinate axes. The coordinates are real numbers.

Data Types:

Point, Line, Circle, REAL

Variables:

$P_1, \ldots, P_j$	: *Point*
$L_1, \ldots, L_k$	: *Line*
$C_1, \ldots, C_l$	: *Circle*
$X_1, \ldots, X_i$	: *REAL*

Constants:

P_0, L_x, L_y *(origin* and *coordinate axes)*

Functions: (examples)

circle(P, Radius)	: $Point \times REAL \rightarrow Circle$
line(P, Degree)	: $Point \times REAL \rightarrow Line$
distance(P_1, P_2)	: $Point \times Point \rightarrow REAL$
intersection(C, L)	: $Circle \times Line \rightarrow Point$

The language introduced so far may be extended in two ways:

1) We may interpret the expressions as the elements of a procedural programming language by additionally introducing the conceptions of *assignment*, *decision operations* and some higher level control structures, such as "*if-then-else*", or "*while-do*", known from programming languages such as Pascal. With this semantics we can define geometric objects by a sequence of computation steps. In other words we may "compute" the objects with a program expressed in the language. Theoretical investigations of programming languages for compass and ruler operations are found in [ENG 83], [HUC 86] and [SCH 75].

2) By introducing some geometric predicates and the logical operators $\wedge$ (conjunction), $\vee$ (disjunction), $\neg$ (negation), as well as universal and existential quantifiers ($\forall$, $\exists$) we may define well formed formulas of first order predicate calculus. With such formulas we can describe a theory of elementary geometry by axioms. Also we can express theorems with these formulas.

Predicates. Geometric constraints are relations between points which may be expressed by predicates on these points. First we define some relations that assign ground terms (i.e. terms containing no variables) to tuples of points. These predicates have the same interpretation as assignments in a procedural programming language, therefore they may be used to interpret a sequence of formulas as a procedural program. Also we may view these terms as Skolem functions that replace existentially quantified variables.

```
p(P, Pos).
d(P1, P2, D).
s(P1, P2, Alpha).
v(P1, P2, [Alpha, D]).
a(P1, P2, P3, Beta).
tr(P1, P2, P3, [Alpha, Beta]).
```

- The predicate *p(P, Pos)* assigns a position *Pos* to the point variable *P*. *Pos* is either a constant (e.g. a pair of coordinates *[10, 0.5]*), or a functional expression, like e.g. *intersection(line(P2, 90), circle(P3, 20))*. We may use any functional expression of type Point, with the restriction that it may not contain variables. Therefore, the positions of the points referenced (here *P2* and *P3*) are assumed to be known.
- The predicate *d(P1, P2 Dist)* expresses that we know the distance between two points. For the value of *Dist* we may again write a constant number, or a function of type REAL, e.g. *length(vector(P3, P4))*, i.e. the distance of two points with already specified position.
- *s(P2, P1, Alpha)* constrains the slope of the line connecting *P1* and *P2* by the angle *Alpha* measured counterclockwise from to the x-axis.
- In fig. 3 an angle Beta is associated with three points by the predicate *a(P1, P2, P3, Beta)*.

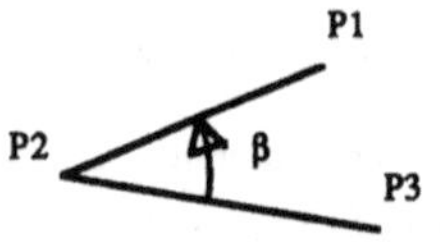

fig.3

- The angle *Beta* is measured counterclockwise from the *line (P2, P3)* to the *line (P2, P1)*. An angle going in the other direction would be written with negative sign: *a(P1, P2, P3, Beta)* ⇔ *a(P3, P2, P1, -Beta)*.
- If we know the value of two angles of a triangle we express this by the predicate *tr(P1, P2, P3, [Alpha, Beta])*. The predicate *tr* stands for triangle, and implies that the value of the third angle is known implicitly: *Gamma = 180°- (Alpha + Beta)*. Note that only angles, but not distances between points are defined by 'tr'.

Congruence relations. In addition to the above predicates which assign constant values to angles distances, positions, etc., the following equations are used for expressing the so called congruence relations:

`d(P1, P2) = d(P3, P4).`	The distance of points *P1* and *P2* is congruent to the distance between points *P3* and *P4*.
`a(P1,P2,P3) = a(P4,P5,P6).`	The two angles are congruent.
`s(P1, P2) = s(P3, P4).`	The *line(P1,P2)* is parallel to *line(P3,P4)*.
`v(P1, P2) = v(P3, P4).`	The *vector(P1,P2)* is equal to *vector(P3,P4)*.
`tr(P1,P2,P3) = tr(P4,P5,P6).`	The two triangles are similar.
`tr(P1,P2,P3) = str(P4,P5,P6).`	The two triangles are similar up to symmetry.
`p(P1) = p(P2).`	Two points are identical

What can be expressed by the predicates? The predicates above suffice for specifying two-dimensional geometric objects and theorems of planar elementary geometry. In comparison, the axiomatic system by A. Tarski (see [SST 83]) only requires two predicates, namely, *d(P1,P2,P3,P4)* for expressing the "*equidistance relation*" and a "*betweenness relation*" *b(P1,P2,P3)* (expressing that point *P2* is between points *P1* and *P3*). Since the predicate "*s*" (for slope) also defines a linear order of points we can express *b(P1,P2,P3)* by the parallelism relation, namely with *s(P1,P2) = s(P2,P3)*. All predicates introduced above could be expressed with equidistance and betweenness relations alone. The fact that we use so many more predicates is justified by the better geometric intuition rendered by individual predicates than by expressing everything with a combination of only two relations, as well as by some properties of the algorithm to be discussed later.

The constraints we talk about here are restricted to relations between points. Each pair of points implicitly defines a line which may be specified by constraining those points. To specify the properties of a circle we would have to constrain some characteristic points (e.g. center point, and a point on the periphery).

Two triangles that are joined by two common points build a quadrilateral. By stating the similarity of joined triangles we can describe the similarity of polygons. In combination with other predicates every property of polygons which may be expressed algebraically may be expressed by the geometric predicates.

5) Geometric axioms and constructions rules

The most important goal of the approach presented in this paper is to express the axioms of a geometric theory in a way, such that a computer may apply them automatically for proving the truth of a theorem, or for finding a geometric construction. We found that a representation of axioms by rewrite-rules makes it easy to apply a simple inference mechanism. Later we show how it is possible to prove the completeness and termination properties of these inferences. In addition to the rewrite-rules we find implication rules that derive relations implied by other relations. The implication rules are simply realized with Prolog-rules, using the built-in reasoning mechanism of the programming language Prolog [CLM 81]. In this paper we cannot give a complete description of the algorithm, but describe the general ideas by typical examples.

Rules for compass and ruler construction. An axiomatic definition of compass and ruler constructions can be found in [ENG 83]. In this approach we apply rules known from constructing with compass and ruler to derive positions of points that are not explicitly specified by a corresponding predicate "*p*". An example: given the positions of two points *P1* and *P2*, the distance between *P1* and a third point *P3* and the distance between *P2* and *P3*, we may construct *P3* by intersecting two circles. We first write the precondition of the rule by a conjunction of predicates, such as defined in the previous section:

```
p(P1, [Pos1]) ∧ p(P2, [Pos2]) ∧ d(P1, P3, R1) ∧ d(P2, P3, R2)
```

The position of the third point *P3* is found by intersecting two circles with centers *P1* and *P2*. This is expressed symbolically by:

```
p(P3, intersection(circle(P1, R1), circle(P2, R2))).
```

We now want to express this construction rule as a rewrite-rule. We introduce the following notation: Conjunctions of predicates are expressed as lists of predicates '[ ]', and the symbol '->' indicates the direction in which the rewrite-rule is applied. The above rule is represented by the following rewrite-rule:

```
[p(P1, [Pos1]), p(P2, [Pos2]), d(P1, P3, R1), d(P2, P3, R2)]
-> [p(P1, [Pos1]), p(P2, [Pos2]),
    p(P3, intersection(circle(P1, R1), circle(P2, R2)))].
```

We find similar rules for intersecting circles and lines, or two lines.

```
[p(P1, [Pos1]), p(P2, [Pos2]), d(P1, P3, R), s(P2, P3, S1)]
-> [p(P1, [Pos1]), p(P2, [Pos2]),
    p(P3, intersection(circle(P1, R), line(P2, S1)))].

[p(P1, [Pos1]), p(P2, [Pos2]), s(P1, P3, S11), s(P2, P3, S12)]
-> [p(P1, [Pos1]), p(P2, [Pos2]),
    p(P3, intersection(line(P1, S11), line(P2, S12)))].
```

All rules are of the form "φ -> Ψ". The *precondition* φ and the *postcondition* Ψ are first order formulas in Skolem form. I.e., they are conjunctions of literals (the predicates on points) with Skolem functions for the point positions, distances, etc. The functions express geometric operations. So far the idea is similar to Hoare's approach [HOA 69] used for program verification, although the notation is different.

All constraints (which are defined by the specification of an interactive user) are expressed by predicates which are stored as 'facts' in the Prolog database. An inference mechanism tries to apply all the rules. A rule can only be applied if the preconditions holds. If the corresponding predicates of the precondition can be matched with facts in the database, the rule fires, and a *transaction* on the database is performed. Those facts on the left side of the rule not occurring on the right side are replaced in the database by the ones occurring only on the right side of the rule. Inserting and retracting facts in the database is realized with the commands *retract* and *assertz*, which are built-in predicates in PROLOG. A transaction must be carried out as an atomic operation, i.e. either it is carried out completely, or not at all, and thus preserves consistency of the database. The algorithm terminates when no rule applies.

The inference mechanism, together with rules like the one described above determine an algorithm for automatically constructing a geometric object from a geometric specification. The result can be regarded as a proof for the correctness of the specification by constraints. Every step of the proof corresponds to a geometric operation. If we write down the sequence of operations during the proof, we get a prescriptive definition of the specified object. The specification of a geometric object is complete and consistent, if for each point P_i, there exists a predicate *p(Pi, POSi)*.

The following 2-D example shows the effect of the construction algorithm. The initial content of the database is (see fig. 4):

```
p(P1, [100,100]).
p(P2, [200,100]).
d(P1, P3, 75).
d(P2, P3, 80).
d(P3, P4, 60).
s(P2, P4, 90).
```

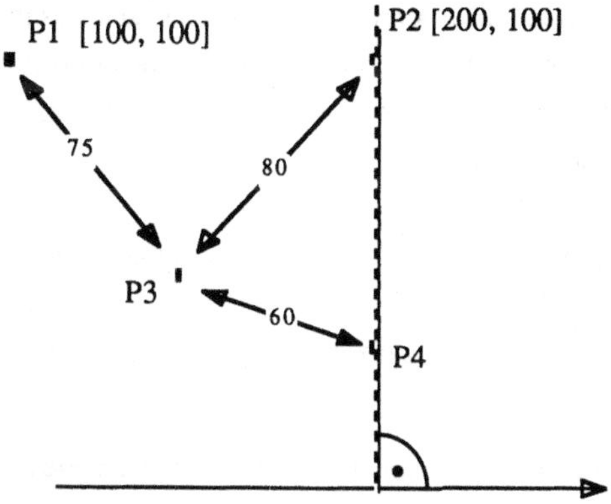

fig. 4 Four constrained points

After running the construction algorithm, the database contains:

```
p(P1, [100,100]).
p(P2, [200,100]).
p(P3, intersection(circle(P1, 75), circle(P2, 80))).
p(P4, intersection(circle(P3, 60), line(P2, -90))).
```

The new facts are asserted in a sequence determined by the algorithm. The expressions specifying point positions therefore may refer to points defined earlier in the construction process (in our example, the symbolic expression for the position of point *P4* refers to points *P2* and *P3*). The result of the algorithm is a symbolic prescription for constructing the points, and therefore has a procedural interpretation.

Axioms of Euclidean geometry: For every pair of triangles, if two sides and the angle between them are congruent, then also the other angles are congruent (fig. 5):

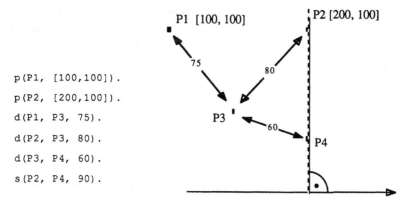

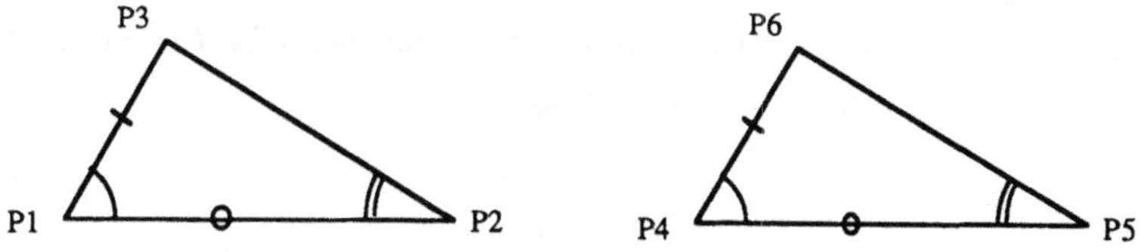

fig. 5

$$d(P1,P2) = d(P4,P5) \wedge d(P1,P3) = d(P4,P6)$$
$$\wedge\ a(P3,P1,P2) = a(P6,P4,P5)$$
$$\supset a(P1,P2,P3) = a(P4,P5,P6)$$

We wish to express this axiom as a rewrite-rule which enables us to employ our inference mechanism. When carrying out the inference the predicates on the left side of the rule are replaced by those on the right side; therefore it is important that no information is lost by applying the rewrite-rule. First we express the above axiom as an equation on clauses.

```
[d(P1,P2) = d(P4,P5), d(P1,P3) = d(P4,P6), a(P3,P1,P2) = a(P6,P4,P5)]
    =
[a(P1,P2,P3) = a(P4,P5,P6), a(P3,P1,P2) = a(P6,P4,P5), d(P1,P2) = d(P4,P5)]
```

By imposing a direction of application the equation may be expressed as a rewrite-rule with the desired property.

r_1:

```
[d(P1,P2) = d(P4,P5), d(P1,P3) = d(P4,P6), a(P3,P1,P2) = a(P6,P4,P5)]
    ->
[a(P1,P2,P3) = a(P4,P5,P6), a(P3,P1,P2) = a(P6,P4,P5), d(P1,P2) = d(P4,P5)]
```

With the following rewrite-rule we introduce a definition for the predicate '*tr*:

d_3:

```
[a(P3,P1,P2) = a(P6,P4,P5), a(P1,P2,P3) = a(P4,P5,P6)]
        -> [tr(P1,P2,P3) = tr(P4,P5,P6)].
```

The following rewrite-rule d_1 is the definition of the predicate *v*:

d_1:

```
[d(P1,P2, Dist), s(P1,P2,  S1)] -> [v(P1,P2, vect(S1, Dist))]
```

We also want to express congruence relations for these predicates, for instance, $v(P1, P2) = v(P3, P4)$, and $tr(P1, P2, P3) = tr(P4, P5, P6)$ which are defined by d_2 and d_3:

d_2:

```
[d(P1,P2) = d(P3,P4), s(P1,P2) = s(P3,P4)]
        -> [v(P1,P2) = v(P3,P4)].
```

Forward vs. backward reasoning. The information associated with the predicates retracted from the database by application of rewrite-rules is still implied by the new predicates. Instead of adding the inverse rule (which would cause termination problems) we define some implications. Such implications can be represented in Prolog by so called "Prolog-rules". The notation of a Prolog rule is as follows: "A :- B." means that A is implied by B, where A is a predicate (the head of the rule), and B is a conjunction of predicates (the body of the rule). The Prolog backtracking mechanism is used to derive the truth of predicates either directly from facts '[ ]', or indirectly by applying Prolog-rules. Together with the rewrite-rule d_3 we define the following implications.

```
a(P1,P2,P3) = a(P4,P5,P6) :-  [tr(P1,P2,P3) = tr(P4,P5,P6)]
a(P2,P3,P1) = a(P5,P6,P4) :-  [tr(P1,P2,P3) = tr(P4,P5,P6)]
a(P3,P1,P2) = a(P6,P4,P5) :-  [tr(P1,P2,P3) = tr(P4,P5,P6)]
```

Another axiom of Euclidean geometry used in our system is expressed by the rewrite-rule r_2.

r_2:

```
[a(P1, P2, P3) = a(P4, P5, P6), s(P2, P3) = s(P5, P6)]
          -> [s(P2, P3) = s(P5, P6), s(P1, P2) = s(P4, P5)]
```

The corresponding implication is:

```
a(P1, P2, P3) = a(P4, P5, P6):-
          [s(P1, P2) = s(P4, P5)], [s(P2, P3) = s(P5, P6)].
```

Here the importance of the order of the arguments becomes clear. If we reversed the order of points in one of the predicates the rule would be in contradiction to its geometric meaning.

It is a general principle of our program to apply the *rewrite-rules* are in one direction (for replacing the predicates in the database with new predicates, and thus bringing them to a certain normal form), and the *implications* in the other direction (for deriving predicates implied by other predicates). The method of applying rewrite-rules is also called *forward reasoning*, whereas the built-in reasoning mechanism of Prolog is called *backward reasoning*. Both reasoning mechanisms are combined in this program. The implications used in combination with the so called construction rules introduced at the beginning of this section may be interpreted as measuring operations. The following implication may be used for measuring the distance between two points:

```
distance(P1, P2, D):-
    p(P1, [X1,Y1]),
    p(P2, [X2,Y2]),
    D is sqrt((X2 - X1) *  (X2 - X1) + (Y2 - Y1) * (Y2 - Y1)).
```

Here forward reasoning is used for construction operations and backward reasoning is used for measuring operations.

6) Theoretical problems

Uniform termination. The idea behind the geometric rewrite-rules was to replace some facts in the database by facts which are in some sense "simpler". To prove that the inference mechanism applying the rules to a given database eventually terminates, we have to find orderings such that the right hand side of a rule is always smaller than the left hand side, and furthermore to prove that the orderings have lower bounds, so that no infinite decreasing sequence is possible.

Definition 6.1: (*well-foundedness*) Given a set of terms T, a partial ordering '$\succ$' on T is *well-founded* (or Noetherian) if there is no infinite descending chain of terms $t_1 \succ t_2 \succ \ldots$

Definition 6.2: (*order-function* τ) For our geometric predicates and for the set of conjunctions C we introduce a function $\tau : C \rightarrow N$ which is defined as follows:

for *literals*:

$$\tau(d(P1, P2, \_)) = 6$$
$$\tau(a(P1, P2, P3, \_)) = 5$$
$$\tau(s(P1, P2, \_)) = 4$$
$$\tau(tr(P1, P2, P3, \_)) = 3$$
$$\tau(v(P1, P2, \_)) = 2$$
$$\tau(p(P1, \_)) = 1$$
$$\tau(false) = 0$$

$$\tau(d(P1, P2) = d(P3, P4)) = 6$$
$$\tau(a(P1, P2, P3) = a(P4, P5, P6)) = 5$$
$$\tau(s(P1, P2) = s(P3, P4)) = 4$$
$$\tau(tr(P1, P2, P3) = tr(P4, P5, P6)) = 3$$
$$\tau(tr(P1, P2, P3) = str(P4, P5, P6)) = 3$$
$$\tau(v(P1, P2) = v(P3, P4)) = 2$$
$$\tau(p(P1) = p(P2)) = 1$$

For conjunctions X, Y: $\tau(X \wedge Y) = \tau(X) + \tau(Y)$, for the empty conjunction: $\tau() = 0$

Lemma 6.1: If for all rules $l \rightarrow r$: $\tau(l) > \tau(r)$ then τ on C is well founded.

Proof of lemma 6.1: For each conjunction X to which a rewrite-rule can be applied τ has some finite value $\tau(X) = n \in N$. Each rule applied to X reduces $\tau(X)$ at least by one (by assumption), 0 is a lower bound for τ, therefore X is reduced in at most n steps to an irreducible conjunction X_o. ♦

Unique termination. With the following example we want to show what happens when different rules may be applied to the same facts in a database (see fig. 5.1). For the points *P1* to *P6* some relations for distance-congruence, angle-congruence, and parallelism are expressed by the facts in the database. When examining the rewrite-rules described in section 5 we can see that rules r_1, and r_2 can both be applied to the initial database. Depending on which of the two rules is applied the database will be in a different state (fig. 5.2, or fig. 5.3). Proceeding from there we may continue applying rules. On the left side, after applying rule d_2 twice, we end with a database state as shown in fig. 5.4. On the right side we may apply rule r_2 twice and then d_2, and the database will have a final state as shown in fig. 5.6.

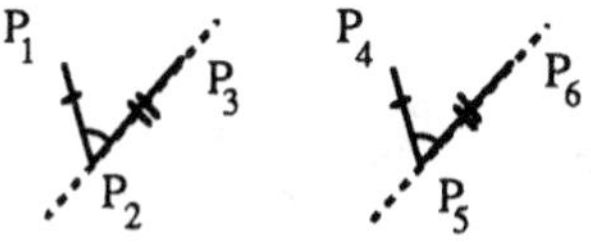

fig. 5.1

d(P1, P2) = d(P4, P5), s(P2, P3) = s(P5, P6),
d(P2, P3) = d(P5, P6), a(P1, P2, P3) = a(P4, P5, P6).

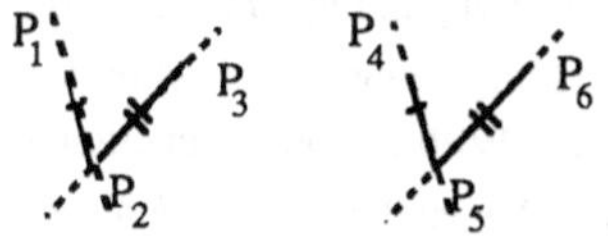

fig. 5.2

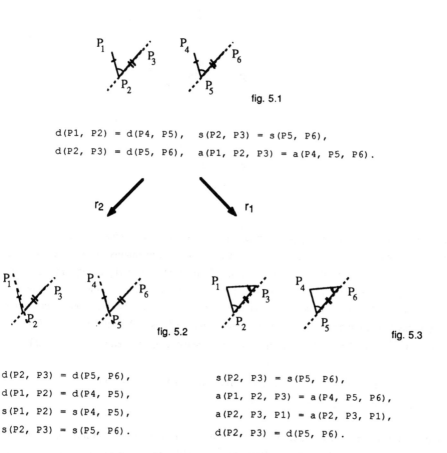

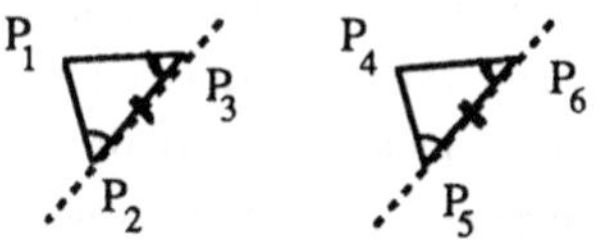

fig. 5.3

d(P2, P3) = d(P5, P6),
d(P1, P2) = d(P4, P5),
s(P1, P2) = s(P4, P5),
s(P2, P3) = s(P5, P6).

s(P2, P3) = s(P5, P6),
a(P1, P2, P3) = a(P4, P5, P6),
a(P2, P3, P1) = a(P2, P3, P1),
d(P2, P3) = d(P5, P6).

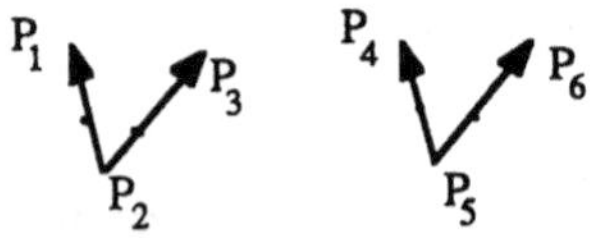

fig. 5.4

```
v(P2, P3) = v(P5, P6),
v(P1, P2) = v(P4, P5).
```

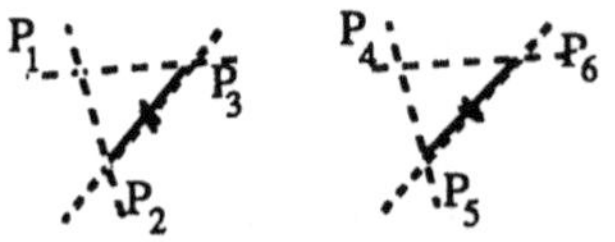

fig. 5.5

```
d(P2, P3) = d(P5, P6),
s(P2, P3) = s(P5, P6),
s(P1, P2) = s(P4, P5),
s(P1, P3) = s(P4, P6).
```

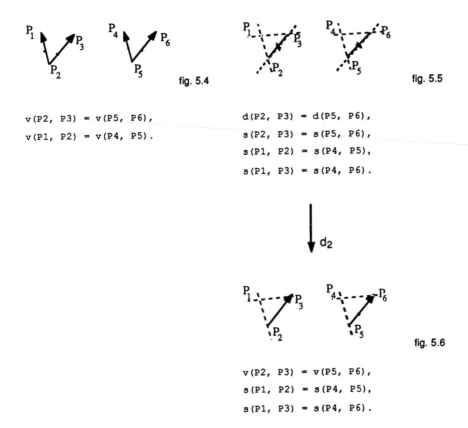

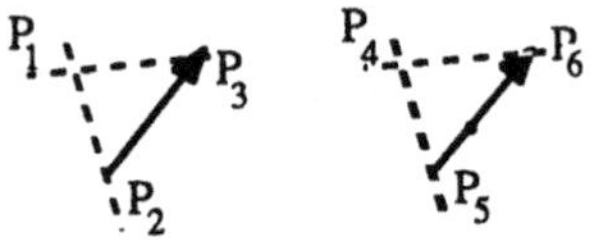

fig. 5.6

```
v(P2, P3) = v(P5, P6),
s(P1, P2) = s(P4, P5),
s(P1, P3) = s(P4, P6).
```

In both cases we can't find rules that apply. The relations in the databases are what we call *irreducible.* Since both states do have the same offspring and the inference mechanism maintains the consistency of the database the two final states are equivalent but obviously not identical. The sensitivity in regard to selection is an undesired property of the rules. The actual situation may be cured by adding a new rewrite-rule that either rewrites the facts in fig. 5.4 to those of fig. 5.6 or the other way round. With the order function τ defined above we may find out the direction. The total value τ for the facts of the database in fig. 5.4 is *4*, and the value of the facts in fig. 5.6 is *10*. To maintain well-foundedness the new reduction rule must be:

```
[v(P2, P3) = v(P5, P6), s(P1, P2) = s(P4, P5), s(P1, P3) = s(P4, P6)]
->  [v(P2, P3) = v(P5, P6), v(P1, P2) = v(P4, P5)]
```

In order to find out whether we need more rules it seems that we need to inspect an infinite number of such examples. We may therefore ask ourselves: "Does there exist a finite set of reduction rules that suffice for obtaining the desired property, namely to bring any database to normal form, independently of the order in which the rules are applied, or do we have to add more and more rules?"

The answers to these question can be given by a theory applied for solving word problems in Algebra, described by Knuth and Bendix in 1965 (see [KNB 70] or [HSI 83]). The theory and its application to this specific type of geometric rewrite-rules, taking into consideration the associativity and commutativity laws of the conjunction, is discussed formally in [BRU 87]. Also the implementation of a Knuth-Bendix completion algorithm for finding a complete (canonical) set of rewrite-rules is described there.

7) Rules for inconsistently and insufficiently specified objects

Inconsistencies. In addition to the rules that infer predicates from other predicates we need rules for detecting inconsistent specifications of geometric objects. One of the reasons for a contradiction may be that some relation is defined twice for the same point(s), for instance, when the position of one point is defined by two facts in the database. This inconsistency is formally expressed by the following rewrite-rule:

```
[p(P1, Pos1), p(P1, Pos2)] -> [false]
```

The same principle holds as well for all other predicates. If a relation which is already implied by other predicates, is explicitly stated by a predicate of its own, this is inconsistent. An example is given below.

```
[p(P1, Pos1), p(P2, Pos2), d(P1, P2, D)] -> [false]
```

Here the distance is already implied by the positions of the two points. Inconsistencies also occur when arguments of some predicates are repeated illegally as expressed by the following rule:

```
[a(P1, P2, P3) = a(P4, P5, P4)] -> [false]
```

The inconsistency rules are also a means of expressing "*negation*" and therefore are essential in the theory on which the algorithm is based. The consequences are discussed in [BRU 87] (see chapter 4).

The inconsistency rules may be practically applied in an interactive system. Inconsistencies are reported to the user to give him a hint for changing the specifications. The user may decide to discard the new constraint, or to cancel some of the constraints specified earlier. We may trace back the inferences carried out, and can find out which of the constraints might be deleted.

The rules make no distinction between redundant and inconsistent information. Redundancy is considered as a special case of inconsistency in the theory. In practice, this means that the fact *'false'* can be interpreted as *"redundant"* or as *"inconsistent"*, depending on the context.

Insufficient specification. Fig. 6 illustrates the geometric specification for 8 points which are related to each other by some predicates stored in the database:

```
tr(1,2,3 [45,90]).
tr(8,1,3, [50,45]).
```

```
tr(4,8,3, [30,40]).
tr(5,6,4, [100,60]).
d(1,2, 25.0).
d(6,7, 60.0).
p(7, [150,50]).
```

If a relation "*tr*" (*triangle*) is stated for three points, and a distance between two of them is constrained, these points are connected rigidly (e.g. *tr(1,2,3, [45,90])* and *d(1,2, 25.0)*). By means of the addition property of triangles we can easily find out which points form a rigid polygon. By coloring the area covered by connected triangles in a uniform manner we can make the rigidity visible. For the above example the algorithm found out that points 1, 2, 3, 4 and 8 are rigidly connected to each other (represented by a *filled area*), the angles between points 4, 5 and 6 are known (*empty area*), point 7 has a fixed position (*fat point*) and its distance to point 6 is specified (*thick line*).

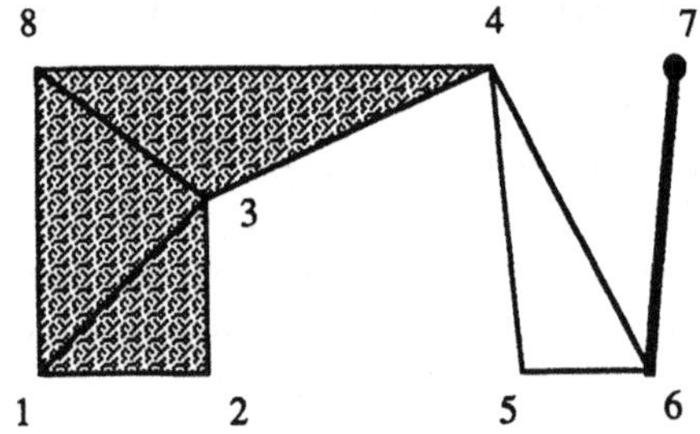

fig. 6

From the drawing we can see immediately that the object consists of two rigid sub-objects with 5 and 2 points respectively. To make the whole object rigid, we need to specify constraints between sub-objects. For instance, we may specify the distance between points 5 and 6, and say that points 1, 2, 5, and 6 are collinear, and points 8, 4, and 7 are collinear.

8) Conclusions

The benefit of a symbolic, geometric approach is the nearness of the language in which it is realized to the language of the interactive CAD user. The predicates and functions used in the algorithm have a direct geometric interpretation. Therefore it is relatively straight forward to develop a user interface that explains the results in words or graphically. In the implemented prototype system the graphical user interface and the numerical evaluation of the symbolical expressions are realized with the language interface of the Prolog interpreter with the procedural programming language Modula-2 (see [MUL 85] and [WIR 83]).

The rewrite-rules arise from equations of formulas. Therefore applying these rules cannot invent new points that don't occur in the description of the problem, although this is sometimes necessary for finding a proof or a construction. Theoretically it is possible to have the computer invent such points, but then the program would be intractable for real-life problems. The intended application of the system is to support the interactive user in finding a solution when constructing geometric objects. Also the system may be used for interactively proving elementary geometry theorems. The strength of the system is that it helps detecting inconsistencies in the definition (in an interactive session such inconsistencies are quite often) and gives hints where to add constraints.

References

[BUL 82] B. Buchberger and R. Loos. Algebraic Simplification. Computing, Suppl. 4, 11, (1982) pp. 11 - 43

[BRU 86] B. Brüderlin. Constructing Three-Dimensional Geometric Objects Defined By Constraints. 1986 Workshop on Interactive 3D Graphics, Conference Proceedings, Chapel Hill, North Carolina, published by ACM Siggraph, 1986

[BRU 87] B. Brüderlin. Rule-Based Geometric Modelling. Ph.D. thesis, ETH Zürich, Switzerland, Verlag der Fachvereine, vdf-Verlag, Zürich 1987 (ISBN 3 7281 1638)

[CHO 84] Shang-Ching Chou. Proving Elementary Geometry Theorems Using Wu's Algorithm. Contemporary Mathematics, Volume 29, 1984, American Mathematical Society, pp 234 - 287

[CHS 86] Shang-Ching Chou and William F. Schelter. Proving Geometry Theorems with Rewrite Rules. Journal of Automated Reasoning 2 (1986) pp. 253 - 373

[CHO 86] Shang-Ching Chou. A Collection of Geometry Theorems Proved Mechanically. Technical Report 50, July 1986. Institute for Computing Science, University of Texas at Austin

[CLM 81] W.F. Clocksin, C.S. Mellish. Programming in Prolog. Springer Verlag, Berlin, Heidelberg, New York 1981

[COP 86] Helder Coelho, Luiz Moniz Pereira. Automated Reasoning in Geometry Theorem Proving with Prolog. Journal of Automated Reasoning 2 (1986)

[COL 75] G.E. Collins. Quantifier Elimination for Real Closed Fields by Cylindrical Algebraic Decomposition. Lecture Notes in Computer Science No. 33, pp. 134 - 183, Springer Verlag, 1975

[ENG 83] E. Engeler. Metamathematik der Elementarmathematik. Springer Verlag, Berlin, Heidelberg, New York 1983

[GEL 59] H. Gelernter. Realization of a Theorem Proving Machine, 1959. Published in "Automation of Reasoning" Vol.1, Springer Verlag, 1983

[GOS 83] James Gosling. Algebraic Constraints. Ph.D. thesis, Carnegie-Mellon University, May 1983

[HIL 71] D. Hilbert. Foundations of Geometry. Open Court Publishing Company, La Salla, Illinois 1971

[HUC 86] Ulrich Huckenbeck. Geometrische Maschinenmodelle (german). PhD. thesis, Universität Würzburg, Germany, 1986

[HSI 83] Jieh Hsiang. Topics in Automated Theorem Proving and Program Generation. Ph.D. Thesis, Univ. of Illinois at Urbana-Champaign, 1983

[HOA 69] C.A.R. Hoare. An Axiomatic Basis for Computer Programming. Communications of the ACM. Vol. 12 No. 10, 1969

[KNB 70] D.E. Knuth, P.B. Bendix. Simple Word Problems in Universal Algebra. Computational Problems in Abstract Algebra. Conference Proceedings, Oxford 1967, J. Leech ed., Pergamon 1970

[KOH 85] H.-P. Ko and M.A. Hussain. ALGE-PROVER. An Algebraic Geometry Theorem Proving Software. Report No. 85CRD139, july 1985. Technical Information Series, General Electric

[LIG 82] R. Light, D. Gossard. Modification of geometric models through variational geometry. CAD vo. 14, No. 4, Butterworth 1982

[MUL 85] C. Muller. Modula -- Prolog, User Manual. Rep. No. 63, July 1985. Inst. für Informatik, ETH Zürich, Switzerland

[NEL 85] G. Nelson. Juno, a constraint-based graphics system. 1985 ACM Siggraph Conference Proceedings

[POP 86] R.J. Popplestone. The Edinburgh Designer System as a Framework for Robotics

[SCH 88] F. Schmid. A Symbolic Approach to Solving Formulas in Projective Geometry. Ph.D. Thesis, ETH, Switzerland To appear, 1988

[SCH 75] Peter Schreiber. Theorie der geometrischen Konstruktionen (german). VEB Verlag der Wissenschaften, Berlin, 1975

[SST 83] W. Schwabhäuser, W. Szmielev, A. Tarski. Metamathematische Methoden in der Geometrie (german). Springer Verlag Berlin, Heidelberg, New York 1983

[SHA 78] Michael Ian Shamos. Computational Geometry. Ph.D. thesis, Yale University, New Haven, Connecticut, 1978

[SUT 63] I. Sutherland. Sketchpad, A Man-Machine Graphical Communication System. Ph.D. thesis, MIT, January 1963

[TAR 51] A. Tarski. A Decision Method for Elementary Algebra and Geometry. Univ. of Calif. Press, Berkeley, 1951

[WIR 83] N. Wirth. Programming in Modula-2. Texts and Monographs in Computer Science. Springer Verlag Berlin, Heidelberg, New York 1983

[WUW 84a] Wu Wen-tsün. Some Recent Advances in Mechanical Theorem Proving of Geometries. Contemporary Mathematics, Volume 29, 1984, American Mathematical Society, pp. 235 - 241

[WUW 84b] Wu Wen-tsün. On the decision Problem and the mechanization of Theorem-Proving in Elementary Geometry. Contemporary Mathematics, Volume 29, 1984, American Mathematical Society, pp. 213 - 23

[WUW 86] Wu Wen-tsün. Basic Principles of Mechanical Theorem Proving in Elementary Geometries. Journal of Automated Reasoning 2 (1986) pp. 221 - 252